Recycling and Recovery of Biomass Materials

Recycling and Recovery of Biomass Materials

Editor

Leonel Jorge Ribeiro Nunes

MDPI • Basel • Beijing • Wuhan • Barcelona • Belgrade • Manchester • Tokyo • Cluj • Tianjin

Editor
Leonel Jorge Ribeiro Nunes
PROMETHEUS - Unidade de
Investigação em Materiais,
Energia e Ambiente para a
Sustentabilidade
Instituto Politécnico de Viana do
Castelo
Viana do Castelo
Portugal

Editorial Office
MDPI
St. Alban-Anlage 66
4052 Basel, Switzerland

This is a reprint of articles from the Special Issue published online in the open access journal *Recycling* (ISSN 2313-4321) (available at: www.mdpi.com/journal/recycling/special_issues/ Recycling_biomass_materials).

For citation purposes, cite each article independently as indicated on the article page online and as indicated below:

LastName, A.A.; LastName, B.B.; LastName, C.C. Article Title. *Journal Name* **Year**, *Volume Number*, Page Range.

ISBN 978-3-0365-1650-9 (Hbk)
ISBN 978-3-0365-1649-3 (PDF)

© 2021 by the authors. Articles in this book are Open Access and distributed under the Creative Commons Attribution (CC BY) license, which allows users to download, copy and build upon published articles, as long as the author and publisher are properly credited, which ensures maximum dissemination and a wider impact of our publications.
The book as a whole is distributed by MDPI under the terms and conditions of the Creative Commons license CC BY-NC-ND.

Contents

About the Editor . . . . **vii**

Leonel J. R. Nunes, Liliana M. E. F. Loureiro, Letícia C. R. Sá, João C. O. Matias, Ana I. O. F. Ferraz and Ana C. P. B. Rodrigues
Energy Recovery of Agricultural Residues: Incorporation of Vine Pruning in the Production of Biomass Pellets with ENplus® Certification
Reprinted from: *Recycling* **2021**, *6*, 28, doi:10.3390/recycling6020028 . . . . **1**

Leonel J. R. Nunes, Abel M. Rodrigues, Liliana M. E. F. Loureiro, Letícia C. R. Sá and João C. O. Matias
Energy Recovery from Invasive Species: Creation of Value Chains to Promote Control and Eradication
Reprinted from: *Recycling* **2021**, *6*, 21, doi:10.3390/recycling6010021 . . . . **21**

Leonel J. R. Nunes, Liliana M. E. F. Loureiro, Letícia C. R. Sá and Hugo F.C. Silva
Thermochemical Conversion of Olive Oil Industry Waste: Circular Economy through Energy Recovery
Reprinted from: *Recycling* **2020**, *5*, 12, doi:10.3390/recycling5020012 . . . . **39**

Lukas Jasiūnas, Thomas Helmer Pedersen and Lasse Aistrup Rosendahl
Biocrude Production via Non-Catalytic Supercritical Hydrothermal Liquefaction of *Fucus vesiculosus* Seaweed Processing Residues
Reprinted from: *Recycling* **2021**, *6*, 45, doi:10.3390/recycling6030045 . . . . **47**

Simone Di Piazza, Mirko Benvenuti, Gianluca Damonte, Grazia Cecchi, Mauro Giorgio Mariotti and Mirca Zotti
Fungi and Circular Economy: *Pleurotus ostreatus* Grown on a Substrate with Agricultural Waste of Lavender, and Its Promising Biochemical Profile
Reprinted from: *Recycling* **2021**, *6*, 40, doi:10.3390/recycling6020040 . . . . **65**

Mateus Manabu Abe, Marcia Cristina Branciforti and Michel Brienzo
Biodegradation of Hemicellulose-Cellulose-Starch-Based Bioplastics and Microbial Polyesters
Reprinted from: *Recycling* **2021**, *6*, 22, doi:10.3390/recycling6010022 . . . . **77**

Carlos Sánchez-Alvarracín, Jessica Criollo-Bravo, Daniela Albuja-Arias, Fernando García-Ávila and M. Raúl Pelaez-Samaniego
Characterization of Used Lubricant Oil in a Latin-American Medium-Size City and Analysis of Options for Its Regeneration
Reprinted from: *Recycling* **2021**, *6*, 10, doi:10.3390/recycling6010010 . . . . **99**

Felipe Arcos and Lina Uribe
Evaluation of the Use of Recycled Vegetable Oil as a Collector Reagent in the Flotation of Copper Sulfide Minerals Using Seawater
Reprinted from: *Recycling* **2021**, *6*, 5, doi:10.3390/recycling6010005 . . . . **121**

Tatyana Ponomareva, Maria Timchenko, Michael Filippov, Sergey Lapaev and Evgeny Sogorin
Prospects of Red King Crab Hepatopancreas Processing: Fundamental and Applied Biochemistry
Reprinted from: *Recycling* **2021**, *6*, 3, doi:10.3390/recycling6010003 . . . . **139**

Nana Kariada Tri Martuti, Isti Hidayah, Margunani Margunani and Radhitya Bayu Alafima
Organic Material for Clean Production in the Batik Industry: A Case Study of Natural Batik
Semarang, Indonesia
Reprinted from: *Recycling* **2020**, *5*, 28, doi:10.3390/recycling5040028 . . . . **155**

Kouwelton Kone, Karl Akueson and Graeme Norval
On the Production of Potassium Carbonate from Cocoa Pod Husks
Reprinted from: *Recycling* **2020**, *5*, 23, doi:10.3390/recycling5030023 . . . . **169**

**Idi Guga Audu, Abraham Barde, Othniel Mintang Yila, Peter Azikiwe Onwualu and Buga
Mohammed Lawal**
Exploring Biogas and Biofertilizer Production from Abattoir Wastes in Nigeria Using a
Multi-Criteria Assessment Approach
Reprinted from: *Recycling* **2020**, *5*, 18, doi:10.3390/recycling5030018 . . . . **175**

Hoi Nguyen Xa, Thanh Nguyen Viet, Khanh Nguyen Duc and Vinh Nguyen Duy
Utilization of Waste Cooking Oil via Recycling as Biofuel for Diesel Engines
Reprinted from: *Recycling* **2020**, *5*, 13, doi:10.3390/recycling5020013 . . . . **199**

About the Editor

Leonel Jorge Ribeiro Nunes

Leonel Jorge Ribeiro Nunes holds a PhD in Industrial Engineering and Management from the University of Beira Interior, a MSc in Hydraulics and Water Resources from the University of Lisbon, a MSc in Geological Engineering from the University NOVA of Lisbon, a degree in Geology from the University of Coimbra, a degree in Geology from the University of Minho, and a degree in Mining Engineering from the Faculty of Engineering of the University of Oporto. He is an Invited Assistant at the Agrarian Higher School of the Polytechnic Institute of Viana do Castelo. He lectures undergraduate and graduate courses of Natural Resources Management, Sustainability, Production Efficiency, Biomass Energy and Biomass Conversion Technologies. He is a full researcher at proMetheus - Research Unit in Materials, Energy and Environment for Sustainability. He has 150 scientific works published in books, chapters, articles, and conference proceedings.

Article

Energy Recovery of Agricultural Residues: Incorporation of Vine Pruning in the Production of Biomass Pellets with ENplus® Certification

Leonel J. R. Nunes [1,*], Liliana M. E. F. Loureiro [2], Letícia C. R. Sá [2], João C. O. Matias [3,4], Ana I. O. F. Ferraz [1] and Ana C. P. B. Rodrigues [1]

[1] PROMETHEUS, Unidade de Investigação em Materiais, Energia e Ambiente para a Sustentabilidade, Escola Superior Agrária, Instituto Politécnico de Viana do Castelo, Rua da Escola Industrial e Comercial de Nun'Alvares, 4900-347 Viana do Castelo, Portugal; aferraz@esa.ipvc.pt (A.I.O.F.F.); acrodrigues@esa.ipvc.pt (A.C.P.B.R.)
[2] YGE—Yser Green Energy SA, Área de Acolhimento Empresarial de Úl/Loureiro, Lote 17, 3720-075 Loureiro OAZ, Portugal; liliana.loureiro@ygenergia.com (L.M.E.F.L.); leticia.reis@ygenergia.com (L.C.R.S.)
[3] GOVCOPP, Unidade de Investigação em Governança, Competitividade e Políticas Públicas, Universidade de Aveiro, Campus Universitário de Santiago, 3810-193 Aveiro, Portugal; jmatias@ua.pt
[4] DEGEIT, Departamento de Economia, Gestão, Engenharia Industrial e Turismo, Universidade de Aveiro, Campus Universitário de Santiago, 3810-193 Aveiro, Portugal
* Correspondence: leonelnunes@esa.ipvc.pt

Citation: Nunes, L.J.R.; Loureiro, L.M.E.F.; Sá, L.C.R.; Matias, J.C.O.; Ferraz, A.I.O.F.; Rodrigues, A.C.P.B. Energy Recovery of Agricultural Residues: Incorporation of Vine Pruning in the Production of Biomass Pellets with ENplus® Certification. *Recycling* **2021**, *6*, 28. https://doi.org/10.3390/recycling6020028

Academic Editors: Junbeum Kim and Eugenio Cavallo

Received: 26 January 2021
Accepted: 21 April 2021
Published: 22 April 2021

Publisher's Note: MDPI stays neutral with regard to jurisdictional claims in published maps and institutional affiliations.

Copyright: © 2021 by the authors. Licensee MDPI, Basel, Switzerland. This article is an open access article distributed under the terms and conditions of the Creative Commons Attribution (CC BY) license (https://creativecommons.org/licenses/by/4.0/).

Abstract: The use of residual biomass of forest and/or agricultural origin is an increasingly common issue regarding the incorporation of materials that, until recently, were out of the typical raw material supply chains for the production of biomass pellets, mainly due to the quality constraints that some of these materials present. The need to control the quality of biomass-derived fuels led to the development of standards, such as ENplus®, to define the permitted limits for a set of parameters, such as the ash or alkali metal content. In the present study, samples of vine pruning, and ENplus®-certified pellets were collected and characterized, and the results obtained were compared with the limits presented in the standard. The values presented from vine pruning approximated the values presented by *Pinus pinaster* wood, the main raw material used in the production of certified pellets in Portugal, except for the values of ash, copper (Cu), and nitrogen (N) contents, with vine pruning being out of the qualifying limits for certification. However, it was found that the incorporation of up to 10% of biomass from vine pruning allowed the fulfillment of the requirements presented in the ENplus® standard, indicating a path for the implementation of circular economy processes in the wine industry.

Keywords: energy recovery of agricultural waste; biomass pellets; circular economy; ENplus®

1. Introduction

The use of biomass as a primary source of energy is currently an established reality, with a developed and regulated market in which products are evaluated according to quality criteria and compliance with parameters defined by regulatory processes such as certification standards [1,2]. The increasingly frequent use of solid biomass-derived fuels, as is the case with biomass pellets, has led to the development of standards regulating the physical–chemical parameters of the final product [3,4]. The development of the demand, supported by a regulatory instrument, conditioned the use of raw materials presenting parameters, mainly of chemical nature, out of normative requirements [5–7].

Initially, different standards appeared in different countries to regulate the criteria based on the regional availability of available raw materials, making it possible, when compared with each other, for the values required for the parameters to differ [8,9].

The methodologies used to perform the laboratory tests were also not uniform in the different standards, making it difficult to directly compare the results obtained for the same product, although certified by different standards [10]. For this reason, the standardization of criteria in a single standard, allowing direct comparison of products that started to have an increasingly wider global dispersion, became a necessity [11].

However, the need for this standard to present ranges of results for the different parameters, which are sufficiently extended to be used in densified materials produced with a diverse range of raw materials in different geographical locations, has already become evident. There is a high diversity of forest species that can be used in the production process and, although often related, they show variance in their chemical composition and physical structure [12,13]. For example, although they belong to the same genus *Pinus*, the species *Pinus radiata*, originally from the American continent, differs significantly from the species *Pinus pinaster*, common on the European Atlantic coasts from Portugal to England, and thus the biomass pellets produced also present significant differences from a chemical and combustibility point of view [14,15].

In this way, the standardization of the qualitative characterization criteria of biomass pellets through a single standard was a decisive step toward the stabilization of the product, since it led to homogenization of the production processes and the selection of a set of raw materials that fit the criteria defined by the new standard, ENplus® [16]. However, this regulation also came to limit the use of raw materials of waste origin reduced the quality of the final product and its market value, since producers opt mostly for products with a higher market value and commercial margins more interesting from a business perspective [17,18].

The use of residues from operations of forest management operations, as well as those resulting from agroindustrial operations, may result in the introduction of a significant volume of low-cost raw materials, provided they are properly characterized and studied so that this introduction takes place in a proportion that does not interfere with the product quality criteria defined by the norms that regulate the sector, as is the case with the ENplus® standard [19].

An example of this type of agroforestry waste is the material resulting from the pruning of vineyards, which is traditionally used as firewood in traditional domestic fireplaces and in bakery ovens all over the Mediterranean [20–22]. Currently, with the advent of the industrialization of processes and the exponential growth of the wine industry, the quantities of residual biomass resulting from such pruning reaches significant volumes. Thus, the incorporation of this residual biomass in industrial pellet production processes can be an opportunity, both from the perspective of reducing raw material costs, as well as from the environmental perspective of reducing the volume of waste, which is otherwise often eliminated by burning the remaining materials [23]. Although the available quantities are not known for certain, it is easy to infer the high potential that these waste products present, mainly due to the volume that can be produced annually. In Portugal, currently, there are about 200,000 hectares of vineyards, which can contribute in a very significant way to the supply of biomass residues to be recovered.

The objective of the present work is to characterize the residual material resulting from the pruning of vineyards in all aspects explained in the ENplus® standard, to make a comparison with the values presented by the pellets with ENplus® certification in such a way as to understand the existing differences, and then to determine the feasibility of incorporating vine pruning in the production of biomass pellets with ENplus® certification, as the only raw material, or partially, depending on the different types of pellets that the standard presents.

2. Materials and Methods

2.1. Sample Acquisition and Preparation

2.1.1. Sampling

For the characterization analysis of the pellets, two bags of biomass pellets were purchased in a large commercial area. The 15 kg bags were produced in Portugal and had the identification of being produced using exclusively *Pinus pinaster* wood. The bags were labeled as being pellets with ENplus® certification, which was confirmed on the website of the entity responsible for certification and available at: https://enplus-pellets.eu/pt/certificacoes-pt-pt/produtor-pt-pt.html (10 January 2021).

The two bags were mixed, and the quantities of material necessary for the characterization tests were subsequently removed. After collection, the pellets were ground to simulate the raw material used in their production. The biomass of vine pruning was collected during December 2020 in vineyards located in the Ponte de Lima region. The material was subsequently cut into portions with dimensions close to those of the pellets (Figure 1 to facilitate drying and grinding).

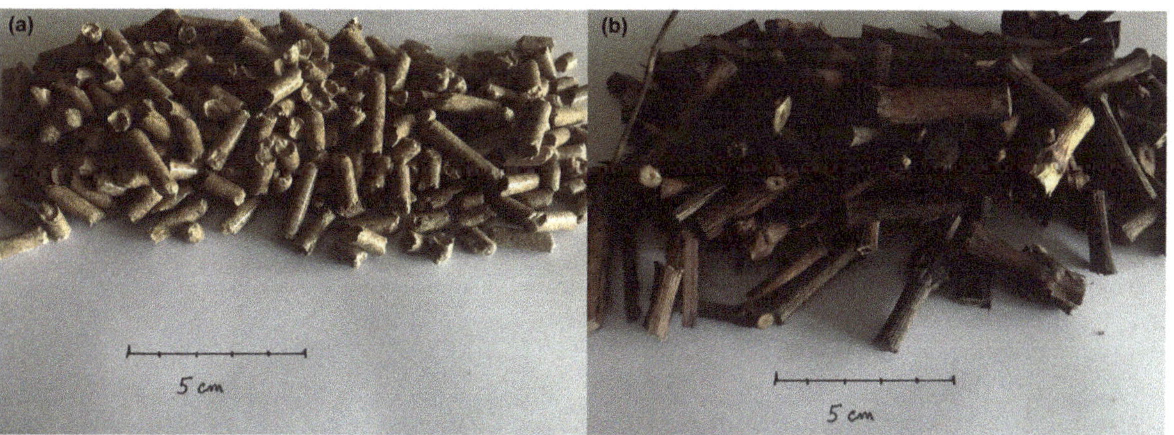

Figure 1. Samples collected for the characterization tests. (**a**) Pellets with ENplus® A1 certification; (**b**) vine pruning already cut to a size close to that of the pellets.

2.1.2. Preparation of Ash for Fusibility Tests

The preparation of samples for the ash fuse tests required a procedure necessary to ensure the production of a sufficient quantity to carry out the tests with replication to ensure statistical representation and treatment. In the present study, we decided to perform all tests in triplicate, to determine an average value used as a comparison with the values from the ENplus® standard, as well as providing the standard deviation of the sample.

The materials collected from vine pruning and from *Pinus pinaster* wood were first reduced and homogenized to a maximum nominal size of 1 mm, placed inside a crucible and taken to a muffle, where they were subjected to a combustion program appropriate to the requirements of the standard in question. In this case, according to the requirements of the ENplus® standard, the program used a heating ramp up to 250 °C over 40 min, and remained at that temperature for 1 h. Then, it reached 815 °C in a period of 1 h, where it remained for 2 h, followed by a cooling period for the samples to be removed from the interior of the muffle.

2.1.3. Sample Preparation for Inductively Coupled Plasma–Optical Emission Spectroscopy (ICP-OES) and Chlorine Determination

The procedure for preparing samples for ICP-OES, both ashes and precombustion material, and for determining the chlorine content, involved the digestion of the materials in two stages. For this purpose, the samples were reduced and homogenized to a maximum nominal size of 2 mm and subsequently mixed with reagents for the digestion of the materials. In the present study, a CEM Mars One microwaves digestion system was used. When the selected program ended, the cooling step to <100 °C began. After this step, we removed the sample holder carousel from the microwave and waited another 15 min before opening the containers and taking the samples. After cooling the samples, a volume of 4% H_3BO_3 was added to carry out the second phase of the digestion. In the second phase of digestion, the procedure began with placing the samples to be analyzed and the standard sample in a microwave where they were subjected to heating to a temperature of 150 °C in 5 min, maintained at this temperature for a period of 15 min, then cooled for 15 min. For methods A and B, each sample was transferred to a 50 mL volumetric flask through a paper filter (taking care to wash the container and lid walls with a 1% HNO_3 solution, as well as the filter paper itself), and making the balloons up to 50 mL before being homogenized. Subsequently, the contents of the flasks were transferred to 60 mL flasks for use in the ICP-OES. In method C, each sample was transferred to a 150 mL beaker, taking care to wash the walls of the container and lid with purified water. Then the digest was diluted to 100 mL, and the beaker was placed in a chloride titrator autosampler where the chlorine content was determined.

2.2. Heating Value

The calorific value indicates the amount of energy released during the combustion of a given amount of biomass. The calorific value of the biomass was determined using a calorimeter. This determination was made at a reference temperature of 30 °C and consisted of combustion of the biomass resulting in liquid water and carbon dioxide as products. The calculated value is defined as a higher calorific value, which includes the energy related to water vaporization (enthalpy of water vaporization; it only making sense to use these values if, during the process, the water vapor is then condensed). In processes in which the water vapor is eliminated, and its calorific value is not used, a lower heating value is used, calculated from the higher heating value and removing the value related to vaporization, i.e., the energy needed to vaporize the water is not considered as heat. A calorimeter is always composed of a combustion chamber where the sample is combusted. To ignite the sample, an electrical impulse is produced between two electrodes. For access to the interior of the combustion chamber, there is a cover which guarantees the tightness of the entire calorimeter. Surrounding the combustion chamber is a thermostatic bath that guarantees homogenization and temperature control through an agitator and a thermometer. The heat exchanges between the thermostated bath and the environment are controlled using a thermal shield. The procedure for determining the calorific value consists of recording temperature changes during the combustion process of a substance. The calorific capacity of a calorimeter indicates the amount of energy required to register a change in a temperature unit. During the combustion of a biomass sample with oxygen at high pressure, the nitrogen present in the atmosphere inside the combustion chamber is oxidized, producing nitrous oxide (NO_2), which will, in turn, be combined with water vapor, producing nitric acid (HNO_3). The heat derived from the formation of HNO_3 does not come from the sample and must be discounted in determining the calorific value. Thus, it is necessary to collect the washing water from the combustion chamber and holder with NaOH 0.1 M to correct the calorific value determined in the combustion. In combustion in an atmosphere rich in oxygen, the sulfur present in the atmosphere is oxidized to SO_3, which, in turn, is combined with water vapor resulting in sulfuric acid (H_2SO_4). In the combustion process, the sulfur should be completely transformed into SO_2, and thus it is necessary to correct the heat derived from the formation of H_2SO_4. For this, it is necessary to know the sulfur content present in the sample. For the determination of the calorific value, the samples should ideally be reduced to a maximum nominal size of 2 mm, and the same sample should be used to calculate the moisture content, the C, H, N, ash, S and Cl contents.

2.3. Elementary Analysis (CHNO)

The content of carbon, hydrogen and nitrogen are important factors in assessing the quality of biomass. Through the carbon content, information can be obtained regarding the amount of CO_2 emission during combustion. The nitrogen content can be used to obtain information regarding NO_x emission. The elemental composition of the samples was analyzed, using an elemental analyzer LECO CHN, in accordance with standard EN 15,104:2011, Solid Biofuels—Determination of the Total Content of Carbon, Hydrogen and Nitrogen—Instrumental Methods. The oxygen content was, thereafter, estimated using weight difference according to Equation (1):

$$w(O) = 100 - w(C) - w(H) - w(N) - w(S), \tag{1}$$

where $w(O)$ is the oxygen content (%), $w(C)$ is the carbon content (%), $w(H)$ is the hydrogen content (%), $w(N)$ is the nitrogen content (%), and $w(S)$ is the sulfur content (%).

2.4. Thermogravimetric Analysis (TGA)

Moisture content is a necessary characteristic for calculation of the various characteristics on a dry basis. The volatile content corresponds to the loss of mass (eliminating the contribution of the loss of mass of moisture) that occurs when biomass, in an inert atmosphere, is heated to a given temperature. The ash content of biomass corresponds to the content of noncombustible (inorganic) content. The quality of the ash depends on the biomass itself, and also on the biomass collection and pretreatment processes. The ash content values vary widely depending on the type of biomass. One of the reasons for using biomass materials with low ash contents is that they have higher calorific values, since there is less noncombustible material. Therefore, for the same biomass, a lower ash content corresponds to a greater calorific value. Another great reason for wanting a low ash content is the treatment and elimination of ash formed in energy conversion processes, such as combustion or pyrolysis. Ash can cause problems, such as slagging or fouling in furnaces, in which deposits of noncombustible inorganic material accumulate. In this way, the ash formed in these processes affects the costs of running and maintaining a factory, as ash is difficult to remove and can obstruct mechanical equipment. Fixed carbon is the portion of a biomass that is consumed in thermal processes, such as combustion. In this way, fixed carbon is a good indicator of the calorific value of biomass. Thermogravimetric analysis (TGA) is a method of thermal analysis in which the physical and chemical properties are determined as a function of temperature or as a function of time. In the present study, an Eltra ThermoStep equipment (Haan, Germany) was used, in which the samples were reduced and homogenized to a maximum nominal size of 1 mm.

2.5. Chemical Analysis by ICP-OES

The analysis of the major elements of the ash of biomass allows for comparison of the formation of ash in the process of thermal conversion according to the metal content. Most of the ash formed in biomass combustion processes is primarily made up of a mixture of oxides of Ca, Si, Mg, and Al. The relationship between these oxides is important because it influences the characteristics of the ash, such as the melting temperature and viscosity. The formation of CaO in combustion processes leads to an increase in the ash melting temperature, which implies a higher ash formation at the same temperature. In some cases, the formation of $CaCO_3$ can occur, which is responsible for the accumulation of material (fouling). SiO_2 is the main component involved in ash formation (slagging). Aluminum (Al) is not used in biological processes by the plant; therefore, its presence in high ash levels may indicate contamination of the biomass with soil. Magnesium (Mg) is a component of chlorophyll, which is present in the green parts of plants. The formation of MgO during the combustion of biomass has the same effect as CaO, i.e., increasing the melting temperature of ash. High concentrations of iron (Fe) indicate that the sample may have been contaminated with dirt. In coal combustion processes, iron causes slagging (ash formation at the base of the furnace). The presence of potassium (K) is usually common in fast-growing plants and its concentration depends on the seasonality of plant growth. In combustion processes, potassium remains volatile, and is an element that contributes to the emission of particles into the environment. Being a volatile element, this is one of the components that causes the formation of fouling (the clogging of pipes). When present in large concentrations, the melting temperature of the ash decreases, increasing the formation of slagging. The presence of sodium (Na) can occur naturally in plants originating from places by the sea. In large concentrations, it may indicate the presence of contamination. The effects of the presence of sodium are analogous to the presence of potassium: fouling, slagging and particle emissions. Sulfur (S) is an element causing air pollution through the formation of SO_2. In addition, vaporized sulfur can lead to the formation of K_2SO_4 particulate emissions. The presence of sulfur decreases the melting temperature of ash and increases the effects of fouling and slagging. The presence of chlorine (Cl) influences the emissions of HCl and KCl particles, which are corrosive substances. Chlorine decreases the melting temperature of the ash by increasing the formation of fouling, as it is a volatile element not found in large quantities at the bottom of the furnaces (slagging). Toxic metals, such as Cd, Pb, Zn, and Cr, are monitored due to their higher concentration in ash. Cd, Pb and Zn are partially volatile and participate in the emission of particles. Ti and Mn are elements typically found in very low concentrations. Their presence in higher concentrations may indicate contamination. For example, titanium (Ti) is a common element in paints and may indicate a mixture of biomass with wood waste with paint remnants. To determine the metal content in trace concentrations, ICP-OES (Inductively Coupled Plasma–Optical Emission Spectroscopy) was used, also called ICP-AES (Inductively Coupled Plasma–Atomic Emission Spectroscopy). ICP produces a high potential difference spark to transform argon gas into plasma. This potential difference produces ionized particles (electrons and ions) that are then accelerated by an applied magnetic field, which causes collisions with the neutral argon particles. These collisions cause a greater degree of ionization-producing plasma. The plasma state is maintained by the continuous collisions induced by the applied magnetic field and can reach temperatures of 10,000 °C. The sample is nebulized to the plasma and its components are immediately ionized. As the elements return to the ground state, electromagnetic radiation is released at specific wave lengths for each element. The wavelength of the electromagnetic radiation released and its intensity are detected by the OES analyzer, which allows for determination of the concentration of an element in a sample by comparison with standards of strictly known concentrations. For each sequence of analysis in an ICP, it is necessary to perform a standard calibration with the elements to be detected. Different dilutions of standards are required depending on the method of analysis to be performed and are listed in Table 1.

The above standards were prepared in 50 mL flasks, according to the following scheme shown in Figure 2.

Subsequently, the 50 mL vials were placed in the ICP-OES autosampler, in ascending order of concentration, depending on the method to be used, and depending on the type of material to be analyzed. When the concentration of a given element was found to be greater than the measurement range, the sample was reanalyzed by performing a 1:2 dilution (5 mL of the sample + 5 mL HNO_3 at 1%), taking care to carry out the same dilution for the digestion blank. For every 10 test tubes, two standards were placed to adjust the values acquired after the analysis (one standard with low concentration and another standard with high concentration).

Table 1. Patterns used to the different sets of elements to be determined using ICP-OES.

Elements to Analyze	Patterns and Respective Dilutions Used in Each Sequence
Al, Ba, Ca, Cd, Co, Cr, Cu, Fe, K, Mg, Mn, Na, Ni, Pb, Zn, Hg	- White pattern (1% HNO_3) - Multielements pattern 0.1 ppm - Multielements pattern 1 ppm - Multielements pattern 10 ppm - Multielements pattern 60 ppm - Multielements pattern 100 ppm
As, S, P, Ti, Si	- White pattern (1% HNO_3) - Pattern As, S, P, Ti 0.01 ppm - Pattern As, S, P, Ti 0.1 ppm - Pattern As, S, P, Ti 1 ppm - Pattern As, S, P, Ti 10 ppm - Pattern As, S, P, Ti 100 ppm - Pattern Si 0.6 ppm - Pattern Si 12 ppm - Pattern Si 80 ppm - Pattern Si 160 ppm
S	- White pattern (1% HNO_3) - Pattern S 0.01 ppm - Pattern S 0.1 ppm - Pattern S 1 ppm - Pattern S 10 ppm

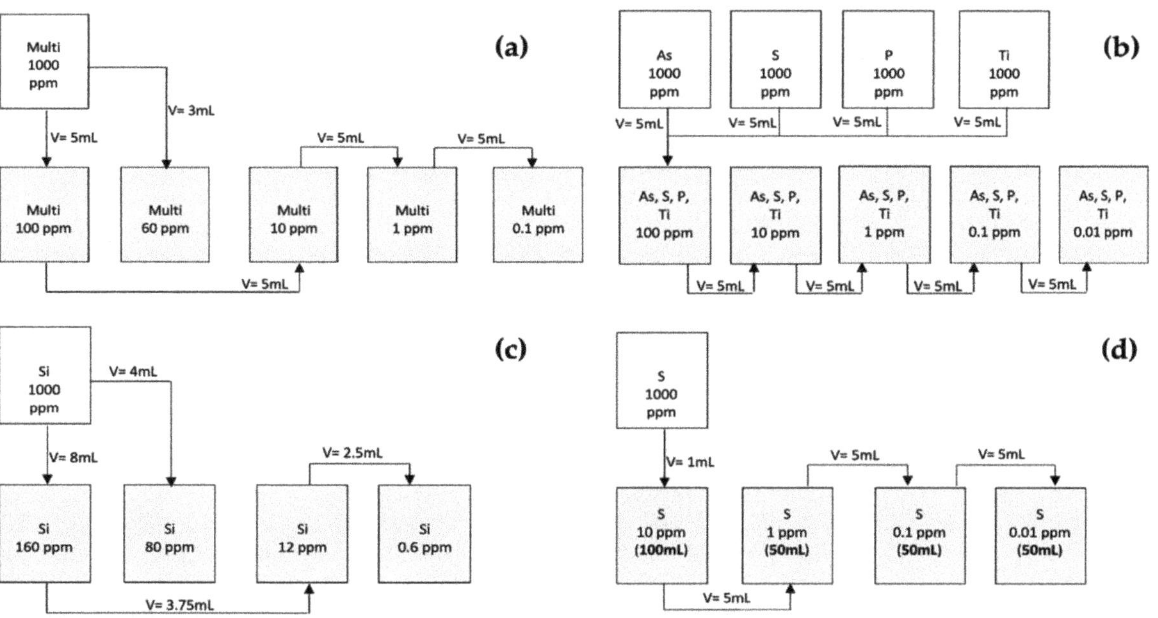

Figure 2. Preparation sequence of the solutions for determining the different elements using ICP-OES. (**a**) Multielements; (**b**) As, S, P, and Ti; (**c**) Si; and (**d**) S.

2.6. Fusibility of the Ashes

The ashes fusibility test can be used for the prediction of the formation of slagging and fouling in thermal conversion processes. These data must be related to the ash content (determined using the TGA) and the content of the different ash components (determined by ICP/OES). The fusibility test can be carried out with an oxidizing atmosphere (air) or reducing atmosphere (60% CO + 40% CO_2). The choice of atmosphere must be related to the combustion conditions of the boiler or burner. If the boiler operates in atmospheres rich in fuel (with an oxygen deficit), the atmosphere will be mostly reducing with incomplete combustion and CO formation. As a general rule, reducing atmospheres cause ash to melt at lower temperatures, thus, causing greater slagging and fouling problems. Therefore, the fuse test must reflect these characteristics and adapt to the customer's combustion process. During the fusibility test, the ash melting behavior was monitored and the following characteristic temperatures were determined.

- Initial temperature: temperature at which the test starts was up to 550 °C.
- Shrinkage temperature: shrinkage to 95% of the area recorded at 550 °C.
- Deformation temperature: temperature at which the first rounding of the vertices of the cylinders occurred.
- Hemisphere temperature: temperature at which the height of the cylinder was equal to half the width (h = L/2).
- Fluidity temperature: temperature at which the height was equal to half the height recorded at the hemisphere temperature (h = h (T_{hem})/2).

In the present study, the samples were converted to ashes according to the procedure described in Section 2.1.2. Subsequently, the ashes were placed in a plastic dish, where two drops of ethyl alcohol were added and, using a spatula, they were homogenized until a uniform paste was obtained. Then, this paste was transferred to the mold where the cylinder was compacted. After being removed from the mold, the cylinders were placed on the zirconia lamella. The samples were then placed inside the chamber of the ash fusibility furnace, which, in this specific case, was a SYLAB IF 2000-G device.

2.7. Determination of Chlorine Content

The determination of chloride content was conducted through a potentiometric titration, where an Ag–AgCl electrode (silver–silver chloride) with an internal KCl electrolyte (potassium chloride) was used. The electrode consisted of a filament of Ag (s) coated with AgCl (s). This filament was, in turn, in contact with an aqueous solution of KCl. A potentiometric titration consists of measuring a signal (potential difference) as titrant is added. The equivalence point is calculated by plotting the first derivative and identifying the volume that corresponds to the maximum of the first derivative. In the potentiometric titration of chlorides, an $AgNO_3$ solution of known concentration is used as the titrant. The oxidation-reduction reaction that occurs in this titration is presented in Equation (1):

$$Cl^-_{(aq.)} + AgNO_{3(aq.)} \rightarrow AgCl_{(s)} + NO^-_{3\,(aq.)}. \tag{2}$$

With the equivalence volume and the concentration value of the $AgNO_3$ solution, it is possible to calculate the concentration of chlorides (Cl^-) using a SI Analytics automatic titrator.

2.8. Fouling and Slagging Rates for Ash in Industrial Furnaces

Measurements of ash fuse temperatures aim to identify the behavior of the different types of compounds that make up the ash that forms during combustion processes, especially in an industrial environment given the quantities of materials that may be involved in the processes and the size of the equipment used. The damage caused by the occurrence of certain types of compounds, for example the elements belonging to the group of alkali metals, or chlorine, can cause considerable losses, since the natural corrosion and incrustation processes, related to the chemistry of combustion processes, can be exponentially enhanced and accelerated forcing technical stops for maintenance and repair of equipment. These phenomena of corrosion and scale that occur inside the combustion equipment, mainly in the furnace areas due to interaction with the bottom ash, but also in the areas where the heat exchange occurs due to the presence of fly ash and gases containing chlorine or sulfur, have been studied with regard the combustion processes for producing thermal energy in an industrial environment. The development of methodologies for the analysis of ash fusibility and its behavior started with the use of coal as an energy source, and all the indices that are currently used were derived from the analysis of coal ash to other solid fuels, such as biomass.

Most of the indices that are applied in the analysis of coal ash are based on the melting temperatures of the ash and its chemical composition, mainly on the ratios of acidic metal oxides, such as SiO_2 and Al_2O_3, in relation to basic metal oxides, such as Fe_2O_3, CaO, MgO, Na_2O and K_2O. These indices present a perspective of the ash fuse temperature, which then allows determination of the probability of the occurrence of corrosion, fouling, and slagging phenomena. For this reason, these indices are still widely used in industrial applications, mainly as tools for predicting potential damage and optimizing preventive maintenance, although their technical limitations are recognized. With the advent of the use of fuels derived from biomass in an industrial environment, as an alternative to the consumption of coal, the same indexes have been adapted for the new fuels. Most biomass materials have a strong presence of alkali metals, with K being the dominant element present in the bottom ash, while Na is in a form that enhances its volatilization and appears more associated with fly ash. Thus, the fouling and slagging indices developed for fuels derived from biomass are fundamentally based on the total levels of alkali metals present in the fuel. The indices used in the present study were based on the following equations.

1. Slagging Index (SI) represents the acid/base ratio used to quantify the tendency for the occurrence of slagging caused by ash from a fuel, and is determined numerically by Equation (3):

$$SI\left(\frac{B}{A}\right) = \frac{Fe_2O_3 + CaO + MgO + K_2O + Na_2O}{SiO_2 + TiO_2 + Al_2O_3}. \tag{3}$$

Equation (2) was initially developed for fossil fuels with low levels of phosphorus (P). Later, with the need to apply these methodologies also to fuels with high levels of P, another relationship developed, which is presented in Equation (3):

$$SI\left(\frac{B}{A} + P\right) = \frac{Fe_2O_3 + CaO + MgO + K_2O + Na_2O + P_2O_5}{SiO_2 + TiO_2 + Al_2O_3}, \tag{4}$$

where the $SI < 0.5$ indicates a low propensity for slagging to occur; $0.5 < SI < 1.0$ indicates an average propensity for slagging to occur; $SI = 1.0$ indicates a high propensity for slagging to occur and $SI > 1.75$ indicates a severe propensity for slagging to occur.

2. Fouling Index (FI) represents the propensity for the formation of fouling in industrial furnaces through the relation presented in Equation (5):

$$FI = \frac{B}{A} \times (K_2O + Na_2O), \tag{5}$$

where $FI < 0.6$ represents a low propensity for the formation of fouling; $0.6 < FI < 40$ represents a high propensity for the formation of fouling and $FI > 40$ represents a severe propensity for the formation of fouling.

3. Alkali Index (AI): this index represents and expresses the quantity of alkaline oxides per unit of energy, and is determined using Equation (6):

$$AI = \frac{1 \times 10^6 \times Ash\ (\%) \times (K_2O + Na_2O)\ (\%)}{HHV\ (kJ/kg)}, \tag{6}$$

where $AI < 0.17$ indicates a low propensity for the occurrence of slagging and fouling phenomena; $0.17 < AI < 0.34$ indicates a high propensity for the occurrence of fouling and slagging phenomena and $AI > 0.34$ indicates a severe propensity for the occurrence of slagging and fouling phenomena.

3. Results

3.1. Samples of Vine Pruning

3.1.1. Heating Value

The values obtained for the high calorific value varied between 18.94 MJ/kg and 18.95 MJ/kg, and the average value calculated for the three samples was 18.95 ± 0.002 MJ/kg with a standard deviation of 0.002 MJ/kg. As expected, the low calorific value also showed very close values, with an average of 17.58 ± 0.002 MJ/kg and a standard deviation of 0.002 MJ/kg.

3.1.2. Elementary Analysis

The results obtained for the elementary analysis are shown in Table 2.

Table 2. Elemental analysis (CHNO).

	C (%)	H (%)	N (%)	O (%)
Sample 1	47.12	6.36	0.710	45.79
Sample 2	46.48	6.42	0.616	46.47
Sample 3	45.23	6.05	0.280	48.42
Average	46.28	6.28	0.540	46.89
Standard deviation	0.96	0.20	0.230	1.37
Confidence	1.09	0.22	0.26	1.55

The value of the C content fluctuated between a minimum of 45.23% and a maximum of 47.12%, with an average value of $46.28 \pm 1.09\%$ and a standard deviation of 0.96%. H varied between a minimum value of 6.05% and a maximum value of 6.42%, with an average value of $6.28 \pm 0.22\%$ and a standard deviation of 0.20%. N had a minimum value of 0.280% and a maximum value of 0.710%, with an average value of $0.540 \pm 0.26\%$ and a standard deviation of 0.230%. O oscillated between the minimum value of 45.79% and the maximum value of 48.42%, with an average value of $46.89 \pm 1.55\%$ and a standard deviation of 1.37%.

3.1.3. Thermogravimetric Analysis

The results obtained for the thermogravimetric analysis are shown below in Table 3.

Table 3. Thermogravimetric analysis (moisture, volatiles, ashes, and fixed carbon content).

	Moisture (%)	Volatiles (%)	Ashes (%)	Fixed Carbon (%)
Sample 1	3.74	77.83	1.42	19.51
Sample 2	3.77	77.84	1.41	19.46
Sample 3	3.51	77.72	1.42	19.78
Average	3.67	77.80	1.42	19.58
Standard deviation	0.14	0.07	0.01	0.17
Confidence	0.16	0.08	0.01	0.20

The moisture of the samples of vine pruning after drying varied between the minimum value of 3.51% and the maximum value of 3.77%, with an average value of $3.67 \pm 0.16\%$ and a standard deviation of 0.14%. The volatile content fluctuated between the minimum value of 77.72% and the maximum value of 77.84%, with an average value of $77.80 \pm 0.08\%$ and a standard deviation of 0.07%. The ash content was between the minimum value of 1.41% and a maximum value of 1.42%, with an average value of $1.42 \pm 0.01\%$ and a standard deviation of 0.01%. The fixed carbon content varied between the minimum value of 19.16% and the maximum value of 19.78%, with an average value of 19.18 ± 0.20% and a standard deviation of 0.17%.

3.1.4. Chemical Analysis by ICP-OES

Major Elements

The results obtained using the ICP-OES for the major elements are shown in Table 4.

Table 4. Major elements obtained using ICP-OES.

	Al (mg/kg)	Ca (mg/kg)	Fe (mg/kg)	Mg (mg/kg)	P (mg/kg)	K mg/kg)	Si (mg/kg)	Na (mg/kg)	Ti (mg/kg)
Sample 1	74.5	6972.5	31.2	1303.7	1101.9	8120.5	188.8	407.3	2.7
Sample 2	53.7	7722.4	38.6	1386.4	1077.5	8444.4	137.5	413.7	3.5
Sample 3	53.4	7641.2	32.3	1387.8	1107.9	8168.2	131.7	444.4	2.8
Average	60.6	7445.4	34.0	1359.3	1095.8	8244.4	152.7	421.8	3.0
Standard deviation	12.1	411.5	4.0	48.1	16.1	174.9	31.5	19.8	0.4
Confidence	13.71	465.69	4.52	54.46	18.26	197.87	35.60	22.45	0.47

The Al content varied between a minimum value of 53.4 mg/kg and a maximum value of 74.5 mg/kg, with an average value of 60.6 ± 13.71 mg/kg and a standard deviation of 12.1 mg/kg. Ca varied between the minimum value of 6972.5 mg/kg and the maximum value of 7722.4 mg/kg, with an average value of 7445.4 ± 465.69 mg/kg and a standard deviation of 411.5 mg/kg. Fe varied between the minimum value of 31.2 mg/kg and the maximum value of 38.6 mg/kg, with an average value of 34.0 ± 4.52 mg/kg and a standard deviation of 4.0 mg/kg. Mg varied between the minimum value of 1303.7 mg/kg and the maximum value 1387.8 mg/kg, with an average value of 1359.3 ± 54.46 mg/kg and a standard deviation of 48.1 mg/kg. P oscillated between a minimum value of 1077.5 mg/kg and a maximum value of 1107.9 mg/kg, with an average value of 1095.8 ± 18.26 mg/kg and a standard deviation of 16.1 mg/kg. K had a minimum value of 8120.4 mg/kg and a maximum value of 8444.4 mg/kg, with an average value of 8244.4 ± 197.87 mg/kg and a standard deviation of 174.9 mg/kg. Si varied between the minimum value of 131.7 mg/kg and the maximum value of 188.8 mg/kg, with an average value of 152.7 ± 35.60 mg/kg and a standard deviation of 31.5 mg/kg. Na showed a minimum value of 407.3 mg/kg and a maximum value of 444.4 mg/kg, with an average value of 421.8 ± 22.45 mg/kg and a standard deviation of 19.8 mg/kg. Ti showed a minimum value of 2.7 mg/kg and a maximum value of 3.5 mg/kg, with an average value of 3.0 ± 0.47 mg/kg and a standard deviation of 0.4 mg/kg.

Minor Elements

The results obtained using the ICP-OES for the minor elements are those shown in Table 5.

Table 5. Minor elements obtained using ICP-OES.

	As (mg/kg)	Cd (mg/kg)	Co (mg/kg)	Cr (mg/kg)	Cu (mg/kg)	Mn (mg/kg)	Ni (mg/kg)	Pb (mg/kg)	Zn (mg/kg)
Sample 1	0.58	0.58	0.40	0.48	23.16	39.64	0.63	0.29	14.51
Sample 2	0.88	0.34	0.20	0.34	26.48	43.44	0.41	0.10	14.53
Sample 3	0.73	0.39	0.32	0.04	25.15	40.95	0.13	0.09	78.40
Average	0.73	0.44	0.31	0.29	24.93	41.34	0.39	0.16	35.81
Standard deviation	0.15	0.13	0.10	0.22	1.67	1.93	0.25	0.11	36.88
Confidence	0.17	0.14	0.11	0.25	1.89	2.18	0.28	0.13	41.73

As had a minimum value of 0.58 mg/kg and a maximum value of 0.88 mg/kg, with an average value of 0.73 $\pm$ 0.17 mg/kg and a standard deviation of 0.15 mg/kg. Cd showed a minimum value of 0.34 mg/kg and a maximum value of 0.58 mg/kg, with an average value of 0.44 $\pm$ 0.14 mg/kg and a standard deviation of 0.13 mg/kg. Co had a minimum value of 0.20 mg/kg and a maximum value of 0.40 mg/kg, with an average value of 0.31 $\pm$ 0.11 mg/kg and a standard deviation of 0.10 mg/kg. Cr had a minimum value of 0.04 mg/kg and a maximum value of 0.48 mg/kg, with an average value of 0.29 $\pm$ 0.25 mg/kg and a standard deviation of 0.22 mg/kg. Cu varied between the minimum value of 23.16 mg/kg and the maximum value of 26.48 mg/kg, with an average value of 24.93 $\pm$ 1.89 mg/kg and a standard deviation of 1.67 mg/kg. Mn varied between a minimum value of 39.64 mg/kg and a maximum value of 43.44 mg/kg, with an average value of 41.34 $\pm$ 2.18 mg/kg and a standard deviation of 1.93 mg/kg. Ni showed a minimum value of 0.13 mg/kg and a maximum value of 0.63 mg/kg, with an average value of 0.39 $\pm$ 0.28 mg/kg and a standard deviation of 0.25 mg/kg. Pb varied between the minimum value of 0.09 mg/kg and the maximum value of 0.29 mg/kg, with an average value of 0.16 $\pm$ 0.13 mg/kg and a standard deviation of 0.11 mg/kg. Zn had a minimum value of 14.51 mg/kg and a maximum value of 78.40 mg/kg, with an average value of 35.81 $\pm$ 41.73 mg/kg and a standard deviation of 36.88 mg/kg. Hg showed a value below the detection limit of the equipment used in the present study and, therefore, was presented as $\leq$0.1 mg/kg.

Determination of the S and Cl Content

The S content varied between a minimum value of 0.0181% and a maximum value of 0.0190%, with an average value of 0.0185 $\pm$ 0.0005% and a standard deviation of 0.00046%. The Cl content varied between a minimum value of 0.0001% and a maximum value of 0.0008%, with an average value of 0.0005 $\pm$ 0.0004% and a standard deviation of 0.0003%.

3.1.5. Analysis of the Ash Fusibility

The temperatures determined for the ash fusibility test are shown in Table 6.

Table 6. Ash fusibility temperature analysis: initial deformation temperature, softening temperature, hemisphere temperature, and flow temperature.

	Initial Deformation Temperature (°C)	Softening Temperature (°C)	Hemisphere Temperature (°C)	Flow Temperature (°C)
Sample 1	888	1476	1572	1581
Sample 2	858	1570	1579	1588
Sample 3	901	1534	1578	1583
Average	882	1527	1576	1584
Standard deviation	22	47	4	4
Confidence	25	54	4	4

The results obtained for the shrinking temperature varied between the minimum value of 858 °C and the maximum value of 901 °C, with an average value of 882 $\pm$ 25 °C and a standard deviation of 22 °C. The strain temperature showed a minimum value of 1476 °C and a maximum value of 1570 °C, with an average value of 1527 $\pm$ 54 °C and a standard deviation of 47 °C. The hemisphere temperature varied between the minimum value of 1572 °C and the maximum value of 1579 °C, with an average value of 1576 $\pm$ 4 °C and a standard deviation of 4 °C. The fluidity temperature varied between the minimum value of 1581 °C and the maximum value of 1588 °C, with an average value of 1584 $\pm$ 4 °C and a standard deviation of 4 °C.

3.2. Characterization of the Raw Material of Pinus Pinaster Wood Pellets with ENPlus Certification

3.2.1. High and Low Heating Value

The high calorific value showed a minimum value of 19.02 MJ/kg and a maximum value of 19.79 MJ/kg, with an average value of 19.35 ± 0.45 MJ/kg and a standard deviation of 0.40 MJ/kg. The low calorific value varied between the minimum value of 17.20 MJ/kg and the maximum value of 18.47 MJ/kg, with an average value of 17.87 ± 0.72 MJ/kg and a standard deviation of 0.64 MJ/kg.

3.2.2. Elemental Analysis

The results obtained for the elemental analysis are presented in Table 7.

Table 7. *Pinus pinaster* wood elemental analysis.

	C (%)	**H (%)**	**N (%)**	**O (%)**
Sample 1	50.04	6.07	0.075	43.81
Sample 2	50.22	5.96	0.062	43.75
Sample 3	50.38	6.17	0.093	43.35
Average	50.21	6.07	0.08	43.64
Standard deviation	0.17	0.10	0.02	0.25
Confidence	0.19	0.12	0.02	0.28

C varied between a minimum value of 50.04% and a maximum value of 50.38%, with an average value of 50.21 ± 0.19% and a standard deviation of 0.17%. H varied between a minimum value of 5.96% and a maximum value of 6.17%, with an average value of 6.07 ± 0.12% and a standard deviation of 0.10%. N varied between the minimum value 0.062% and the maximum value of 0.093%, with an average value of 0.080 ± 0.02% and a standard deviation of 0.020%. O varied between the minimum value of 43.35% and the maximum value of 43.81%, with an average value of 43.64 ± 0.28% and a standard deviation of 0.25%.

3.2.3. Thermogravimetric Analysis (TGA)

The results obtained for the thermogravimetric analysis are presented in Table 8.

Table 8. *Pinus pinaster* wood thermogravimetric analysis.

	Moisture (%)	**Volatiles (%)**	**Ashes (%)**	**Fixed Carbon (%)**
Sample 1	6.42	80.65	0.72	18.64
Sample 2	6.39	81.19	0.67	18.14
Sample 3	6.44	80.76	0.63	18.61
Average	6.42	80.87	0.67	18.46
Standard deviation	0.02	0.28	0.04	0.28
Confidence	0.03	0.32	0.01	0.31

The humidity present in the Pinus pinaster wood, as received, varied between a minimum value of 6.31% and a maximum value of 6.44%, with an average value of 6.42 ± 0.03% and a standard deviation of 0.02. The volatile content varied between a minimum value of 80.65% and a maximum value of 81.19%, with an average value of 80.87 ± 0.32% and a standard deviation of 0.28%. The ash content varied between a minimum value of 0.63% and a maximum value of 0.72%, with an average value of 0.67 ± 0.01% and a standard deviation of 0.04%. The fixed carbon content varied between a minimum value of 18.14% and a maximum value of 18.64%, with an average value of 18.46 ± 0.31% and a standard deviation of 0.28%.

3.2.4. ICP-OES Chemical Analysis of Pinus Pinaster Wood

Major Elements

The results obtained for the chemical analysis are presented in Table 9.

Table 9. *Pinus pinaster* major elements (mg/kg) obtained using ICP-OES.

	Al (mg/kg)	Ca (mg/kg)	Fe (mg/kg)	Mg (mg/kg)	P (mg/kg)	K (mg/kg)	Si (mg/kg)	Na(mg/kg)	Ti (mg/kg)
Sample 1	277.73	1614.60	256.60	638.34	73.27	672.11	896.04	365.07	17.82
Sample 2	317.93	1539.69	273.16	614.03	81.04	737.10	1010.71	421.34	19.12
Sample 3	326.06	1781.17	275.37	683.48	79.92	762.32	1073.16	523.96	20.73
Average	307.24	1645.15	268.37	645.28	78.08	723.84	993.31	436.79	19.22
Standard deviation	25.88	123.61	10.26	35.24	4.20	46.55	89.84	80.56	1.46
Confidence	29.28	139.87	11.61	39.88	4.75	52.67	101.66	91.17	1.65

The Al content varied between a minimum value of 277.73 mg/kg and a maximum value of 326.06 mg/kg, with an average value of 307.24 ± 29.28 mg/kg and a standard deviation of 25.88 mg/kg. Ca varied from a minimum value of 1539.69 mg/kg to a maximum value of 1781.17 mg/kg, with an average value of 1645.15 ± 139.87 mg/kg and a standard deviation of 123.61 mg/kg. Fe varied from a minimum of 256.60 mg/kg to a maximum of 275.37 mg/kg, with an average value of 268.37 ± 11.61 mg/kg and a standard deviation of 10.26 mg/kg. Mg varied from a minimum of 614.03 mg/kg to a maximum of 683.48 mg/kg, with an average value of 645.28 ± 39.88 mg/kg and a standard deviation of 35.24 mg/kg. P ranged from a minimum value of 73.27 mg/kg to a maximum value of 81.04 mg/kg, with an average value of 78.08 ± 4.75 mg/kg and a standard deviation of 4.20 mg/kg. K ranged from a minimum value of 672.11 mg/kg to a maximum value of 762.32 mg/kg, with an average value of 723.84 ± 52.67 mg/kg and a standard deviation of 46.55 mg/kg. Si fluctuated between the minimum value of 896.04 mg/kg and a maximum value of 1073.16 mg/kg, with an average value of 993.31 ± 101.66 mg/kg and a standard deviation of 89.84 mg/kg. Na varied from a minimum value of 365.07 mg/kg to a maximum value of 523.96 mg/kg, with an average value of 436.70 ± 91.17 mg/kg and a standard deviation of 80.56 mg/kg. Ti varied from a minimum value of 17.82 mg/kg to a maximum value of 20.73 mg/kg, with an average value of 19.22 ± 1.65 mg/kg and a standard deviation of 1.46 mg/kg.

Minor Elements

The results of the determination of the minor elements carried out using the ICP-OES to *Pinus pinaster* wood are shown in Table 10.

Table 10. *Pinus pinaster* minor elements (mg/kg) obtained using ICP-OES.

	As (mg/kg)	Cd (mg/kg)	Co (mg/kg)	Cr (mg/kg)	Cu (mg/kg)	Mn (mg/kg)	Ni (mg/kg)	Pb (mg/kg)	Zn (mg/kg)
Sample 1	0.89	0.35	0.44	1.94	1.89	36.84	1.20	0.27	8.07
Sample 2	1.02	0.32	0.14	2.20	3.84	39.51	0.98	0.32	7.83
Sample 3	0.90	0.34	0.07	1.84	4.92	38.96	1.05	1.55	8.33
Average	0.94	0.34	0.22	1.99	3.55	38.44	1.08	0.71	8.08
Standard deviation	0.07	0.01	0.20	0.19	1.53	1.41	0.11	0.72	0.25
Confidence	0.08	0.02	0.22	0.21	1.73	1.59	0.12	0.82	0.29

The As content varied between the minimum value of 0.89 mg/kg and the maximum value of 1.02 mg/kg, with an average value of 0.94 ± 0.08 mg/kg and a standard deviation of 0.07 mg/kg. Cd varied between the minimum value of 0.32 mg/kg and the maximum value of 0.35 mg/kg, with an average value of 0.34 ± 0.02mg/kg and a standard deviation of 0.01 mg/kg. Co varied between a minimum value of 0.07 mg/kg and a maximum value of 0.44 mg/kg, with an average value of 0.22 ± 0.22 mg/kg and a standard deviation of 0.20 mg/kg. Cr varied between a minimum value of 1.84 mg/kg and a maximum value of 2.20 mg/kg, with an average value of 1.99 ± 0.21 mg/kg and a standard deviation of 0.19 mg/kg. Cu varied between a minimum value of 1.89 mg/kg and a maximum value of 4.92 mg/kg, with an average value of 3.55 ± 1.73 mg/kg and a standard deviation of 1.53 mg/kg. Mn varied between the minimum value of 36.84 mg/kg and the maximum value of 39.51 mg/kg, with an average value of 38.44 ± 1.59 mg/kg and a standard deviation of 1.41 mg/kg. Ni varied between a minimum value of 0.98 mg/kg and a maximum value of 1.20 mg/kg, with an average value of 1.08 ± 0.12 mg/kg and a standard deviation of 0.11 mg/kg. Pb varied between a minimum value of 0.27 mg/kg and a maximum value of 1.55 mg/kg, with an average value of 0.71 ± 0.82 mg/kg and a standard deviation of 0.72 mg/kg. Zn varied between a minimum value of 8.07 mg/kg and a maximum value of 8.33 mg/kg, with an average value of 8.08 ± 0.29 mg/kg and a standard deviation of 0.25 mg/kg. Hg showed a value below the detection limit of the equipment used in the present study, and, therefore, it was presented as ≤0.1 mg/kg.

Determination of the S and Cl Content

The S content determined using the ICP-OES varied between a minimum value of 0.0039% and a maximum value of 0.0055%, with an average value of 0.0045 ± 0.001% and a standard deviation of 0.0008%. The chlorine content varied between a minimum value of 0.012% and a maximum value of 0.021%, with an average value of 0.020 ± 0.005% and a standard deviation of 0.005%.

3.2.5. Ash Fusibility

The temperatures determined for the ash fusibility test obtained by burning *Pinus pinaster* wood are summarized in Table 11.

Table 11. Ash fusibility temperature analysis: initial deformation temperature, softening temperature, hemisphere temperature, and flow temperature.

	Initial Deformation Temperature (°C)	Softening Temperature (°C)	Hemisphere Temperature (°C)	Flow Temperature (°C)
Sample 1	867	1212	1226	1234
Sample 2	869	1215	1227	1243
Sample 3	901	1214	1227	1239
Average	879	1214	1227	1239
Standard deviation	19	2	1	5
Confidence	22	2	1	5

The results obtained for the shrinking temperature varied between a minimum temperature of 867 °C and a maximum temperature of 901 °C, with an average temperature of 879 ± 22 °C and a standard deviation of 19 °C. The deformation temperature varied between a minimum temperature of 1212 °C and a maximum temperature of 1215 °C, with an average temperature of 1214 ± 2 °C and a standard deviation of 2 °C. The temperature of the hemisphere varied between a minimum temperature of 1226 °C and a maximum temperature of 1227 °C, with an average temperature of 1227 ± 1 °C and a standard deviation of 1 °C. The fluidity temperature varied between a minimum temperature of 1234 °C and a maximum temperature of 1243 °C, with an average temperature of 1239 ± 5 °C and a standard deviation of 5 °C.

3.3. Determination of the Fouling and Slagging Propensity Indices

The values determined for the fouling and slagging indices, namely through the calculation of the Slagging indexes (B/A and B/A + P), Fouling index (FI), and Alkali index (AI), are shown in Table 12.

Table 12. The Fouling and Slagging indexes for samples of vine pruning and *Pinus pinaster*.

	Vine Pruning		*Pinus pinaster*	
Slagging Index (B/A)	1.94	Severe	1.89	Severe
Slagging Index (B/A + P)	2.05	Severe	1.96	Severe
Fouling Index (FI)	0.28	Low	0.28	Low
Alkali Index (AI)	20.48	Severe	4.78	Severe

For the slagging index (B/A), values of 1.94 and 1.89 were obtained, respectively, for samples of vine pruning and for *Pinus pinaster* wood. For index slagging (B/A + P), the values were 2.05 and 1.96, a value of 0.28 was obtained both for samples of vine pruning and for wood samples of *Pinus pinaster*. For the alkali index, the value for samples of vine pruning was 20.48, while, for *Pinus pinaster* wood, the value was 4.78.

4. Discussion

The analysis of the physical–chemical parameters of the pellets acquired with ENplus® certification proved that the referred pellets fulfilled all the requirements presented by the standard. Table 13 shows the results obtained, both for *Pinus pinaster* wood pellets and for the biomass resulting from vine pruning, as well as the parameters defined by the ENplus® standard [17].

Table 13. Parameters defined by the ENplus® standard and the results obtained for *Pinus pinaster* wood pellets and the biomass of vine pruning.

		ENplus®			*Pinus pinaster*	**Vine Pruning**
		A1 *	**A2 ***	**B ***	**ENplus® A1**	
Moisture	%		≤10		6.42	3.67
Ashes	%	≤0.7	≤1.2	≤2	0.62	1.42
LHV	MJ/kg		≥16.50		17.87	17.58
N	%	≤0.3	≤0.5	≤1.0	0.08	0.536
S	%	≤0.04		≤0.05	0.0045	0.0185
Cl	%	≤0.02		≤0.03	0.02	0.0005
Softening Temperature	°C	≥1200		≥1100	1215	1570
As	mg/kg		≤1		0.94	0.73
Cd	mg/kg		≤0.5		0.34	0.44
Cr	mg/kg		≤10		1.99	0.29
Cu	mg/kg		≤10		3.55	24.93
Pb	mg/kg		≤10		0.71	0.16
Hg	mg/kg		≤0.1		≤0.01	≤0.01
Ni	mg/kg		≤10		1.08	0.39
Zn	mg/kg		≤100		8.08	35.81

* The categories A1, A2 and B correspond to different levels of quality of pellets with ENPlus® certification, and are related to the different types of raw materials used in their production.

As can be seen in the results presented in Table 13, the values obtained for the biomass of vine pruning are in agreement with the requirements of the ENplus® standard, with the exception of the parameters corresponding to the ash content, where the average value obtained was 1.42%, the N content, where the average value obtained was 0.536% and the Cu content, where the average value obtained was 24.93 mg/kg.

Despite being outside the requirements presented by the standard for ash content, making it impossible to use for the production of pellets in categories A1 and A2, the ash content determined had significantly lower values compared with the characterizations carried out in other previous studies, such as the study by Zanetti et al. (2017) where the values varied between 3.3% and 5.5% [24].

This difference may be related to the fact that there may be differences between the compositions of the different varieties, and also with the influence of the types of soils and the method of harvesting during pruning, since the method used to collect the samples for the present study was that of manual collection, directly from the grapevine. Therefore, there was no contamination with the soil through mixing with inert materials such as earth, stones, or sand. Regarding the N content, the value presented is in line with the values presented by Zanetti et al. (2017), which varied between 0.560% and 0.640%. That is, values close to 0.536% were determined for the samples analyzed in the present study. The Cu content had a value significantly higher than that presented in the study referred to above, which reported values ranging between 13.1 and 16.3 mg/kg. The value of 24.93 mg/kg may indicate the presence of the remains of a Bordeaux mixture. A combination of copper sulfate, lime and water used as fungicide and bactericide in the vineyards, which when mixed properly, provides a long-lasting protection against diseases. However, this statement requires confirmation through further analysis and monitoring of the vineyard where the samples were collected.

The occurrence of a set of elements, namely those belonging to the group of alkali metals, such as Na or K, with levels that may indicate a low melting temperature of ash, has been described in several previous works. An example of this is the work of Niue et al., (2010), which presented the analysis of residual biomasses of agricultural origin, such as capsicum stalks, cotton stalks and wheat stalks, with the ash prepared by calcining the material at 400, 600, and 815 °C [25]. In that study, there was a tendency for the occurrence of low ash melting temperatures, closely related to the occurrence of high levels of alkali metals and also to other elements, such as Ca, which is a recognized melting agent [25,26].

Other works, such as those presented by Ma et al., (2016), Wang et al., (2017), Rizvi et al., (2015) and Li et al., (2019), analyzed the relationships between the contents of various elements with the melting temperatures of the ash and with the behavior of these melting materials and their chemical and structural reorganization [27–30]. Another way of assessing the impact of ash fusion was described in several studies, which resorted to the transposition of analysis methodologies that are common in coal science through the adaptation of indices that relate the different constituents to each other, allowing a qualitative assessment of the potential for the occurrence of fouling or slagging phenomena. Examples of this include the works of Yao et al., (2017), Lee et al., (2018), Yao et al., (2020), Ruscio et al., (2016), and Yao et al., (2020) [31–35]. The results obtained in the determination of fouling and slagging prediction indices in the present study, described in Section 2.8, indicated a severe tendency for the occurrence of slagging processes, both for vine pruning and for wood pellets of *Pinus pinaster*. This is most likely due to the presence of Na_2O, which contribute to the sintering of bottom ash, while K_2O contributes in a greater proportion to the potential occurrence of fouling processes but, due to the relationships with other present compounds, shows a low tendency for the occurrence of this phenomenon.

This is an issue of increasing importance, particularly for large-scale uses, where the amounts of these compounds involved can gain significant weight, since the potential damage caused may lead to unplanned stops in energy conversion equipment. In the case of smaller equipment, such as domestic equipment, this problem is not important and there is an excellent possibility for reducing production costs by the inclusion of residual materials [36,37]. The high deformation temperatures presented by the samples of vine pruning, indicate that their use is not a problem, as they were significantly higher than those observed for the *Pinus pinaster* wood samples.

From the results obtained for the characterization of the samples of vine pruning, with the exception of the situations described in the previous paragraphs, there is a strong probability of making mixtures to incorporate biomass from vine pruning for the production of pellets with ENplus® certification. Table 14 presents the calculated results for the main physical–chemical parameters resulting from the mixtures between percentages of biomass from vine pruning with *Pinus pinaster* wood. As can be seen, for an incorporation of 50% of biomass from vine pruning, the Cu content precludes certification in any of the types (A1, A2, and B). However, for the incorporation of 25%, Cu has values within the permitted ranges, including for type A1, which is not achieved due to the ash content, but is still higher than 0.7%, and which only allows type A2 certification. For incorporations of 10%, all parameters fall within type A1.

Table 14. Mixtures of *Pinus pinaster* wood and vine pruning (PP% + VP%).

		ENPlus®			Mixtures		
		A1	A2	B	50% + 50%	75% + 25%	90% + 10%
Moisture	%		≤10		5.04	5.73	6.14
Ashes	%	≤0.7	≤1.2	≤2	1.02	0.82	0.70
LHV	MJ/kg		≥16.50		17.72	17.79	17.84
N	%	≤0.3	≤0.5	≤1.0	0.31	0.19	0.12
S	%	≤0.04		≤0.05	0.012	0.008	0.006
Cl	%	≤0.02		≤0.03	0.01	0.01	0.02
Softening Temperature	°C	≥1200		≥1100	1393	1304	1251
As	mg/kg		≤1		0.83	0.89	0.92
Cd	mg/kg		≤0.5		0.39	0.36	0.35
Cr	mg/kg		≤10		1.14	1.56	1.82
Cu	mg/kg		≤10		14.24	8.90	5.69
Pb	mg/kg		≤10		0.44	0.57	0.66
Hg	mg/kg		≤0.1		≤0.01	≤0.01	≤0.01
Ni	mg/kg		≤10		0.73	0.90	1.01
Zn	mg/kg		≤100		21.94	15.01	10.85

The ash content values may vary significantly and, therefore, the incorporation of residual biomass should lead to the incorporation in type B productions, as defined by the standard for the use of residual materials in type B products. Preferably, these materials should be applied for domestic use, where the quantities are not continuously high and where the potential negative effects of corrosive and fouling phenomena are more easily controllable.

The definitive validation of these considerations still lacks reference to the production of pellets since the present work dealt with characterization of parameters related to the chemical properties of the materials. However, due to the fact that densification is a purely physical process, it does not appear that there were constraints in the parameters related to the quality of the pellets, such as the final moisture, durability, fines content, dimensions or density at bulk.

5. Conclusions

The current perspective of mitigating the effects of climate change has led to an increasing demand for alternative forms of energy, which can be used to replace sources of fossil origin, such as oil, coal or natural gas. Biomass appears as a very interesting possibility, and has proven to be viable as demonstrated by several large-scale tests carried out in coal-fired power plants with the use of wood pellets in cocombustion with mineral coal, through the use of wood chips produced from forest residues for the production of steam in industrial units in the textile sector, or in the heating of agricultural and aviary greenhouses. On a smaller scale, and with greater proximity, the use of fuels derived from biomass for domestic uses, mainly for heating residential spaces but also for heating spaces of a commercial and small industrial nature (as is the case, for example, with bakeries), has positively and increasingly adhered to the use of these solid fuels. However, due to the quality requirements imposed by the regulatory instruments, the production of these materials has not incorporated a wide range of residual materials resulting from activities of an agroindustrial nature, which includes the wine sector, a major producer of waste. Regarding vine pruning, the characterization studies carried out over the past few years have shown that these materials present properties, both energetic, physical and chemical, that enable their incorporation in the production processes of pellets of biomass or briquettes, which is already a current practice. However, given the volumes produced annually, the existence of constraints related to the quality of the final products, namely the ash content or the copper content, has not allowed the incorporation of the majority of the waste produced, leading wine producers to frequently resort to less environmentally acceptable practices, such as burning the leftovers or simply abandoning them in piles.

Author Contributions: Conceptualization, L.J.R.N., A.I.O.F.F., A.C.P.B.R., J.C.O.M. and L.M.E.F.L.; methodology, L.J.R.N., A.I.O.F.F., J.C.O.M. and L.M.E.F.L.; validation, L.J.R.N., L.C.R.S., J.C.O.M. and L.M.E.F.L.; formal analysis, L.J.R.N., A.I.O.F.F., A.C.P.B.R., J.C.O.M. and L.M.E.F.L.; investigation, L.J.R.N. and L.C.R.S.; resources, L.M.E.F.L.; data curation, L.J.R.N., A.I.O.F.F., A.C.P.B.R., L.M.E.F.L., J.C.O.M. and L.C.R.S.; writing—original draft preparation, L.J.R.N., A.I.O.F.F., A.C.P.B.R. and L.M.E.F.L.; writing—review and editing, L.J.R.N., A.I.O.F.F., A.C.P.B.R., J.C.O.M. and L.M.E.F.L.; supervision, L.J.R.N., A.I.O.F.F., A.C.P.B.R., J.C.O.M. and L.M.E.F.L. All authors have read and agreed to the published version of the manuscript.

Funding: This work is a result of the project TECH—Technology, Environment, Creativity and Health, Norte-01-0145-FEDER-000043, supported by the Norte Portugal Regional Operational Program (NORTE 2020), under the PORTUGAL 2020 Partnership Agreement, through the European Regional Development Fund (ERDF). L.J.R.N. was supported by proMetheus, Research Unit on Energy, Materials and Environment for Sustainability-UIDP/05975/2020, funded by national funds through FCT—Fundação para a Ciência e Tecnologia.

Data Availability Statement: The data presented in this study are available on request from the corresponding author. The data are not publicly available because the research is not yet concluded, and the data will be updated.

Acknowledgments: The authors would like to acknowledge the companies YGE (Yser Green Energy SA) and AFS (Advanced Fuel Solutions SA), both in Portugal, for the execution of the laboratory tests.

Conflicts of Interest: The authors declare no conflict of interest.

References

1. Karkania, V.; Fanara, E.; Zabaniotou, A. Review of sustainable biomass pellets production–A study for agricultural residues pellets' market in Greece. *Renew. Sustain. Energy Rev.* **2012**, *16*, 1426–1436. [CrossRef]
2. García-Maraver, A.; Popov, V.; Zamorano, M. A review of European standards for pellet quality. *Renew. Energy* **2011**, *36*, 3537–3540. [CrossRef]
3. Nunes, L.; Matias, J.; Catalão, J. Mixed biomass pellets for thermal energy production: A review of combustion models. *Appl. Energy* **2014**, *127*, 135–140. [CrossRef]
4. Mostafa, M.E.; Hu, S.; Wang, Y.; Su, S.; Hu, X.; Elsayed, S.A.; Xiang, J. The significance of pelletization operating conditions: An analysis of physical and mechanical characteristics as well as energy consumption of biomass pellets. *Renew. Sustain. Energy Rev.* **2019**, *105*, 332–348. [CrossRef]
5. Smith, A.; Stirling, A.; Berkhout, F. The governance of sustainable socio-technical transitions. *Res. Policy* **2005**, *34*, 1491–1510. [CrossRef]
6. Moser, F.; Karavezyris, V.; Blum, C. Chemical leasing in the context of sustainable chemistry. *Environ. Sci. Pollut. Res.* **2015**, *22*, 6968–6988. [CrossRef] [PubMed]
7. Nunes, L.; Matias, J.; Catalão, J. Analysis of the use of biomass as an energy alternative for the Portuguese textile dyeing industry. *Energy* **2015**, *84*, 503–508. [CrossRef]
8. Trienekens, J.H. Agricultural value chains in developing countries; a framework for analysis. *Int. Food Agribus. Manag. Rev.* **2011**, *14*, 51–83.
9. Nunes, L.; Matias, J.; Catalão, J. Energy recovery from cork industrial waste: Production and characterisation of cork pellets. *Fuel* **2013**, *113*, 24–30. [CrossRef]
10. Zhang, Z.; Zhang, Y.; Zhou, Y.; Ahmad, R.; Pemberton-Pigott, C.; Annegarn, H.; Dong, R. Systematic and conceptual errors in standards and protocols for thermal performance of biomass stoves. *Renew. Sustain. Energy Rev.* **2017**, *72*, 1343–1354. [CrossRef]
11. Shang, L.; Nielsen, N.P.K.; Stelte, W.; Dahl, J.; Ahrenfeldt, J.; Holm, J.K.; Arnavat, M.P.; Bach, L.S.; Henriksen, U.B. Lab and bench-scale pelletization of torrefied wood chips—process optimization and pellet quality. *Bioenergy Res.* **2014**, *7*, 87–94. [CrossRef]
12. Christoforou, E.; Fokaides, P.A. Solid biofuels in trading form in global markets. In *Advances in Solid Biofuels*; Springer: Berlin/Heidelberg, Germany, 2019; pp. 57–68.
13. Nunes, L.J.; Godina, R.; Matias, J.C.; Catalão, J.P. Economic and environmental benefits of using textile waste for the production of thermal energy. *J. Clean. Prod.* **2018**, *171*, 1353–1360. [CrossRef]
14. Nunes, L.; Matias, J.C.; Catalao, J.P. Wood pellets as a sustainable energy alternative in Portugal. *Renew. Energy* **2016**, *85*, 1011–1016. [CrossRef]

15. Viana, H.d.S.; Rodrigues, A.M.; Godina, R.; Matias, J.d.O.; Nunes, L.J.R. Evaluation of the physical, chemical and thermal properties of Portuguese maritime pine biomass. *Sustainability* **2018**, *10*, 2877. [CrossRef]
16. Jelonek, Z.; Drobniak, A.; Mastalerz, M.; Jelonek, I. Assessing pellet fuels quality: A novel application for reflected light microscopy. *Int. J. Coal Geol.* **2020**, *222*, 103433. [CrossRef]
17. García, R.; Gil, M.; Rubiera, F.; Pevida, C. Pelletization of wood and alternative residual biomass blends for producing industrial quality pellets. *Fuel* **2019**, *251*, 739–753. [CrossRef]
18. Prvulovic, S.; Gluvakov, Z.; Tolmac, J.; Tolmac, D.a.; Matic, M.; Brkic, M. Methods for determination of biomass energy pellet quality. *Energy Fuels* **2014**, *28*, 2013–2018. [CrossRef]
19. Braghiroli, F.L.; Passarini, L. Valorization of biomass residues from forest operations and wood manufacturing presents a wide range of sustainable and innovative possibilities. *Curr. For. Rep.* **2020**, *6*, 172–183. [CrossRef]
20. Bottalico, F.; Pesola, L.; Vizzarri, M.; Antonello, L.; Barbati, A.; Chirici, G.; Corona, P.; Cullotta, S.; Garfì, V.; Giannico, V. Modeling the influence of alternative forest management scenarios on wood production and carbon storage: A case study in the Mediterranean region. *Environ. Res.* **2016**, *144*, 72–87. [CrossRef]
21. Spinelli, R.; Magagnotti, N.; Nati, C. Benchmarking the impact of traditional small-scale logging systems used in Mediterranean forestry. *For. Ecol. Manag.* **2010**, *260*, 1997–2001. [CrossRef]
22. Velázquez-Martí, B.; Fernández-González, E.; López-Cortés, I.; Salazar-Hernández, D.M. Quantification of the residual biomass obtained from pruning of trees in Mediterranean olive groves. *Biomass Bioenergy* **2011**, *35*, 3208–3217. [CrossRef]
23. San-Miguel-Ayanz, J.; Moreno, J.M.; Camia, A. Analysis of large fires in European Mediterranean landscapes: Lessons learned and perspectives. *For. Ecol. Manag.* **2013**, *294*, 11–22. [CrossRef]
24. Zanetti, M.; Brandelet, B.; Marini, D.; Sgarbossa, A.; Giorio, C.; Badocco, D.; Tapparo, A.; Grigolato, S.; Rogaume, C.; Rogaume, Y. Vineyard pruning residues pellets for use in domestic appliances: A quality assessment according to the EN ISO 17225. *J. Agric. Eng.* **2017**, *48*, 99–108. [CrossRef]
25. Niu, Y.; Tan, H.; Wang, X.; Liu, Z.; Liu, H.; Liu, Y.; Xu, T. Study on fusion characteristics of biomass ash. *Bioresour. Technol.* **2010**, *101*, 9373–9381. [CrossRef]
26. Du, S.; Yang, H.; Qian, K.; Wang, X.; Chen, H. Fusion and transformation properties of the inorganic components in biomass ash. *Fuel* **2014**, *117*, 1281–1287. [CrossRef]
27. Ma, T.; Fan, C.; Hao, L.; Li, S.; Song, W.; Lin, W. Fusion characterization of biomass ash. *Thermochim. Acta* **2016**, *638*, 1–9. [CrossRef]
28. Wang, Q.; Han, K.; Wang, J.; Gao, J.; Lu, C. Influence of phosphorous based additives on ash melting characteristics during combustion of biomass briquette fuel. *Renew. Energy* **2017**, *113*, 428–437. [CrossRef]
29. Rizvi, T.; Xing, P.; Pourkashanian, M.; Darvell, L.; Jones, J.; Nimmo, W. Prediction of biomass ash fusion behaviour by the use of detailed characterisation methods coupled with thermodynamic analysis. *Fuel* **2015**, *141*, 275–284. [CrossRef]
30. Li, F.; Li, Y.; Fan, H.; Wang, T.; Guo, M.; Fang, Y. Investigation on fusion characteristics of deposition from biomass vibrating grate furnace combustion and its modification. *Energy* **2019**, *174*, 724–734. [CrossRef]
31. Yao, X.; Xu, K.; Yan, F.; Liang, Y. The influence of ashing temperature on ash fouling and slagging characteristics during combustion of biomass fuels. *BioResources* **2017**, *12*, 1593–1610. [CrossRef]
32. Lee, Y.-J.; Choi, J.-W.; Park, J.-H.; Namkung, H.; Song, G.-S.; Park, S.-J.; Lee, D.-W.; Kim, J.-G.; Jeon, C.-H.; Choi, Y.-C. Techno-economical method for the removal of alkali metals from agricultural residue and herbaceous biomass and its effect on slagging and fouling behavior. *ACS Sustain. Chem. Eng.* **2018**, *6*, 13056–13065. [CrossRef]
33. Yao, X.; Zhao, Z.; Li, J.; Zhang, B.; Zhou, H.; Xu, K. Experimental investigation of physicochemical and slagging characteristics of inorganic constituents in ash residues from gasification of different herbaceous biomass. *Energy* **2020**, *198*, 117367. [CrossRef]
34. Ruscio, A.; Kazanc, F.; Levendis, Y.A. Comparison of fine ash emissions generated from biomass and coal combustion and valuation of predictive furnace deposition indices: A review. *J. Energy Eng.* **2016**, *142*, E4015007. [CrossRef]
35. Yao, X.; Zhao, Z.; Xu, K.; Zhou, H. Determination of ash forming characteristics and fouling/slagging behaviours during gasification of masson pine in a fixed-bed gasifier. *Renew. Energy* **2020**, *160*, 1420–1430. [CrossRef]
36. Feldmeier, S.; Wopienka, E.; Schwarz, M.; Schö, C.; Pfeifer, C. Applicability of fuel indexes for small-scale biomass combustion technologies, part 1: Slag formation. *Energy Fuels* **2019**, *33*, 10969–10977. [CrossRef]
37. Royo, J.; Canalís, P.; Quintana, D.; Díaz-Ramírez, M.; Sin, A.; Rezeau, A. Experimental study on the ash behaviour in combustion of pelletized residual agricultural biomass. *Fuel* **2019**, *239*, 991–1000. [CrossRef]

recycling

MDPI

Article

Energy Recovery from Invasive Species: Creation of Value Chains to Promote Control and Eradication

Leonel J. R. Nunes [1,*], Abel M. Rodrigues [2,3], Liliana M. E. F. Loureiro [4], Letícia C. R. Sá [4] and João C. O. Matias [5,6]

1 PROMETHEUS, Unidade de Investigação em Materiais, Energia e Ambiente para a Sustentabilidade, Escola Superior Agrária, Instituto Politécnico de Viana do Castelo, Rua da Escola Industrial e Comercial de Nun'Alvares, 4900-347 Viana do Castelo, Portugal
2 INIAV—Instituto Nacional de Investigação Agrícola e Veterinária, Av. da República, Quinta do Marquês (Edifício Sede), 2780-157 Oeiras, Portugal; abel.rodrigues@iniav.pt
3 MARETEC—Marine, Environment & Technology Center, Secção de Ambiente e Energia, Departamento de Engenharia Mecânica, Instituto Superior Técnico, Av. Rovisco Pais, N°. 1, 1049-001 Lisboa, Portugal
4 YGE—Yser Green Energy SA, Área de Acolhimento Empresarial de Úl/Loureiro, Lote 17, 3720-075 Loureiro Oaz, Portugal; liliana.loureiro@ygenergia.com (L.M.E.F.L.); leticia.sa@ygenergia.com (L.C.R.S.)
5 GOVCOPP, Unidade de Investigação em Governança, Competitividade e Políticas Públicas, Campus Universitário de Santiago, Universidade de Aveiro, 3810-193 Aveiro, Portugal; jmatias@ua.pt
6 DEGEIT, Departamento de Economia, Gestão, Engenharia Industrial e Turismo, Campus Universitário de Santiago, Universidade de Aveiro, 3810-193 Aveiro, Portugal
* Correspondence: leonelnunes@esa.ipvc.pt

Abstract: The use of biomass as an energy source presents itself as a viable alternative, especially at a time when the mitigation of climate change requires that all possibilities of replacing fossil fuels be used and implemented. The use of residual biomass also appears as a way to include in the renewable energy production system products that came out of it, while allowing the resolution of environmental problems, such as large volumes available, which are not used, but also by the elimination of fuel load that only contributes to the increased risk of rural fires occurrence. Invasive species contribute to a significant part of this fuel load, and its control and eradication require strong investments, so the valorization of these materials can allow the sustainability of the control and eradication processes. However, the chemical composition of some of these species, namely *Acacia dealbata*, *Acacia melanoxylon*, *Eucalyptus globulus*, *Robinia pseudoacacia* and *Hakea sericea*, presents some problems, mainly due to the nitrogen, chlorine and ash contents found, which preclude exclusive use for the production of certified wood pellets. In the case of *Eucalyptus globulus*, the values obtained in the characterization allow the use in mixtures with *Pinus pinaster*, but for the other species, this mixture is not possible. From a perspective of local valorization, the use of materials for domestic applications remains a possibility, creating a circular economy process that guarantees the sustainability of operations to control and eradicate invasive species.

Keywords: invasive forest species; wood pellets; circular economy; sustainability; value chain

check for updates

Citation: Nunes, L.J.R.; Rodrigues, A.M.; Loureiro, L.M.E.F; Sá, L.C.R.; Matias, J.C.O. Energy Recovery from Invasive Species: Creation of Value Chains to Promote Control and Eradication. *Recycling* **2021**, *6*, 21. https://doi.org/10.3390/recycling6010021

Received: 8 February 2021
Accepted: 6 March 2021
Published: 13 March 2021

Publisher's Note: MDPI stays neutral with regard to jurisdictional claims in published maps and institutional affiliations.

Copyright: © 2021 by the authors. Licensee MDPI, Basel, Switzerland. This article is an open access article distributed under the terms and conditions of the Creative Commons Attribution (CC BY) license (https://creativecommons.org/licenses/by/4.0/).

1. Introduction

The use of biomass as an energy source is increasingly presented as a current alternative, in the permanent search for more sustainable forms of energy, which can somehow replace traditional sources of fossil origin, such as oil and coal [1]. However, the use of biomass as an energy source is not a recent application, since this is the oldest energy source that man learned to use, from the moment that they discovered how to control fire for their own benefit, passing this on to be part of daily life situations [2]. Since time immemorial, human populations have started to have in their routine the acquisition of biomass fuels, through their collection and storage, thus being available to supply needs such as space heating, cooking, lighting, and even protection by keeping wild animals away [3].

Currently, biomass remains the most widespread source of energy among the most remote populations, essentially due to its availability and ease of use, giving it the epithet "energy of the poor", since they are usually the most disadvantaged populations, mainly in Africa, Southeast Asia and some regions of Latin America, which this more rudimentary use continues to occur [4]. However, in the more developed countries there is also an increase in the use of biomass as a source of primary energy because of, in addition to the traditional consumption associated with the heating of residential spaces in the form of firewood consumption, the consumption of fuels derived from biomass, such as wood pellets and briquettes, both for heating, but also for more industrial applications, such as the production of industrial steam, and even the production of electric energy [5].

These more industrial uses, however, have led to an increasing standardization of fuel quality criteria in order to optimize their use, defining a set of characteristics, namely its heating value, but also the maximum limits of certain chemical constituents, such as the content of sulfur, chlorine and alkali metals, due to their behavior during combustion, and contribution to the occurrence of corrosive, fouling and slagging phenomena [6,7]. For this reason, the use of biomass is currently very limited to selected types which meet a set of quality requirements, leaving a set of forms of biomass considered to be residual in the supply chains, as are the materials resulting from operations forest management and agricultural activity [8,9].

Within this huge group of residual biomasses resulting from forest management operations, there are numerous tree and shrub species, which, since they have no commercial and/or industrial application, are abandoned on forest land, even after cutting, contributing to the increase in fuel load and consequent increase in the risk of rural fires [10]. Some of these species, in turn, are even exotic species, with invasive behavior, which due to their aggressiveness and competitiveness vis-à-vis native species, are conquering space and replacing native flora, making it possible to identify these situations all over the world [11]. This substitution of native species by invasive species has very negative effects that go far beyond the loss of biodiversity and changes in ecosystems, since they also hinder the development of productive forests by competing directly with the installed species [12].

In Portugal, the phenomenon of the expansion of invasive species has acquired very worrying proportions, mainly with a group of species of the genus *Acacia*, from which the *Acacia dealbata* and the *Acacia melanoxylon* stand out, but also with other species, namely the *Eucalyptus globulus*, the *Robinia pseudoacacia* and *Hakea sericea* [13]. This group of species has progressed almost exponentially, already covering extensive areas. However, the problem is not limited to these species, there are also problematic situations with *Acacia longifolia*, *Cortaderia selloana*, *Arundo donax* and *Ailanthus altissima*, among others, with which different means have been employed in order to try to eradicate, or at least control, the progression of these species [14]. The most serious situation in Portugal is that of the uncontrolled expansion of *Acacia dealbata*. This species has grown in area by more than 400% since the 70s of the 20th century, being currently the invasive species that occupies the largest area in the national territory (Figure 1). However, other species exhibit similar behaviors, making the situation even more serious.

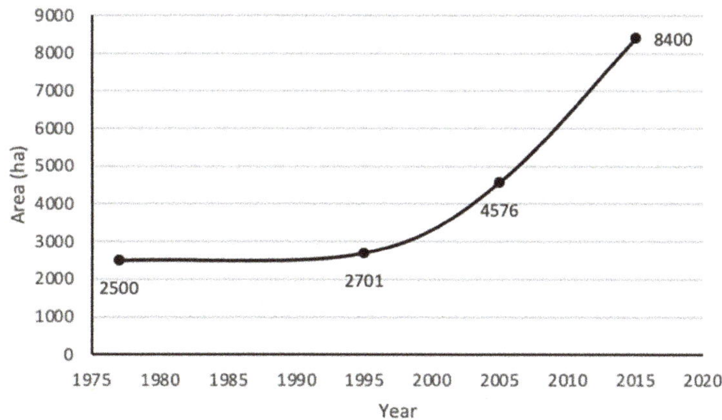

Figure 1. Evolution of the area occupied by *Acacia dealbata* (adapted from [14–16]).

The creation of value chains, with the objective of promoting the use of these species, presents itself as a possibility that allows the balance of the costs generated with the operations of control of the invaders [17]. These value chains, which already exist today, encourage the forwarding of these residual materials to biomass thermoelectric plants, which give them a low financial value, often scarcely enough to pay for the different tasks of forest management, such as cutting, filling and transport, and giving no value to the combustible material, claiming its low density, high moisture content and, mainly, due to the low energy properties that some species have [18]. However, there are some species that may have a good potential for energy recovery, and that, even if they are not used as the sole source of raw material for the production of fuels derived from biomass, can be incorporated with traditional species, namely the *Pinus pinaster*, or other resinous species [19].

This perspective of valorization of residual biomasses, originating from invasive species, can be very helpful in combating, almost always unevenly, the dispersion of these species, contributing to the share of costs with the control and eradication operations [20]. In this way, the creation of these value chains may play a decisive role in the preservation of indigenous biodiversity, while contributing to the reduction in the risk of occurrence of rural fires, since the need to supply valuable raw materials promotes permanent pressure on invasive species, controlling their growth and dispersion, limiting the accumulation of fuel load [21]. In this article, the potential use of these species in the production of wood pellets is discussed, both individually and in mixtures with *Pinus pinaster* wood, in order to justify the creation of value chains that promote pressure on invasive species, ensuring the sustainability of control and eradication operations for these species. The present work has as its main objective the characterization of a set of species, namely, *Acacia dealbata*, *Acacia melanoxylon*, *Robinia pseudoacacia*, *Eucalyptus globulus* and *Hakea sericea*, and their subsequent comparison with the dominant species most used in production of solid fuels derived from biomass, such as wood pellets, which is *Pinus pinaster*.

2. Results

2.1. Thermogravimetric Analysis

The results obtained in the thermogravimetric analysis are shown in Figure 2.

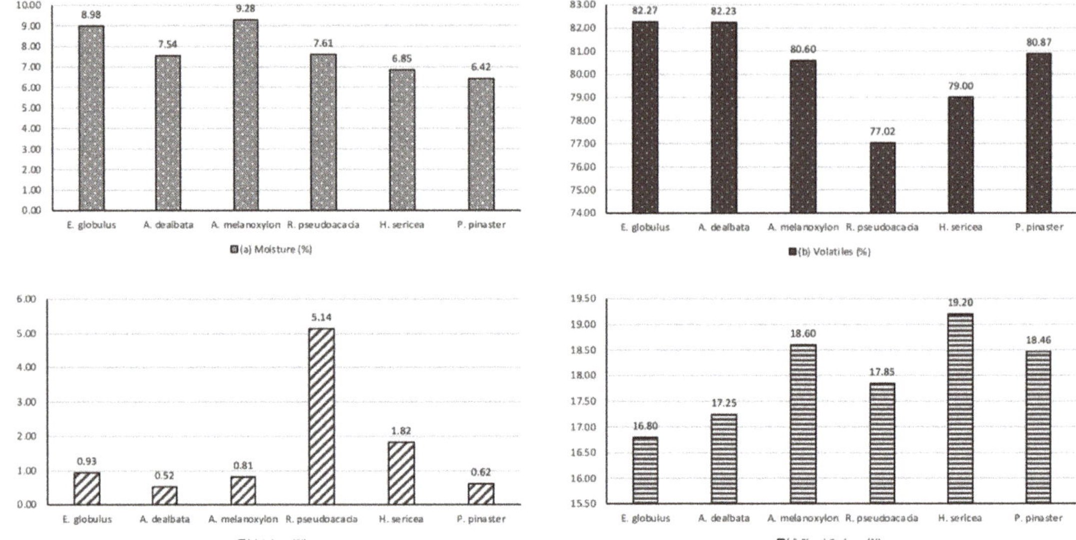

Figure 2. Results of the thermogravimetric analysis.

Moisture is a less important characteristic because it depends directly on the time and type of drying performed. For drying described in Section 2.1. Sampling and preparation, the results varied between 6.42% of *Pinus pinaster* and 9.28% of *Acacia melanoxylon*. The remaining species show very close values, between 6.85% and 8.98%. The volatile content varied between a minimum value of 77.02%, for *Robinia pseudoacacia*, and a maximum value of 82.27%, for *Eucalyptus globulus*. The remaining species varied between 79.00% and 82.23%. The ash content varied between a minimum value of 0.52% for *Acacia dealbata* and a maximum value of 5.14% for *Robinia pseudoacacia*. The remaining species varied between a minimum value of 0.62% and a maximum value of 1.82. The fixed carbon content showed a minimum value of 16.80%, for *Eucalyptus globulus*, and a maximum value of 19.20%, for *Hakea sericea*. The remaining species showed values between 17.25% and 18.60%.

2.2. Elemental Analysis CHNO

The results obtained in the elemental analysis are shown in Figure 3.

The carbon content varied between a minimum value of 47.00% for *Acacia dealbata* and a maximum value of 60.00% for *Hakea sericea*. The remaining species varied between 47.30% and 50.21%. The hydrogen content varied between a minimum value of 5.61%, for *Acacia melanoxylon*, and a maximum value of 6.07%, for *Pinus pinaster*. The remaining species varied between 5.67% and 5.92%. The nitrogen content varied between a minimum value of 0.080%, for *Pinus pinaster*, and a maximum value of 0.711%, for *Hakea sericea*. The remaining species varied between 0.099% and 0.582%. The oxygen content varied between a minimum value of 33.40% for *Hakea sericea* and a maximum value of 47.07% for *Acacia melanoxylon*. The remaining species varied between 43.64% and 46.93%.

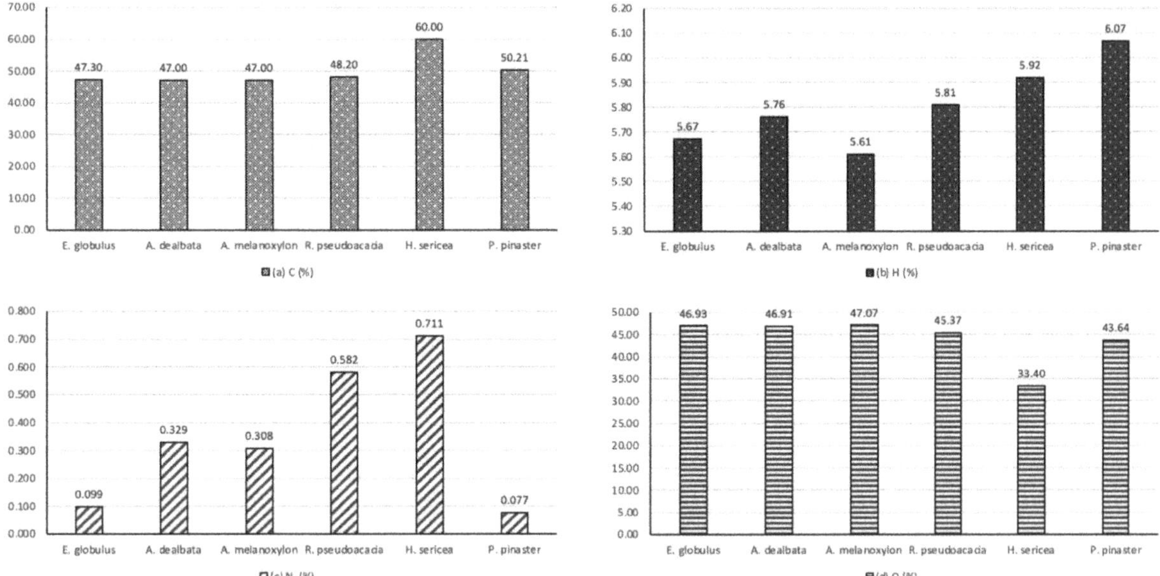

Figure 3. Results of the elemental analysis CHNO.

2.3. Determination of Sulfur and Chlorine Content

The results of the sulphur and chlorine content determination are shown in Figure 4.

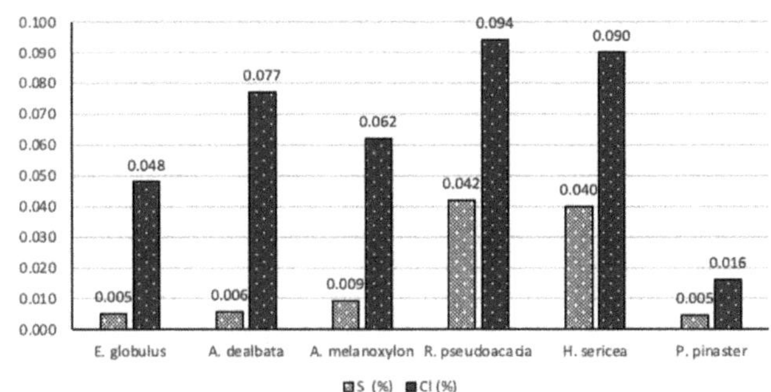

Figure 4. Sulphur and chlorine content.

The sulfur content varied between a minimum value of 0.005%, for *Eucalyptus globulus* and *Pinus pinaster*, and a maximum value of 0.042%, for *Robinia pseudoacacia*. The remaining species varied between 0.006% and 0.040%. The chlorine content varied between a minimum value of 0.016% for *Pinus pinaster* and a maximum value of 0.094% for *Robinia pseudoacacia*. The remaining species varied between 0.048% and 0.090%.

2.4. Determination of the High and Low Heating Value

The results of the heating value determination are shown in Figure 5.

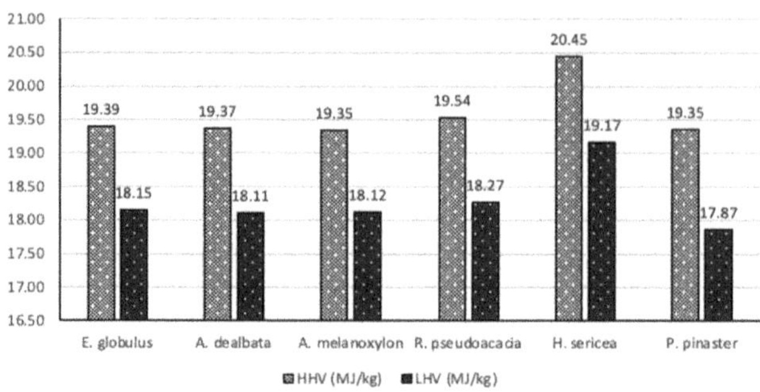

Figure 5. Higher heating value (HHV) and lower heating value (LHV) results.

The high heating value varied from the minimum value of 19.35 MJ/kg, for *Acacia melanoxylon* and for *Pinus pinaster*, to the maximum value of 20.45 MJ/kg, for *Hakea sericea*. The remaining species varied between 19.37 MJ/kg and 19.54 MJ/kg. The low heating value varied from the minimum value of 17.87 MJ/kg, for *Pinus pinaster*, to the maximum value of 19.17 MJ/kg, for *Hakea sericea*. The remaining species varied between 18.11 MJ/kg and 18.27 MJ/kg.

2.5. Determination of the Content of Major Elements

The results of the content of major elements are shown in Figure 6.

The Al content varied between a minimum value of 8.05 mg/kg for *Acacia dealbata* and a maximum value of 307.24 mg/kg for *Pinus pinaster*. The remaining species varied between 19.08 mg/kg and 215.18 mg/kg. The Ca content varied between a minimum value of 1645.15 mg/kg, for *Pinus pinaster*, and a maximum value of 21,917.90 mg/kg, for *Robinia pseudoacacia*. The remaining species ranged from 2214.41 mg/kg to 4830.83 mg/kg. The Fe content varied between a minimum of 21.51 mg/kg for *Acacia dealbata* and a maximum of 268.37 mg/kg for *Pinus pinaster*. The remaining species varied between 38.77 mg/kg and 187.90 mg/kg. The Mg content varied between a minimum of 564.19 mg/kg for *Acacia dealbata* and a maximum of 1474.89 mg/kg for *Robinia pseudoacacia*. The remaining species varied between 645.28 mg/kg and 1120.13 mg/kg. The P content varied from a minimum value of 78.08 mg/kg, for *Pinus pinaster*, up to a maximum value of 480.22 mg/kg. The remaining species varied between 124.93 mg/kg and 312.28 mg/kg. The K content varied from a minimum value of 723.84 mg/kg, for *Pinus pinaster*, up to a maximum value of 3848.58 mg/kg, for *Robinia pseudoacacia*. The remaining species ranged from 1427.42 mg/kg to 3003.84 mg/kg. The Si content varied from a minimum value of 5.13 mg/kg, for *Acacia melanoxylon*, to a maximum value of 993.31 mg/kg, for *Pinus pinaster*. The remaining species varied between 31.47 mg/kg and 856.28 mg/kg. The Na content varied between a minimum value of 436.79 mg/kg, for *Pinus pinaster*, and a maximum value of 1984.70 mg/kg, for *Hakea sericea*. The remaining species varied between 738.66 mg/kg and 1106.17 mg/kg The Ti content varied between a minimum of 1.22 mg/kg for *Acacia dealbata* and a maximum of 19.22 mg/kg for *Pinus pinaster*. The remaining species ranged from 1.84 mg/kg to 15.68 mg/kg.

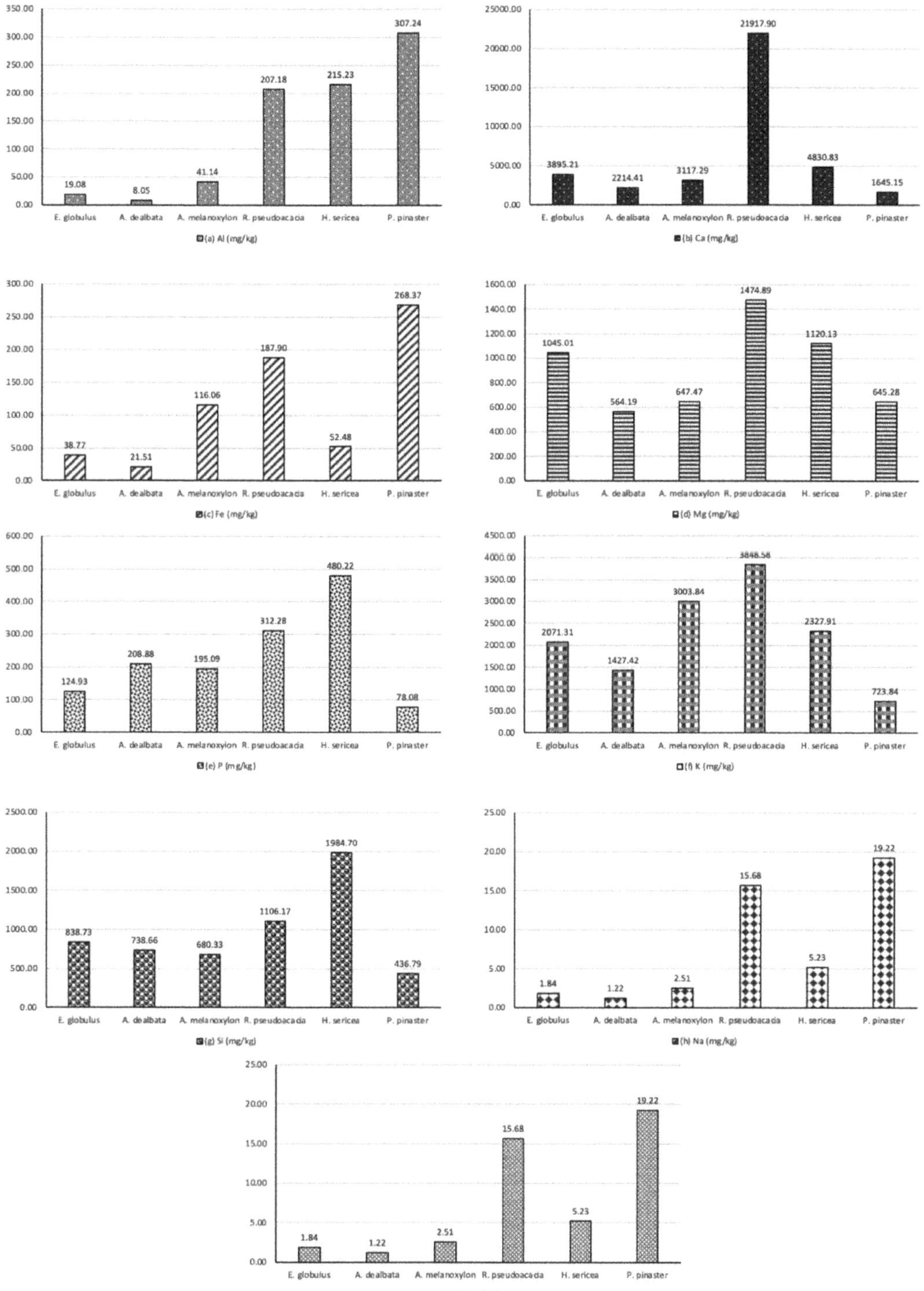

Figure 6. Content of major elements.

2.6. Determination of the Content of Minor Elements

The results of the content of minor elements are shown in Figure 7.

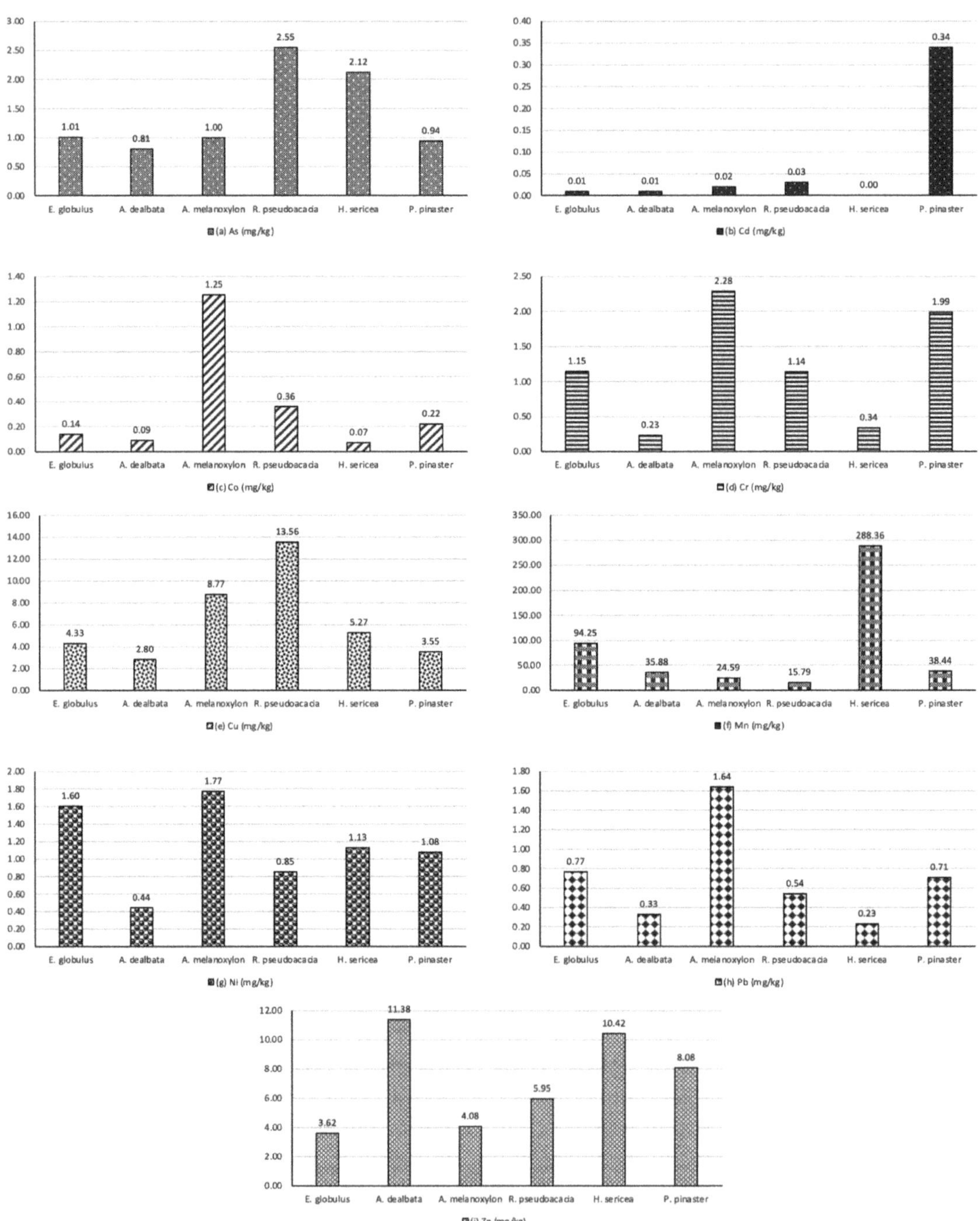

Figure 7. Content of minor elements.

The As content varied between the minimum value of 0.81 mg/kg, for *Acacia dealbata*, and the maximum value of 2.55 mg/kg, for *Robinia pseudoacacia*. The remaining species ranged from 0.94 mg/kg to 2.12 mg/kg. The Cd content varied between the minimum value of 0.00 mg/kg, for *Hakea sericea*, up to a maximum value of 0.34 mg/kg, for *Pinus pinaster*. The remaining species varied between 0.01 mg/kg and 0.03 mg/kg. The Co content varied between the minimum value of 0.07 mg/kg, for *Hakea sericea*, and the maximum value of 1.25 mg/kg, for *Acacia melanoxylon*. The remaining species varied between 0.09 mg/kg and 0.36 mg/kg. The Cr content varied between the minimum value of 0.23 mg/kg, for *Acacia dealbata*, and the maximum value of 2.28 mg/kg, for *Acacia melanoxylon*. The remaining species varied between 0.34 mg/kg and 1.99 mg/kg. The Cu content varied between the minimum value of 2.80 mg/kg, for *Acacia dealbata*, and the maximum value of 13.56 mg/kg, for *Robinia pseudoacacia*. The remaining species varied between 3.55 mg/kg and 8.77 mg/kg. The Mn content varied between the minimum value of 15.79 mg/kg, for *Robinia pseudoacacia*, and the maximum value of 288.36 mg/kg, for *Hakea sericea*. The remaining species varied between 24.59 mg/kg and 94.25 mg/kg. The Ni content varied between the minimum value of 0.44 mg/kg, for *Acacia dealbata*, and the maximum value of 1.77 mg/kg, for *Acacia melanoxylon*. The remaining species varied between 0.85 mg/kg and 1.60 mg/kg. The Pb content varied between the minimum value of 0.23 mg/kg, for *Hakea sericea*, and the maximum value of 1.64 mg/kg, for *Acacia melanoxylon*. The remaining species varied between 0.33 mg/kg and 0.77 mg/kg. The Zn content varied between the minimum value of 3.62 mg/kg, for *Eucalyptus globulus*, and the maximum value of 11.38 mg/kg. The remaining species varied between 4.08 mg/kg and 10.42 mg/kg.

2.7. Ash Fusibility

The results of the ash fusibility temperatures are shown in Figure 8.

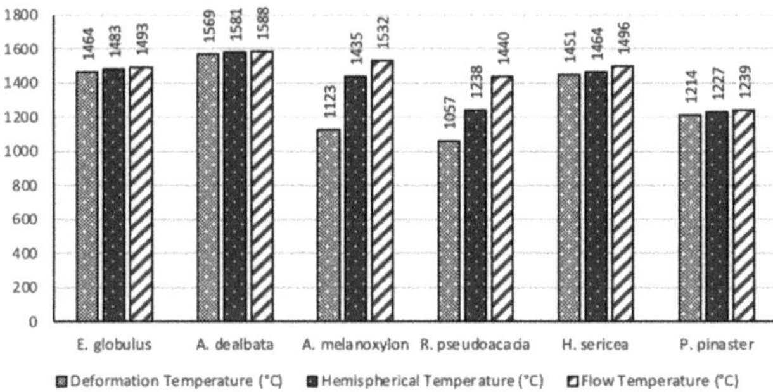

Figure 8. Ash fusibility temperatures.

Deformation temperatures ranged from the minimum value of 1057 °C, for *Robinia pseudoacacia*, and the maximum value of 1569 °C, for *Acacia dealbata*. The remaining species varied between 1123 °C and 1464 °C. Hemispherical temperatures varied between the minimum value of 1227 °C, for *Pinus pinaster*, and the maximum value of 1581 °C, for *Acacia dealbata*. The remaining species varied between 1238 °C and 1483 °C. The flow temperatures varied between the minimum value of 1239 °C, for *Pinus pinaster*, and the maximum value of 1588 °C, for *Acacia dealbata*.

3. Discussion

There are many works available where characterizations of the most diverse types of biomass are presented, in order to assess their combustibility and their physical-chemical properties. An example of this is the review work presented by Cai et al. (2017), where several lignocellulosic biomasses are characterized, including several residual biomasses of agricultural origin, such as rice husk, rice stalk, cotton stalk, wheat straw or corn stalk, but also residual biomasses of forest origin, like pine or poplar [22]. The objective of this study by Cai et al. (2017) fits that of many others, such as those presented by García et al. (2012), Chiang et al. (2012), Wilson et al. (2011), Fang et al. (2015) or Patel and Gami (2012), always from the perspective that using lignocellulosic biomass-derived biofuels can reduce reliance on fossil fuels and contribute to climate change mitigation [23–27]. Many works are also available regarding the behavior of different forms of biomass when subjected to thermochemical conversion processes, such as torrefaction, pyrolysis or gasification, such as the work of Chen et al. (2014), which addresses the non-oxidative and oxidative torrefaction characterization and SEM observations of fibrous and ligneous biomass [28], or the work of Neves et al. (2011), where biomass pyrolysis is addressed, regarding models, mechanisms, kinetics and some information on product yields and properties [29]. In fact, this perspective of energy recovery through the use of energy densification technologies foresees, at the outset, a need to improve, or in some way, correct, the less positive characteristics, such as low heating value, low density or high content of ashes. However, the most frequent uses presuppose the direct combustion of materials [30–32].

The comparative analysis of forms of residual biomass and their potential framing with premium raw materials, certified by the international standards in force, as is the case of ENplus®, has also been addressed by extensive literature. For example, the work presented by Agar et al. (2018) addresses the production of pellets from agricultural and forest biomass [33], while de Souza et al. (2020) addressed the possibility of producing pellets from eucalyptus biomass and coffee growing wastes residues with acceptable properties for commercialization standards, which includes the ENplus® standard [34]. In other words, the possibility of integrating different forms of biomass has been a necessity for a long time, since it would allow, in case of success and compliance with the quality criteria, the reduction in the cost of the acquisition of raw materials at the same time, which presents a solution for the disposal/reuse of a set of waste, which until now has not been subject to any type of recovery [13].

Thus, the classification of the properties of any residual raw material, as is the case of the species selected in the present study, with the parameters defined with one of the standards used internationally for the certification of fuels derived from biomass, as is the case of ENplus® standard, allows, in a simple and accessible way, the validation or exclusion of the use of a certain material (species), or at least, it allows indicating whether it is possible to incorporate a certain percentage of these materials in any way [35]. Table 1 shows the values allowed for the main parameters indicated for the raw materials used in the production of wood pellets by the ENplus® standard. This standard divides products into three categories according to the origin of the raw materials, with categories A2 and B destined to products resulting from the processing of waste materials, which include the species used in the present study [36]. However, in the case of incorporating percentages of residual biomass with premium raw materials, and if the parameters defined in the standard are met, the final products can be included in category A1, which has the highest added value [37].

The results obtained by the characterization of these biomass species, which are summarized in Table 2, present a relatively different framework. In the table, the values marked in italics represent the values that meet the requirements of categories A2 or B, and in bold, the values that do not fit into any of the categories. The remaining values are within the limits imposed by the ENplus® A1 standard. As expected, *Pinus pinaster* fully complies with all the requirements presented by the ENplus® standard, while none of the other species fully comply with all parameters. However, some of the species present values very close to the permitted limits, so that their incorporation with, for example, *Pinus pinaster*, appears as possible, and thus meet the requirements defined by the standard ENplus®.

Table 1. Limit values for properties defined by the ENplus ® standard.

		ENplus ®		
		A1	**A2**	**B**
Moisture	%		≤10	
Ash	%	≤0.7	≤1.2	≤2
LHV	MJ/kg		≥16.50	
N	%	≤0.3	≤0.5	≤1.0
S	%		≤0.04	≤0.05
Cl	%		≤0.02	≤0.03
Deformation Temp.	°C		≥1200	≥1100
As	mg/kg		≤1	
Cd	mg/kg		≤0.5	
Cr	mg/kg		≤10	
Cu	mg/kg		≤10	
Pb	mg/kg		≤10	
Hg	mg/kg		≤0.1	
Ni	mg/kg		≤10	
Zn	mg/kg		≤100	

Table 2. Comparative analysis of the results obtained with the parameters defined by the ENplus ® standard.

		P.p.	*E.g.*	*A.d.*	*A.m.*	*R.p.*	*H.s.*
Moisture	%	6.42	8.98	7.54	9.28	7.61	6.85
Ash	%	0.62	*0.93*	0.52	*0.81*	**5.14**	*1.82*
LHV	MJ/kg	17.87	18.15	18.11	18.12	18.27	19.17
N	%	0.08	0.099	*0.329*	*0.308*	*0.582*	*0.711*
S	%	0.0045	0.005	0.006	0.009	*0.042*	0.040
Cl	%	0.016	**0.05**	**0.08**	**0.06**	**0.09**	**0.09**
Deformation Temp.	°C	1214	1464	1569	*1123*	**1057**	1451
As	mg/kg	0.94	**1.01**	0.81	1.00	**2.55**	**2.12**
Cd	mg/kg	0.34	0.00	0.00	0.00	0.03	0.00
Cr	mg/kg	1.99	1.15	0.23	2.28	1.14	0.34
Cu	mg/kg	3.55	4.33	2.80	8.77	**13.56**	5.27
Pb	mg/kg	0.71	0.77	0.33	1.64	0.00	0.23
Hg	mg/kg	<0.001	<0.001	<0.001	<0.001	<0.001	<0.001
Ni	mg/kg	1.08	1.60	0.44	1.77	0.85	1.13
Zn	mg/kg	8.08	3.62	11.38	4.08	5.95	10.42

As can be seen in the results presented in Table 2, the incorporation of these materials in the production of certifiable wood pellets, only seems possible for the biomass of *Eucalyptus globulus*, which would easily dilute the values above the requirements defined in the standard, allowing the incorporation of a percentage of 25%, which allows the production of wood pellets of category A1. An incorporation of 50% of *Eucalyptus globulus* biomass would only allow for category B certification, since the chlorine content would always be close to the upper limit of 0.03% indicated in the standard for this category. The remaining species present values that are too high in some parameters, namely in nitrogen content, with values ranging between 0.308% and 0.711%, mainly for *Hakea sericea* and *Robinia pseudoacacia*, but also due to the accumulation of other parameters outside the limits defined, as for the ash content, with *Robinia pseudoacacia* reaching 5.14%, and in the chlorine content, where it reaches 0.09%, together with Hakea sericea. However, the use of these residual biomasses remains possible, especially if the objective is not to produce certified materials, but rather to be used less domestically, and to support the local biomass recovery.

The local valorization of residual biomasses, as those analyzed in the present study, as well as others with similar properties, can always be a solution. Usually, given the lower requirement of the proximity markets, where the most evaluated requirement is the fuel cost, in detriment to the quality and combustibility requirements, such products can be used. The possibility of using these residual biomasses in thermochemical conversion processes, namely in the production of biochar, not as an energy product, but rather as a soil amendment product and as a carbon sequestration methodology, must be evaluated, with regard to the creation of value chains for residual biomass. This perspective of creating value chains, which aim to promote the maintenance of actions to control and eradicate invasive species, contributes systematically to the revitalization of ecosystems. This positive condition is effective when the pressure caused on the populations of invasive species decrease their strength, allowing native species to develop and return to occupy their space.

4. Materials and Methods

4.1. Sampling and Material Preparation

The species were selected for their availability and abundance and were collected in areas where their proliferation has been noted for the speed of propagation and occupation of space. Thus, the samples were collected in the locations shown in Table 3. Thus, samples of five invasive species were collected, namely, *Acacia dealbata*, *Acacia melanoxylon*, *Eucalyptus globulus*, *Robinia pseudoacacia* and *Hakea sericea*. Samples of *Pinus pinaster* were also collected to serve as a point of comparison. All samples were collected in the form of adult tree trunks, and in the case of invasive species, the all-in method was chosen. That is, none of the constituent parts, such as branches or leaves, have been discarded.

Table 3. Location of sample collection points used in the present study.

Species	Location
Acacia dealbata	Casal do Rei (Seia—Portugal)
Acacia melanoxylon	Loureiro (Oliveira de Azeméis—Portugal)
Eucalyptus globulus	Loureiro (Oliveira de Azeméis—Portugal)
Robinia pseudoacacia	Albernoa (Aljustrel—Portugal)
Hakea sericea	Casal do Rei (Seia—Portugal)
Pinus pinaster	Vale de Cambra (Portugal)

In the sample preparation procedure, the sequence used in a wood pellet production unit was followed, using exclusively *Pinus pinaster* as a raw material, so the *Pinus pinaster* wood was previously debarked before proceeding with the drying. The rest of the wood was not debarked, since the industrial debarking process used is optimized to operate only with *Pinus pinaster* logs, and if the other mentioned species were included in the process, this operation would not be carried out efficiently. Then, the size of the collected samples was reduced to a granulometry equivalent to that normally used in the industrial process, that is, to a G30 size woodchips, which was subjected to drying in a laboratory oven at 100 °C for a period of 12 h. After drying, the material of all samples was ground again, until the dimension normally used in the industrial process of wood pellet production was reached, with a d_{50} within the range [1.13–3.86]. Subsequently, the laboratory characterization of the samples followed with the thermogravimetric analysis, the elemental analysis, the calorimetric analysis, the chemical analysis and the analysis of the fusibility of the ashes. All samples were collected and analyzed in triplicate and the results presented are the average values for each species.

4.2. Thermogravimetric Analysis (TGA)

The thermogravimetric analyzer used was an ELTRA THERMOSTEP model. One gram of each sample was introduced into crucibles and placed inside an oven, along with an empty reference crucible. As temperature increased, crucibles were weighted on a precision scale. Moisture, volatiles, and fixed carbon content were determined in this order throughout the heating process. Lastly, the final residue represents the ash content.

4.3. Elemental Analysis (CHNO)

Elemental analysis was performed in a LECO CHN628. The operational principle consists of weighing a sample in tin foil that is later placed in the autoloader. The sample is then introduced into the primary furnace containing only pure oxygen, which results in fast and complete combustion. Carbon, hydrogen, and nitrogen present in the sample are oxidized to CO_2, H_2O, and NO_x, respectively, and are swept by the oxygen carrier gas through a secondary furnace for further oxidation and particulate removal. Detection of H_2O and CO_2 occurs through separate, optimized, non-dispersive infrared cells, while the NO_x gases are reduced to N. Lastly, N_2 is detected when the gas passes through a thermal conductivity cell. After the analysis is complete, moisture content obtained through thermogravimetric analysis is introduced into the software and the CHN contents are automatically calculated. Following that, it is possible to estimate the oxygen content on a dry basis ($w_{O,db}$) from Equation (1), as follows:

$$w_{O,db}\,(\%) = 100 - w_{C,db} - w_{H,db} - w_{N,db} - w_{S,db} - w_{Cl,db}, \tag{1}$$

where $w_{C,db}$ is the carbon content on a dry basis (%), $w_{H,db}$ is the hydrogen content on a dry basis (%), $w_{N,db}$ is the nitrogen content on a dry basis (%), $w_{S,db}$ is the sulfur content on a dry basis (%), and $w_{Cl,db}$ is the chlorine content on a dry basis (%).

4.4. Determination of Chlorine Content

Chloride titration was the method chosen to determine the chlorine content, and the equipment used was a TITROLINE 7000 titrator from SI Analytics. For this procedure, sample preparation involves previous digestion of the sample, performed in a SINEO MDS-6G microwave, since titration requires a liquid sample. Chlorine content determination is achieved by potentiometric titration. This method consists of measuring the potential difference while the titrant, in this case, $AgNO_3$, is added. Equation (2), as follows, presents the redox reaction that occurs:

$$Cl^-_{(aq.)} + AgNO_{3(aq.)} \rightarrow AgCl_{(S)} + NO^-_{3\,(aq.)}, \tag{2}$$

As next step, a software creates a spreadsheet with the potential difference and titrant volume variation over time. First derivative can then be calculated through Equation (3), as follows, and the equivalence point can be determined as the volume corresponding to the maximum of the first derivative:

$$f'(x) = \frac{\Delta U}{\Delta V} = \frac{U_i - U_{i-1}}{V_i - V_{i-1}} f'(x) = \frac{\Delta U}{\Delta V} = \frac{U_i - U_{i-1}}{V_i - V_{i-1}}, \tag{3}$$

where ΔU is the potential difference variation (mV) and ΔV is the volume variation (mL).

Chlorine content on a dry basis ($w_{Cl,db}$) is then determined by Equation (4), in compliance with the European standard EN15289:

$$w_{Cl,db}\,(\%) = \frac{(C - C_0) \times V}{m} \times 100 \times \frac{100}{100 - M_{ad}} w_{Cl,db}(\%) = \frac{(C - C_0) \times V}{m} \times d \times \frac{100}{100 - M_{ad}}, \tag{4}$$

where C is the concentration of chloride in the solution (mg/L), C_0 is the concentration of chloride in the blank solution (mg/L), V is the volume of the solution (l), m is the mass of the test portion used in the digestion (mg), and M_{ad} is the moisture content in the analysis test sample (%).

4.5. Heating Value

Heating value, also known as the heating value, defines the energy content of biomass fuel [38]. This parameter can be described in two ways: higher heating value (HHV), or gross heating value, refers to the heat released from fuel combustion along with the vaporization energy from water. On the other hand, lower heating value (LHV) or net heating value is based on steam as the product, which means its vaporization energy is not considered heat [39]. The heating value of biomass, both higher and lower, can be determined experimentally by employing an adiabatic bomb calorimeter. The model used in this project was the 6400 Automatic Isoperibol Calorimeter by PARR INSTRUMENT. After each procedure, the equipment provides the corrected temperature increase that is later used for the determination of the heating value. Due to the nitrogen and oxygen-rich atmosphere inside the calorimeter, nitric acid and sulphuric acid are formed, respectively, and the heat of formation of both acids must be disregarded. For HNO_3, the wash water for the pump was titrated with NaOH (0.1 M), and Equation (5) was applied, while for H_2SO_4, knowing the sulfur content Equation (6) can be applied:

$$Q_{N,S} = 1.43 \times V_{NaOH}, \tag{5}$$

where $Q_{N,S}$ is the heat contribution relative to nitric acid formation (cal) and V_{NaOH} is the volume of NaOH used in the titration of the wash water of the pump (ml).

$$Q_{S,add} = 13.61 \times w_{S,db}, \tag{6}$$

where $Q_{S,add}$ is the additional contribution relative to sulfur dioxide formation and $w_{S,db}$ is the sulfur content on a dry basis (%).

With this information, Equation (7) can be applied to obtain the gross heating value, or high heating value, at a constant volume, $q_{V,gr}$ (J/g), as follows:

$$q_{V,gr} = \left(\frac{\varepsilon \times \theta - Q_{thread} - Q_{N,S}}{m} - Q_{S,add} \right) \times 4.1868, \tag{7}$$

where ε is the heating capacity of the calorimeter (previously determined) (cal/°C), θ is the corrected temperature increase (°C), Q_{thread} is the heat contribution relative to the thread combustion (cal), $Q_{N,S}$ is the heat contribution relative to nitric acid formation (cal), $Q_{S,add}$ is the additional contribution relative to sulfur dioxide formation (cal), and m is the mass of the sample (g).

Equation (8) was used to calculate the gross heating value at constant volume on a dry basis, $q_{V,gr,db}$ (J/g), as follows:

$$q_{V,gr,db} = q_{V,gr} \times \frac{100}{100 - M_{ad}} \quad q_{V,gr,db} = q_{V,gr} \times \frac{100}{100 - M_{ad}}, \tag{8}$$

where $q_{V,gr,db}$ is the gross heating value at constant volume (J/g) and M_{ad} is the moisture content in the analysis test sample (%). Lastly, the net heating value at constant pressure on a dry basis, $q_{p,net,db}$ (J/g), can be calculated through Equation (9), as follows:

$$q_{p,net,db} = q_{V,gr,db} - 212.2 \times w_{H,db} - 0.8 \times \left(w_{O,db} + w_{N,db} \right), \tag{9}$$

where $q_{V,gr,db}$ is the gross heating value at constant volume on a dry basis (J/g), $w_{H,db}$ is the hydrogen content on a dry basis (%), $w_{O,db}$ is the oxygen content on a dry basis (%), and $w_{N,db}$ is the nitrogen content on a dry basis (%). According to the European standard EN14918, $(w_{O,db} + w_{N,db})$ is obtained from Equation (10), as follows:

$$\left(w_{O,db} + w_{N,db} \right) = 100 - w_{A,db} - w_{C,db} - w_{H,db} - w_{S,db} \tag{10}$$

where $w_{A,db}$ is the ash content on a dry basis (%), $w_{C,db}$ is the carbon content on a dry basis (%), $w_{H,db}$ is the hydrogen content on a dry basis (%), and $w_{S,db}$ is the sulfur content on a dry basis (%).

4.6. Chemical Analysis by ICP-OES

Inductively coupled plasma atomic emission spectroscopy (ICP-AES), also known as inductively coupled plasma optical emission spectrometry (ICP-OES), is an analytical technique, which produces excited atoms and ions that emit electromagnetic radiation at different wavelengths and is used for the determination of trace elements. The main advantages are its multi-element capability, broad dynamic range, and effective background correction [40]. For the preparation of the samples, microwave digestion was once again necessary to ensure that the capillaries did not get obstructed. The model used for the analysis was a THERMO SCIENTIFIC (iCAP 6000 series). A peristaltic pump delivered the digested samples to an analytical nebulizer and introduced them into the plasma flame that breaks down the samples into charged ions, releasing radiation with specific wavelengths. In the end, a software generates a spreadsheet with the results. Equations (11) and (12) were used, as follows, to calculate the content of each element in the sample on a dry basis ($w_{i,db}$), in compliance with standards EN15289, EN15290 and EN15297:

$$w_{i,db} = \left(\frac{mg}{kg}\right) = \frac{(C - C_{i,0}) \times V}{m} \times \frac{100}{100 - M_{ad}} \tag{11}$$

where C_i is the concentration of the element in the diluted sample digest (mg/L), $C_{i,0}$ is the concentration of the element in the solution of the blank experiment (mg/L), V is the volume of the diluted sample digest solution (mL), m is the mass of the test portion used (g), and M_{ad} is the moisture content in the analysis test sample (%).

$$w_{S,db}\,(\%) = \frac{(C - C_0) \times V}{m} \times 0.3338 \times 100 \times \frac{100}{100 - M_{ad}} \tag{12}$$

where C_1 is the concentration of sulfate in the solution (mg/L), $C_{i,0}$ is the concentration of sulfate in the solution of the blank experiment (mg/L), V is the volume of the diluted sample digest solution (mL), m is the mass of the test portion used (g), 0.3338 is the stoichiometric ratio of the relative molar masses of sulfur and sulfate, and M_{ad} is the moisture content in the analysis test sample (%).

4.7. Fusibility of the Ashes

This determination makes it possible to estimate the behavior of biomass ash when combustion is carried out, for example, in boilers or burners. During the combustion process, ashes may occur, such as slagging and fouling. Slagging is the deposit of ash in the bottom and walls of the furnaces, while fouling is the deposit of ash in the air flow zones (which can cause clogging of pipes). The formation of slagging and fouling depends on the ash content of a biomass, as well as on the elemental composition of the ash. The determination of ash fusibility consists of monitoring the behavior of ash melting along a temperature ramp. Thus, this test allows to predict the formation of slagging and fouling in thermal conversion processes. These data must be related to the ash content (determined using the TGA) and the content of the different ash components (determined by ICP/OES). The fusibility test can be carried out with an oxidizing atmosphere (air) or reducing atmosphere (60% CO + 40% CO_2). The choice of atmosphere must be related to the combustion conditions of the boiler or burner. If the boiler operates in atmospheres rich in fuel (with oxygen deficit), its atmosphere will be mostly reducing, with incomplete combustion and CO formation. As a general rule, reducing atmospheres cause ash to melt at lower temperatures, thus causing greater slagging and fouling problems. Therefore, the fuse test must reflect these characteristics and adapt to the customer's combustion process. During the fusibility test, the ash melting behavior is monitored and the following characteristic temperatures are determined:

- Initial temperature: temperature at which the test starts up to 550 °C;
- Shrinking temperature: shrinkage to 95% of the area recorded at 550 °C;
- Deformation temperature: temperature at which the first rounding of the vertices of the cylinders occurs;
- Hemisphere temperature: temperature at which the height of the cylinder is equal to half the width;

- Fluid temperature: temperature at which the height is equal to half the height recorded at the hemisphere temperature.

In the present study, the samples, converted to ashes, were subsequently placed in a plastic dish, where two drops of ethyl alcohol are added, and using a spatula they are homogenized until a uniform paste is obtained. Then this paste is transferred to the mold, where the cylinder is compacted. After being removed from the mold, the cylinders are placed on the zirconia lamella. The samples are then placed inside the chamber of the ash fusibility furnace, which in this specific case was a SYLAB IF 2000-G device.

5. Conclusions

The use of residual biomass in the production of wood pellets is an opportunity for the circularization of the local economy associated with forest management operations. It also presents itself as an opportunity for the sustainability of operations to control invasive species, as it contributes to the creation of value chains for residual products that until now had no added value. The incorporation of these materials in the production of certified wood pellets presents some difficulties, since these materials do not meet the chemical requirements imposed by the standards that regulate the quality of the final products. However, the recovery of residual biomass from actions to control and eradicate invasive species can be a reality, especially for uses that do not imply product certification, and especially when the recovery of materials in industrial environments is not involved, where adverse effects, such as corrosion, fouling or slagging, can result in serious damage and unforeseen maintenance to combustion equipment.

Author Contributions: Conceptualization, L.J.R.N., J.C.O.M. and A.M.R.; methodology, L.J.R.N. and J.C.O.M.; validation, L.J.R.N., L.C.R.S. and L.M.E.F.L.; formal analysis, L.J.R.N., J.C.O.M. and A.M.R.; investigation, L.J.R.N., L.C.R.S. and L.M.E.F.L.; data curation, L.J.R.N., J.C.O.M., A.M.R., L.C.R.S. and L.M.E.F.L.; writing—original draft preparation, L.J.R.N., J.C.O.M. and A.M.R.; writing—review and editing, L.J.R.N., J.C.O.M. and A.M.R.; supervision, L.J.R.N., J.C.O.M. and A.M.R. All authors have read and agreed to the published version of the manuscript.

Funding: This work is a result of the project TECH—Technology, Environment, Creativity and Health, Norte-01-0145-FEDER-000043, supported by Norte Portugal Regional Operational Program (NORTE 2020), under the PORTUGAL 2020 Partnership Agreement, through the European Regional Development Fund (ERDF). L.J.R.N. was supported by proMetheus, Research Unit on Energy, Materials and Environment for Sustainability—UIDP/05975/2020, funded by national funds through FCT—Fundação para a Ciência e Tecnologia.

Acknowledgments: The authors would like to acknowledge the companies YGE—Yser Green Energy SA, and AFS—Advanced Fuel Solutions SA, both in Portugal, for the execution of the laboratory tests.

Conflicts of Interest: The authors declare no conflict of interest.

References

1. Karkania, V.; Fanara, E.; Zabaniotou, A. Review of sustainable biomass pellets production–A study for agricultural residues pellets' market in Greece. *Renew. Sustain. Energy Rev.* **2012**, *16*, 1426–1436. [CrossRef]
2. Lackner, K.S.; Sachs, J. A robust strategy for sustainable energy. *Brook. Pap. Econ. Act.* **2005**, *2005*, 215–284. [CrossRef]
3. Saidur, R.; Abdelaziz, E.; Demirbas, A.; Hossain, M.; Mekhilef, S. A review on biomass as a fuel for boilers. *Renew. Sustain. Energy Rev.* **2011**, *15*, 2262–2289. [CrossRef]
4. Sovacool, B.K. The political economy of energy poverty: A review of key challenges. *Energy Sustain. Dev.* **2012**, *16*, 272–282. [CrossRef]
5. Nejat, P.; Jomehzadeh, F.; Taheri, M.M.; Gohari, M.; Majid, M.Z.A. A global review of energy consumption, CO_2 emissions and policy in the residential sector (with an overview of the top ten CO_2 emitting countries). *Renew. Sustain. Energy Rev.* **2015**, *43*, 843–862. [CrossRef]
6. Xiao, R.; Chen, X.; Wang, F.; Yu, G. The physicochemical properties of different biomass ashes at different ashing temperature. *Renew. Energy* **2011**, *36*, 244–249. [CrossRef]
7. Jenkins, B.; Baxter, L.; Miles, T., Jr.; Miles, T. Combustion properties of biomass. *Fuel Process. Technol.* **1998**, *54*, 17–46. [CrossRef]
8. Teixeira, P.; Lopes, H.; Gulyurtlu, I.; Lapa, N.; Abelha, P. Evaluation of slagging and fouling tendency during biomass co-firing with coal in a fluidized bed. *Biomass Bioenergy* **2012**, *39*, 192–203. [CrossRef]

9. Vamvuka, D.; Kakaras, E. Ash properties and environmental impact of various biomass and coal fuels and their blends. *Fuel Process. Technol.* **2011**, *92*, 570–581. [CrossRef]

10. Tedim, F.; Leone, V.; Xanthopoulos, G. A wildfire risk management concept based on a social-ecological approach in the European Union: Fire Smart Territory. *Int. J. Disaster Risk Reduct.* **2016**, *18*, 138–153. [CrossRef]

11. Schooler, S.S.; Cook, T.; Prichard, G.; Yeates, A.G. Disturbance-mediated competition: The interacting roles of inundation regime and mechanical and herbicidal control in determining native and invasive plant abundance. *Biol. Invasions* **2010**, *12*, 3289–3298. [CrossRef]

12. Ehrenfeld, J.G. Ecosystem consequences of biological invasions. *Annu. Rev. Ecol. Evol. Syst.* **2010**, *41*, 59–80. [CrossRef]

13. Nunes, L.J.; Raposo, M.A.; Meireles, C.I.; Pinto Gomes, C.J.; Ribeiro, N.; Almeida, M. Control of Invasive Forest Species through the Creation of a Value Chain: Acacia dealbata Biomass Recovery. *Environments* **2020**, *7*, 39. [CrossRef]

14. Nunes, L.J.; Meireles, C.I.; Pinto Gomes, C.J.; Almeida Ribeiro, N. Historical development of the portuguese forest: The introduction of invasive species. *Forests* **2019**, *10*, 974. [CrossRef]

15. Fernandes, M.; Devy-Vareta, N.; Rangan, H. Plantas exóticas invasoras e instrumentos de gestão territorial. O caso paradigmático do género Acacia em Portugal. *Rev. De Geogr. E Ordenam. Do Territ.* **2013**, *1*, 83–107. [CrossRef]

16. Ferreira, S.; Monteiro, E.; Brito, P.; Vilarinho, C. Biomass resources in Portugal: Current status and prospects. *Renew. Sustain. Energy Rev.* **2017**, *78*, 1221–1235. [CrossRef]

17. de Jonge, V.N.; Pinto, R.; Turner, R.K. Integrating ecological, economic and social aspects to generate useful management information under the EU Directives 'ecosystem approach'. *Ocean Coast. Manag.* **2012**, *68*, 169–188. [CrossRef]

18. Stupak, I.; Asikainen, A.; Jonsell, M.; Karltun, E.; Lunnan, A.; Mizaraitė, D.; Pasanen, K.; Pärn, H.; Raulund-Rasmussen, K.; Röser, D. Sustainable utilisation of forest biomass for energy—possibilities and problems: Policy, legislation, certification, and recommendations and guidelines in the Nordic, Baltic, and other European countries. *Biomass Bioenergy* **2007**, *31*, 666–684. [CrossRef]

19. Sikkema, R.; Junginger, M.; Van Dam, J.; Stegeman, G.; Durrant, D.; Faaij, A. Legal harvesting, sustainable sourcing and cascaded use of wood for bioenergy: Their coverage through existing certification frameworks for sustainable forest management. *Forests* **2014**, *5*, 2163–2211. [CrossRef]

20. El-Sayed, A.; Suckling, D.; Wearing, C.; Byers, J. Potential of mass trapping for long-term pest management and eradication of invasive species. *J. Econ. Entomol.* **2006**, *99*, 1550–1564. [CrossRef]

21. Stickler, C.M.; Nepstad, D.C.; Coe, M.T.; McGrath, D.G.; Rodrigues, H.O.; Walker, W.S.; SOARES-FILHO, B.S.; Davidson, E.A. The potential ecological costs and cobenefits of REDD: A critical review and case study from the Amazon region. *Glob. Chang. Biol.* **2009**, *15*, 2803–2824. [CrossRef]

22. Cai, J.; He, Y.; Yu, X.; Banks, S.W.; Yang, Y.; Zhang, X.; Yu, Y.; Liu, R.; Bridgwater, A.V. Review of physicochemical properties and analytical characterization of lignocellulosic biomass. *Renew. Sustain. Energy Rev.* **2017**, *76*, 309–322. [CrossRef]

23. García, R.; Pizarro, C.; Lavín, A.G.; Bueno, J.L. Characterization of Spanish biomass wastes for energy use. *Bioresour. Technol.* **2012**, *103*, 249–258. [CrossRef]

24. Chiang, K.-Y.; Chien, K.-L.; Lu, C.-H. Characterization and comparison of biomass produced from various sources: Suggestions for selection of pretreatment technologies in biomass-to-energy. *Appl. Energy* **2012**, *100*, 164–171. [CrossRef]

25. Wilson, L.; Yang, W.; Blasiak, W.; John, G.R.; Mhilu, C.F. Thermal characterization of tropical biomass feedstocks. *Energy Convers. Manag.* **2011**, *52*, 191–198. [CrossRef]

26. Fang, J.; Gao, B.; Chen, J.; Zimmerman, A.R. Hydrochars derived from plant biomass under various conditions: Characterization and potential applications and impacts. *Chem. Eng. J.* **2015**, *267*, 253–259. [CrossRef]

27. Patel, B.; Gami, B. Biomass characterization and its use as solid fuel for combustion. *Iran. J. Energy Environ.* **2012**, *3*, 123–128. [CrossRef]

28. Chen, W.-H.; Lu, K.-M.; Lee, W.-J.; Liu, S.-H.; Lin, T.-C. Non-oxidative and oxidative torrefaction characterization and SEM observations of fibrous and ligneous biomass. *Appl. Energy* **2014**, *114*, 104–113. [CrossRef]

29. Neves, D.; Thunman, H.; Matos, A.; Tarelho, L.; Gómez-Barea, A. Characterization and prediction of biomass pyrolysis products. *Prog. Energy Combust. Sci.* **2011**, *37*, 611–630. [CrossRef]

30. Viana, H.; Vega-Nieva, D.; Torres, L.O.; Lousada, J.; Aranha, J. Fuel characterization and biomass combustion properties of selected native woody shrub species from central Portugal and NW Spain. *Fuel* **2012**, *102*, 737–745. [CrossRef]

31. Nunes, L.; Matias, J.; Catalão, J. Biomass combustion systems: A review on the physical and chemical properties of the ashes. *Renew. Sustain. Energy Rev.* **2016**, *53*, 235–242. [CrossRef]

32. Girón, R.; Ruiz, B.; Fuente, E.; Gil, R.; Suárez-Ruiz, I. Properties of fly ash from forest biomass combustion. *Fuel* **2013**, *114*, 71–77. [CrossRef]

33. Agar, D.A.; Rudolfsson, M.; Kalén, G.; Campargue, M.; Perez, D.D.S.; Larsson, S.H. A systematic study of ring-die pellet production from forest and agricultural biomass. *Fuel Process. Technol.* **2018**, *180*, 47–55. [CrossRef]

34. de Souza, H.J.P.L.; Arantes, M.D.C.; Vidaurre, G.B.; Andrade, C.R.; Carneiro, A.d.C.O.; de Souza, D.P.L.; de Paula Protásio, T. Pelletization of eucalyptus wood and coffee growing wastes: Strategies for biomass valorization and sustainable bioenergy production. *Renew. Energy* **2020**, *149*, 128–140. [CrossRef]

35. Nunes, L.; Matias, J.C.; Catalao, J.P. Wood pellets as a sustainable energy alternative in Portugal. *Renew. Energy* **2016**, *85*, 1011–1016. [CrossRef]

36. Duca, D.; Riva, G.; Pedretti, E.F.; Toscano, G. Wood pellet quality with respect to EN 14961-2 standard and certifications. *Fuel* **2014**, *135*, 9–14. [CrossRef]
37. Quinteiro, P.; Tarelho, L.; Marques, P.; Martín-Gamboa, M.; Freire, F.; Arroja, L.; Dias, A.C. Life cycle assessment of wood pellets and wood split logs for residential heating. *Sci. Total Environ.* **2019**, *689*, 580–589. [CrossRef] [PubMed]
38. Friedl, A.; Padouvas, E.; Rotter, H.; Varmuza, K. Prediction of heating values of biomass fuel from elemental composition. *Anal. Chim. Acta* **2005**, *544*, 191–198. [CrossRef]
39. Yin, C.-Y. Prediction of higher heating values of biomass from proximate and ultimate analyses. *Fuel* **2011**, *90*, 1128–1132. [CrossRef]
40. Uvegi, H.; Chaunsali, P.; Traynor, B.; Olivetti, E. Reactivity of industrial wastes as measured through ICP-OES: A case study on siliceous Indian biomass ash. *J. Am. Ceram. Soc.* **2019**, *102*, 7678–7688. [CrossRef]

 *recycling*

Article

Thermochemical Conversion of Olive Oil Industry Waste: Circular Economy through Energy Recovery

Leonel J. R. Nunes [1,2,]* **, Liliana M. E. F. Loureiro** [3]**, Letícia C. R. Sá** [3] **and Hugo F.C. Silva** [4]

[1] PROMETHEUS—Unidade de Investigação em Materiais, Energia e Ambiente para a Sustentabilidade, Escola Superior Agrária, Instituto Politécnico de Viana do Castelo, Rua da Escola Industrial e Comercial de Nun'Alvares, 4900-347 Viana do Castelo, Portugal

[2] GOVCOPP—Unidade de Investigação em Governança, Competitividade e Políticas Públicas, DEGEIT—Departamento de Economia, Gestão, Engenharia Industrial e Turismo, Universidade de Aveiro, Campus Universitário de Santiago, 3810-193 Aveiro, Portugal

[3] YGE—Yser Green Energy SA, Área de Acolhimento Empresarial de Úl/Loureiro, Lote 17, 3720-075 Loureiro OAZ, Portugal; liliana.loureiro@ygenergia.com (L.M.E.F.L.); leticia.sa@ygenergia.com (L.C.R.S.)

[4] AFS—Advanced Fuel Solutions SA, Área de Acolhimento Empresarial de Úl/Loureiro, Lote 17, 3720-075 Loureiro OAZ, Portugal; h.silva@adfuelsolutions.com

* Correspondence: leonelnunes@esa.ipvc.pt; Tel.: +351-258-909-740

Received: 6 April 2020; Accepted: 31 May 2020; Published: 1 June 2020

Abstract: The demand for new sources of energy is one of the main quests for humans. At the same time, there is a growing need to eliminate or recover a set of industrial or agroforestry waste sources. In this context, several options may be of interest, especially given the amounts produced and environmental impacts caused. Olive pomace can be considered one of these options. Portugal, as one of the most prominent producers of olive oil, therefore, also faces the problem of dealing with the waste of the olive oil industry. Olive pomace energy recovery is a subject referenced in many different studies and reports since long ago. However, traditional forms of recovery, such as direct combustion, did not prove to be the best solution, mainly due to its fuel properties and other characteristics, which cause difficulties in its storage and transportation as well. Torrefaction and pyrolysis can contribute to a volume reduction, optimizing storage and transportation. In this preliminary study, were carried out torrefaction and pyrolysis tests on olive pomace samples, processed at 300 °C, 400 °C, and 500 °C, followed by laboratory characterization of the materials. It was verified an improvement in the energy content of the materials, demonstrating that there is potential for the use of these thermochemical conversion technologies for the energy recovery of olive pomace.

Keywords: olive pomace; thermochemical conversion; energy recovery; circular economy; biomass waste

---

1. Introduction

Olive oil production is an activity of vital economic importance in Mediterranean countries [1]. Portugal is a reference in modern olive growing, and is expected during next decade to be the third largest producer of olive oil in the world and the seventh largest in terms of occupied area [2]. This growth is sustained by the evolution in the country and, in particular, in Alentejo, Southern Portugal, with the contribution of Alqueva dam that provided the irrigation for intensive crops [3]. For example, in 2018, which saw adverse weather conditions, the results presented in SIAZ—Olive Oil and Table Olives Information System-https://www.gpp.pt—indicated a 30% drop in oil production compared to the previous campaign. Despite this, production was 4% higher than the average production in the last four campaigns. Currently, as the ninth producer in the world, Portugal has 361 thousand hectares of olive groves, and 135 thousand tons of oil extracted.

Olive oil production is associated with the generation of residues, olive pomace, and red waters, which present potentially toxic compositions and are harmful to ecosystems when improperly discharged into the environment [4]. Due to the high levels of oils and fats, the chemical deficiency of oxygen, total solids, polyphenols, and red waters, in order to be able to be sent either to emissaries or directly to the receiving environment, need extremely efficient treatments, with significant investment and operating costs that can make the mills unfeasible. Within these residues, olive pomace has viable forms of recovery already known and in use [5–7]. This olive pomace, resulting from the recently introduced two-stage centrifugation system, is a semi-solid, moderately acidic residue formed by pieces of olive stone, olive pulp, and water [8]. Its composition varies according to the variety of olives, fruit ripeness, climatic conditions, and cultivation practices [9]. In general, it consists of high amounts of water (60–70%), residual oil retained in the pulp (2.5–3%), inorganic compounds, and appreciable amounts of lignin, cellulose, and hemicellulose, as well as other organic matter including proteins, polyalcohols, fatty acids, sugars, polyphenols, and pigments [9,10]. This organic load is responsible for its phytotoxic and antimicrobial characteristics, as well as its high moisture content, making it difficult to handle, store, and transport [11,12].

In recent years, many efforts have been made to find a way to efficiently recover olive pomace in order to add commercial value to this waste [13,14]. Within the forms of recovery already studied and mentioned in the literature, bioconversion of these residues into fertilizers can be highlighted, including application in the production of animal feed, the use as a substrate for the production of bioethanol/biomethane and biohydrogen, and bioconversion in some biopolymers or enzymes for other industries [15].

Despite all the advantages that are described in the literature for these conversion technologies, as capable of processing residual biomass mainly due to its ability to homogenize different types of raw materials with different qualities, these technologies still present restrictions, namely with regard to its scalability, and the fact that large production units must operate continuously. Many recent developments have been achieved, mainly with regard to process control and stability. However, it cannot yet be considered a mature technology, so a large investment in R&D is still needed.

Torrefaction and pyrolysis are thermochemical conversion technologies that present as main difference the temperature range where occur, being, respectively, 220 °C to 320 °C for torrefaction, and 320 °C to 600 °C for pyrolysis [16]. Both processes take place at atmospheric pressure, in an oxygen-poor environment, where residence times vary according to the purpose for which the process is intended, and in the case of roasting, the final yield tends towards the solid fraction, whereas in pyrolysis, this yield tends normally to the liquid fraction [17]. The use of these technologies for the processing of agroforestry residues is already widely documented in the bibliography, with some previous work already dedicated specifically to the thermo-conversion of olive pomace, mainly because these processes allow a very significant volume reduction, at the same time that confer properties such as hydrophobicity, thus allowing a more efficient storage without the risk of reaction with water or withstanding biological activity [18–21].

This work aims to evaluate the potential of using olive pomace as a renewable energy source in the perspective of circular economy, where waste is reused and incorporated into a new production cycle. For this, the research was divided into different stages, starting with the characterization of olive pomace, namely with regard to its original properties that will serve as a standard, followed by torrefaction and pyrolysis in a muffle at different temperatures, respectively, 300 °C, 400 °C, and 500 °C, followed by the determination of a set of characteristics, which allow the evaluation of the potential to improved energy recovery processes, optimizing circular economy in olive oil industrial sector, assuming definitively the potential that these materials present, provided that the initial disadvantages as a fuel are overcome, for example, with the use of technologies such as torrefaction or pyrolysis.

2. Materials and Methods

2.1. Samples Collection and Preparation

In this study, several analyses were carried out to characterize the evolution of the olive pomace properties resulting from the torrefaction and pyrolysis processes. Thus, different techniques were used, such as thermogravimetric analysis, elemental analysis, and heating value. The olive pomace samples analyzed, were collected in an olive oil extraction plant that operates in a two stages process, in the region of Vila Real (Northern Portugal), in the 2019 campaign.

Olive pomace samples were dried at a temperature of 90 °C for 6 h. Samples were weighed to be approximately 500 g and wrapped in aluminum foil to ensure an atmosphere with low oxygen content, as presented in Figure 1. Tests were carried out in duplicate, placing two samples at the same time in each batch, to guarantee their reproducibility.

Figure 1. Assemblage of the samples in aluminum foil for torrefaction and pyrolysis tests.

For torrefaction and pyrolysis, the experimental protocol described by Ribeiro et al. (2018) [22] was applied, which proposed using a high-temperature oven for thermochemical processes used already in other previous studies, such as that one presented by Sá et al. (2020) [23]. The samples were torrefied and pyrolyzed in the muffle, which was programmed according to temperature (°C) and residence time (minutes) as presented in Table 1. The selection of parameters was done in a sequential manner, taking into account the results obtained for each one of the tests carried out. Initial and processed materials are presented in Figure 2.

Table 1. First series of conversion tests.

Residence Time (Minutes)	Test at 300 °C	Test at 400 °C	Test at 500 °C
30	Room Temperature to 180 °C	Room Temperature to 180 °C	Room Temperature to 180 °C
60	180 °C to 300 °C	180 °C to 400 °C	180 °C to 500 °C
90	300 °C	400 °C	500 °C
Time enough to safely open the muffle and collect the material	300 °C to Room Temperature	400 °C to Room Temperature	500 °C to Room Temperature

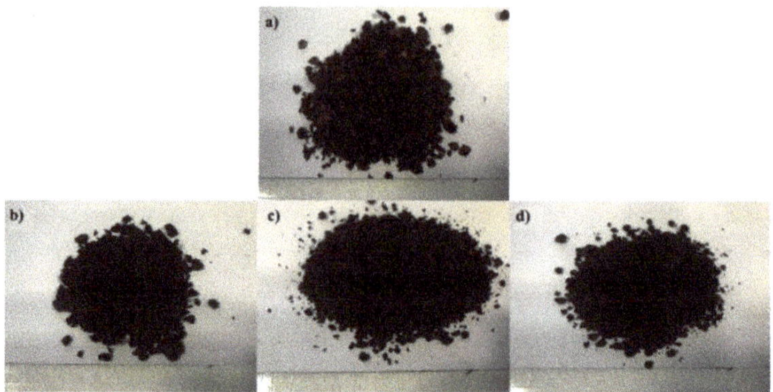

Figure 2. (**a**) Dried sample; (**b**) 300 °C; (**c**) 400 °C; and (**d**) 500 °C.

2.2. Determination of Elemental Composition

To determine the elemental composition was used a CHN analyzer in accordance with the procedure described in the standard EN 15104: 2011—Solid Biofuels—Determination of Total Content of Carbon, Hydrogen and Nitrogen—Instrumental Methods. After obtaining the results for carbon, hydrogen and nitrogen content, the following equation was used to calculate the oxygen content:

$$w(O) = 100 - w(C) - w(H) - w(N) \tag{1}$$

where $w(O)$ is the oxygen content (%), $w(C)$ is the carbon content (%), $w(H)$ is the hydrogen content (%), and $w(N)$ is the nitrogen content (%).

2.3. Thermogravimetric Analysis (TGA)

Thermogravimetric analysis (TGA) was conducted in accordance with the procedures described in the standards EN 14775: 2009—Solid Biofuels—Determination of Ash Content, EN 15148: 2009—Solid Biofuels—Determination of Volatiles Content and EN 14774-3: 2009—Solid Biofuels—Determination of Moisture Content. The entire procedure requires ground materials before being introduced into the melting pots, which were loaded with approximately 1 g of material.

2.4. Determination of Heating Value

To determine HHV, the equation deduced by Parikh et al. (2005) was used as follows [24]:

$$HHV_{(db)} = 0.3536FC + 0.1559V - 0.0078A \tag{2}$$

where $HHV_{(db)}$ is the High Heating Value (MJ.kg^{-1}) on a dry basis, FC is the Fixed Carbon (%) content, V is the Volatile content (%), and A is the Ash content (%).

3. Results and Discussion

3.1. Elemental Analysis

The results obtained in the analysis of the elemental composition are shown in Table 2. As already mentioned, all analyses were performed in duplicate, so the value presented is the average, and the same applies to all the following results.

Table 2. Ultimate analysis for the different process temperatures.

Elements	Original Dry Sample	$T_{300\,°C}$	$T_{400\,°C}$	$T_{500\,°C}$
C (%)	56.2	63.9	77.3	86.7
H (%)	6.8	7.2	3.8	3.4
N (%)	1.2	1.6	2.5	1.2
O (%)	35.8	27.3	16.4	8.7
S (%)	<0.01	<0.01	<0.01	<0.01

The determined sulfur (S) content proved to be lower than the lower detection limit of the equipment used in the analysis, so it was considered, for the purposes of calculating other parameters, to be equal to the referred limit value, which was 0.01% on a dry basis. However, in relation to the S content present in olive pomace, a more rigorous determination of the levels that can be found in these residues will be necessary, since there are very different opinions among other works on the subject [25–30].

In the elemental analysis, there is a decrease in the levels of hydrogen and oxygen, which is understood as normal, due to the fact that with the increase in temperature there is a devolatilization of hydrogenated and oxygenated compounds, mainly related to the depolymerization of hemicellulose, when the process takes place at 300 °C, and lignin and cellulose, when the temperature is increased to 400 °C and 500 °C.

The nitrogen content remains practically unchanged during the different tests at different temperatures, with a small concentration at 400 °C. Nitrogen is eliminated from the solid yield when 500 °C is reached, most likely due to the formation of some compounds, such as light nitriles (acetonitrile, propanenitrile), long chain nitriles and amides, pyrrolic and pyrrolidinic compounds, and diketopiperazines (DKPs), as stated in several previous research works [31–33], and that are emitted at this stage.

3.2. Thermogravimetric Analysis

The results obtained in the analysis of the proximate composition are shown in Table 3.

Table 3. Proximate composition for the different process temperatures.

	Original Dry Sample	Test at 300 °C	Test at 400 °C	Test at 500 °C
Fixed Carbon (%)	18.84	26.12	75.49	82.14
Volatiles (%)	79.85	75.59	21.08	14.36
Ashes (%)	1.31	1.30	3.45	3.51
Moisture (%)	3.51	1.52	0.91	0.57

The TGA results highlight the evolution of fixed carbon, which increases significantly from 18.84% to 82.14%. This fact indicates, first, a directly proportional increase in heating value, but it also opens the door to the creation of products with more added value, which may even justify other studies. Concerning the ashes, there is also an increase in the content, from the initial 1.31%, to 3.51% at 500 °C, which is due to the fact that the loss of mass leads to this concentration of the mineral fraction, which is not lost through the volatilization of oxygenated compounds. Volatile content also presents an expected behavior, since it decreases throughout the process, being minimal in the material produced at 500 °C, with a value of 14.36%.

3.3. Heating Value

The results obtained for the calculation of High Heating Value (HHV) are shown in Table 4.

Table 4. High Heating Value (HHV) calculated for the different process temperatures.

	Original Dry Sample	Test at 300 °C	Test at 400 °C	Test at 500 °C
HHV (MJ.kg^{-1})	19.10	21.01	29.95	31.26
Mass loss	-	37%	76%	78%

The determination of HHV by the method described by Parikh et al. (2005) [24] allows the evaluation of the energy content of a fuel by knowing other properties, such as fixed carbon, volatile and ash content. Thus, the initial heating value of the thermally untreated sample, of 19.20 MJ.kg^{-1}, increases up to 31.26 MJ.kg^{-1}. These values are in accordance to those presented in previous studies, such as the presented by Guizani et al. (2016), where a two-step process based on torrefaction and combustion is proposed for olive pomace recovery and where is concluded that torrefaction improves strongly the combustion properties [20].

The mass loss verified during the tests was, respectively, 37%, 76%, and 78%. This result can be interesting if associated with the energy recovery is also the environmental issue of reducing the volume of waste generated. Previous studies such as those from Bridgeman et al. (2008) found yields for torrefaction of *Phalaris arundinacea*, processed in a range of temperatures comprised from 230 to 290 °C, with mass losses of 7% to 32% [34]. Bergman and Kiel (2005) report typical values of 20% for mass loss [35]. Felfli et al. (2005) found mass losses from 6% to 46%, respectively for temperatures from 220 °C to 270 °C, for wooden briquettes processed for a period of 60 min [36]. Concerning higher temperature tests can be found several in the literature available [37–39]. However, once in these studies the technology used was fast high temperature pyrolysis, the main objective was to obtain a gaseous yield the highest possible, being in the majority of the cases reached 95% [37]. When comparing the results obtained in the tests conducted in the present study with those found in the referred literature, mass loss results for torrefaction temperature range can be considered similar. For the pyrolysis temperature range, above 300 °C, it was not possible to find comparable results, once the procedures used to the obtention of the samples were considerably different from the one used here.

4. Conclusions

Torrefaction and pyrolysis are thermochemical conversion processes that can be the starting point for the development of other conversion technologies. That is, can act as pre-processing technologies before the use of other processes for the production of energy and for a more efficient methodology to reduce waste volumes. It is already possible to find many cases in the literature in reference to the use of these materials in biomass gasification and liquefaction processes, where olive pomace can be included.

The results obtained in this preliminary study showed a significant increase of 11.5%, 57%, and 63% for the heating value, respectively, for the tests performed at 300 °C, 400 °C, and 500 °C, with mass losses of 37%, 76%, and 78%. In this way, the options defined by the tests at 400 °C and 500 °C are potentially interesting, since mass loss, although significant, may allow the process to become viable. However, it is necessary to carry out further tests, namely, using a torrefaction unit on a pilot-industrial scale, in order to confirm the feasibility of the processes.

It is possible to see that there is a strong probability that these technologies can contribute to the resolution of a set of problems, namely, through the elimination and recovery of a set of waste, such as those resulting from the olive oil industry, which are responsible for great environmental concerns, but also because can be an alternative to the production of renewable fuels that allow, for example, the replacement of others with fossil origin, guaranteeing the continuity of its operation, and thus the existence of backup units that ensure the stability of the energy supply, when other renewable sources, more intermittent, are unable to assure it.

Author Contributions: Conceptualization, L.J.R.N.; methodology, L.J.R.N. and H.F.C.S.; validation, L.J.R.N., L.C.R.S. and L.M.E.F.L.; formal analysis, L.J.R.N.; investigation, L.J.R.N., L.C.R.S., L.M.E.F.L.; resources, H.F.C.S.;

data curation, L.J.R.N., L.C.R.S. and L.M.E.F.L.; writing—original draft preparation, L.J.R.N.; writing—review and editing, L.J.R.N. and H.F.C.S.; supervision, L.J.R.N.; funding acquisition, H.F.C.S. All authors have read and agreed to the published version of the manuscript.

Funding: This research received no external funding.

Acknowledgments: The authors would like to acknowledge the Portuguese companies YGE—Yser Green Energy SA, and AFS—Advanced Fuel Solutions SA, both in Portugal, for the execution of the laboratory tests.

Conflicts of Interest: The authors declare no conflict of interest.

References

1. Visioli, F.; Bellosta, S.; Galli, C. Oleuropein, the bitter principle of olives, enhances nitric oxide production by mouse macrophages. *Life Sci.* **1998**, *62*, 541–546. [CrossRef]
2. Czekaj, M.; Hernández, P.; Fonseca, A.; Rivera, M.; Żmija, K.; Żmija, D. Uncovering Production Flows from Small Farms: Results from Poland and Portugal Case Studies. *Rocz. Nauk. Stowarzyszenia Ekon. Rol. I Agrobiz.* **2019**, *21*, 49–61. [CrossRef]
3. Martins, N.; Jiménez-Morillo, N.T.; Freitas, F.; Garcia, R.; da Silva, M.G.; Cabrita, M.J. Revisiting 3D van Krevelen diagrams as a tool for the visualization of volatile profile of varietal olive oils from Alentejo region, Portugal. *Talanta* **2020**, *207*, 120276. [CrossRef]
4. Azbar, N.; Bayram, A.; Filibeli, A.; Muezzinoglu, A.; Sengul, F.; Ozer, A. A review of waste management options in olive oil production. *Crit. Rev. Environ. Sci. Technol.* **2004**, *34*, 209–247. [CrossRef]
5. Aliakbarian, B.; Casazza, A.A.; Perego, P. Valorization of olive oil solid waste using high pressure–high temperature reactor. *Food Chem.* **2011**, *128*, 704–710. [CrossRef]
6. Haddadin, M.S.; Abdulrahim, S.M.; Al-Khawaldeh, G.Y.; Robinson, R.K. Solid state fermentation of waste pomace from olive processing. *J. Chem. Technol. Biotechnol. Int. Res. Process Environ. Clean Technol.* **1999**, *74*, 613–618. [CrossRef]
7. Roig, A.; Cayuela, M.L.; Sánchez-Monedero, M. An overview on olive mill wastes and their valorisation methods. *Waste Manag.* **2006**, *26*, 960–969. [CrossRef]
8. Cucci, G.; Lacolla, G.; Caranfa, L. Improvement of soil properties by application of olive oil waste. *Agron. Sustain. Dev.* **2008**, *28*, 521–526. [CrossRef]
9. López-Piñeiro, A.; Albarrán, A.; Nunes, J.R.; Barreto, C. Short and medium-term effects of two-phase olive mill waste application on olive grove production and soil properties under semiarid Mediterranean conditions. *Bioresour. Technol.* **2008**, *99*, 7982–7987. [CrossRef]
10. Tsantila, N.; Karantonis, H.C.; Perrea, D.N.; Theocharis, S.E.; Iliopoulos, D.G.; Antonopoulou, S.; Demopoulos, C.A. Antithrombotic and antiatherosclerotic properties of olive oil and olive pomace polar extracts in rabbits. *Mediat. Inflamm.* **2007**, *2007*, 36204. [CrossRef]
11. Morillo, J.; Antizar-Ladislao, B.; Monteoliva-Sánchez, M.; Ramos-Cormenzana, A.; Russell, N. Bioremediation and biovalorisation of olive-mill wastes. *Appl. Microbiol. Biotechnol.* **2009**, *82*, 25–39. [CrossRef] [PubMed]
12. Ayed, L.; Asses, N.; Chammem, N.; Othman, N.B.; Hamdi, M. Advanced oxidation process and biological treatments for table olive processing wastewaters: Constraints and a novel approach to integrated recycling process: A review. *Biodegradation* **2017**, *28*, 125–138. [CrossRef] [PubMed]
13. Galanakis, C.M. Recovery of high added-value components from food wastes: Conventional, emerging technologies and commercialized applications. *Trends Food Sci. Technol.* **2012**, *26*, 68–87. [CrossRef]
14. Fiol, N.; Villaescusa, I.; Martínez, M.; Miralles, N.; Poch, J.; Serarols, J. Sorption of Pb (II), Ni (II), Cu (II) and Cd (II) from aqueous solution by olive stone waste. *Sep. Purif. Technol.* **2006**, *50*, 132–140. [CrossRef]
15. Rafatullah, M.; Sulaiman, O.; Hashim, R.; Ahmad, A. Adsorption of methylene blue on low-cost adsorbents: A review. *J. Hazard. Mater.* **2010**, *177*, 70–80. [CrossRef]
16. Nunes, L.J. A Case Study about Biomass Torrefaction on an Industrial Scale: Solutions to Problems Related to Self-Heating, Difficulties in Pelletizing, and Excessive Wear of Production Equipment. *Appl. Sci.* **2020**, *10*, 2546. [CrossRef]
17. Nunes, L.; Matias, J.; Catalão, J. Torrefied Biomass Pellets: An alternative fuel for coal power plants. In Proceedings of the 2016 13th International Conference on the European Energy Market (EEM), Porto, Portugal, 6–9 June 2016; pp. 1–5.

18. Brachi, P.; Chirone, R.; Miccio, M.; Ruoppolo, G. Fluidized bed torrefaction of biomass pellets: A comparison between oxidative and inert atmosphere. *Powder Technol.* **2019**, *357*, 97–107. [CrossRef]
19. Barskov, S.; Zappi, M.; Buchireddy, P.; Dufreche, S.; Guillory, J.; Gang, D.; Hernandez, R.; Bajpai, R.; Baudier, J.; Cooper, R. Torrefaction of biomass: A review of production methods for biocoal from cultured and waste linocellulosic feedstocks. *Renew. Energy* **2019**, *142*, 624–642. [CrossRef]
20. Guizani, C.; Haddad, K.; Jeguirim, M.; Colin, B.; Limousy, L. Combustion characteristics and kinetics of torrefied olive pomace. *Energy* **2016**, *107*, 453–463. [CrossRef]
21. Cellatoğlu, N.; İlkan, M. Effects of torrefaction on carbonization characteristics of solid olive mill residue. *BioResources* **2016**, *11*, 6286–6298. [CrossRef]
22. Ribeiro, J.M.C.; Godina, R.; Matias, J.C.d.O.; Nunes, L.J.R. Future perspectives of biomass torrefaction: Review of the current state-of-the-art and research development. *Sustainability* **2018**, *10*, 2323. [CrossRef]
23. Sá, L.C.; Loureiro, L.M.; Nunes, L.J.; Mendes, A.M. Torrefaction as a pretreatment technology for chlorine elimination from biomass: A case study using Eucalyptus globulus Labill. *Resources* **2020**, *9*, 54. [CrossRef]
24. Parikh, J.; Channiwala, S.; Ghosal, G. A correlation for calculating HHV from proximate analysis of solid fuels. *Fuel* **2005**, *84*, 487–494. [CrossRef]
25. Miranda, T.; Arranz, J.; Montero, I.; Román, S.; Rojas, C.; Nogales, S. Characterization and combustion of olive pomace and forest residue pellets. *Fuel Process. Technol.* **2012**, *103*, 91–96. [CrossRef]
26. Eliche-Quesada, D.; Leite-Costa, J. Use of bottom ash from olive pomace combustion in the production of eco-friendly fired clay bricks. *Waste Manag.* **2016**, *48*, 323–333. [CrossRef]
27. Miranda, T.; Nogales, S.; Román, S.; Montero, I.; Arranz, J.I.; Sepúlveda, F.J. Control of several emissions during olive pomace thermal degradation. *Int. J. Mol. Sci.* **2014**, *15*, 18349–18361. [CrossRef]
28. Miranda, T.; Román, S.; Arranz, J.; Rojas, S.; González, J.; Montero, I. Emissions from thermal degradation of pellets with different contents of olive waste and forest residues. *Fuel Process. Technol.* **2010**, *91*, 1459–1463. [CrossRef]
29. Muscolo, A.; Papalia, T.; Settineri, G.; Romeo, F.; Mallamaci, C. Three different methods for turning olive pomace in resource: Benefits of the end products for agricultural purpose. *Sci. Total Environ.* **2019**, *662*, 1–7. [CrossRef]
30. Lanfranchi, M.; Giannetto, C.; De Pascale, A. Economic analysis and energy valorization of by-products of the olive oil process:"Valdemone DOP" extra virgin olive oil. *Renew. Sustain. Energy Rev.* **2016**, *57*, 1227–1236. [CrossRef]
31. Debono, O.; Villot, A. Nitrogen products and reaction pathway of nitrogen compounds during the pyrolysis of various organic wastes. *J. Anal. Appl. Pyrolysis* **2015**, *114*, 222–234. [CrossRef]
32. Wei, L.; Wen, L.; Yang, T.; Zhang, N. Nitrogen transformation during sewage sludge pyrolysis. *Energy Fuels* **2015**, *29*, 5088–5094. [CrossRef]
33. Choi, S.-S.; Ko, J.-E. Analysis of cyclic pyrolysis products formed from amino acid monomer. *J. Chromatogr. A* **2011**, *1218*, 8443–8455. [CrossRef] [PubMed]
34. Bridgeman, T.; Jones, J.; Shield, I.; Williams, P. Torrefaction of reed canary grass, wheat straw and willow to enhance solid fuel qualities and combustion properties. *Fuel* **2008**, *87*, 844–856. [CrossRef]
35. Bergman, P.C.; Kiel, J.H. Torrefaction for biomass upgrading. In Proceedings of the 14th European Biomass Conference, Paris, France, 17–21 October 2005; pp. 17–21.
36. Felfli, F.F.; Luengo, C.A.; Beat n, P.A. Wood briquette torrefaction. *Energy Sustain. Dev.* **2005**, *9*, 19–22. [CrossRef]
37. Kabakcı, S.B.; Aydemir, H. Pyrolysis of olive pomace and copyrolysis of olive pomace with refuse derived fuel. *Environ. Prog. Sustain. Energy* **2014**, *33*, 649–656. [CrossRef]
38. Volpe, R.; Messineo, A.; Millan, M.; Volpe, M.; Kandiyoti, R. Assessment of olive wastes as energy source: Pyrolysis, torrefaction and the key role of H loss in thermal breakdown. *Energy* **2015**, *82*, 119–127. [CrossRef]
39. Ounas, A.; Aboulkas, A.; Bacaoui, A.; Yaacoubi, A. Pyrolysis of olive residue and sugar cane bagasse: Non-isothermal thermogravimetric kinetic analysis. *Bioresour. Technol.* **2011**, *102*, 11234–11238. [CrossRef]

© 2020 by the authors. Licensee MDPI, Basel, Switzerland. This article is an open access article distributed under the terms and conditions of the Creative Commons Attribution (CC BY) license (http://creativecommons.org/licenses/by/4.0/).

Article

Biocrude Production via Non-Catalytic Supercritical Hydrothermal Liquefaction of *Fucus vesiculosus* Seaweed Processing Residues

Lukas Jasiūnas [1,*], Thomas Helmer Pedersen [2] and Lasse Aistrup Rosendahl [2]

1 Department of Organic Chemistry, Kaunas University of Technology, LT-50254 Kaunas, Lithuania
2 Department of Energy Technology, Aalborg University, 9220 Aalborg Øst, Denmark; thp@et.aau.dk (T.H.P.); lar@et.aau.dk (L.A.R.)
* Correspondence: lukas.jasiunas@ktu.lt

Abstract: The potential of using cold water brown macroalgae *Fucus vesiculosus* for biocrude production via non-catalytic supercritical hydrothermal liquefaction (HTL) was studied. Demineralization, residue neutralization, and high value-added product (alginate and fucoidan) extraction processes were carried out before using the biomass for HTL biocrude production. Acid leaching was carried out using three demineralization agents: distilled water, dilute citric acid solution, and the diluted acidic aqueous by-product from a continuous HTL pilot facility. Alginate was extracted via H_2SO_4 and $NaCO_3$ bathing, and fucoidan was extracted using $CaCl_2$. Experimental data show that none of the leaching agents was greatly efficient in removing inorganics, with citric acid leaching with extensive neutralization reaching the highest ash removal efficiency of 47%. The produced 6 sets of biocrudes were characterized by elemental and thermogravimetric analyses. Short (10-min retention) HTL and the extent of leaching residue neutralization were also investigated. Highest biocrude yields were recorded when liquefying non-neutralized citric acid leaching, alginate, and fucoidan extraction residues. On the other hand, thermochemical conversions of short retention time HTL, full neutralization extent, and baseline (dried raw macroalgae) biomass performed worse. Specifically, the highest biocrude yield of 28.2 ± 2.5 wt.% on dry ash-free feedstock basis was recorded when liquefying alginate extraction residues. Moreover, the highest energy recovery of 52.8% was recorded when converting fucoidan extraction residues.

Keywords: residue valorization; hydrothermal liquefaction; biorefinery; macroalgae; value-added products

Citation: Jasiūnas, L.; Pedersen, T.H.; Rosendahl, L.A. Biocrude Production via Non-Catalytic Supercritical Hydrothermal Liquefaction of *Fucus vesiculosus* Seaweed Processing Residues. *Recycling* **2021**, 6, 45. https://doi.org/10.3390/recycling6030045

Academic Editor: Leonel Jorge Ribeiro Nunes

Received: 30 April 2021
Accepted: 25 June 2021
Published: 4 July 2021

Publisher's Note: MDPI stays neutral with regard to jurisdictional claims in published maps and institutional affiliations.

Copyright: © 2021 by the authors. Licensee MDPI, Basel, Switzerland. This article is an open access article distributed under the terms and conditions of the Creative Commons Attribution (CC BY) license (https://creativecommons.org/licenses/by/4.0/).

1. Introduction

The transportation sector is engaging with innovation to address societal concerns over climate change. Numerous upcoming technologies are posed to significantly diminish our dependence on fossil fuels. Among the rapidly developing technologies are hydrogen-, electricity-, and electrofuel-based alternatives. Given the imminent transitional period, intermediary fuels will undoubtedly play a critical role to gradually transform the current well-established infrastructure of liquid fuels. This is where advanced biofuels come into play and supply for this demand. Biomass, given adequate management, sustainable cultivation, and timely integration, should be the key precursor for several types of fuels—a significant fraction of future energy portfolio. Due to the wide abundance and short life cycles, biomass promises potential for a more sustainable world, one where we are able to lower anthropogenic CO_2 emissions drastically.

Macroalgae, also known as seaweed, constitute numerous large multicellular algae species. These seabed dwelling plants grow in coastal marine areas, and can be harvested at depths less than 50 m below sea level. The environments in such ecosystems are conveniently next to invariant in terms of temperatures and salinity, facilitating continuous growth all year round, albeit not constant in growth rate [1]. Seasonal solar irradiance variance plays a major role in dictating growth rates and the chemical composition of the resultant macroalgal biomass. Previous studies have shown that brown seaweed (e.g., *Saccharina*, *Undaria*, *Ecklonia*, or *Sargassum*) are characterized by growth rates of 3.3–11.3 kg dry weight/m^2 per year [1]. This corresponds to harvest potentials between 2–10 dry tons/ha per year in Danish waters [2]. In fact, brown seaweeds can have a maximum energy yield of more than 45% throughout a single growing period. Such a value is significantly greater when compared to yields of most types of terrestrial biomass (e.g., lignocelluloses: 20–25%, energy crops: 30–35%). Such high productivity rates show a high potential for growing this biomass commercially [3].

Despite stagnant conditions locally, macroalgae are known to vary greatly in terms of chemical composition. Energy storage carbohydrate (e.g., laminaran and mannitol) fractions depend heavily on harvest seasonality, as the plants accumulate and release the compounds throughout the lighter and darker seasons, respectively [1]. Ash content can also vary greatly [1,4]. For instance, brown seaweeds harvested early in spring typically contain high amounts of alginate, proteins, and ash but low concentrations of the other types of carbohydrates [5]. However, upon receiving more light, the photosynthetic activity of the algae surges—the plants produce higher amounts of sugars, whereas the relative amounts of alginate, proteins, and ash drop [5].

Alginate, a linear polysaccharide abundant in free hydroxyl and carboxyl groups, and fucoidan, a fucose-containing sulfated polysaccharide, are amongst many algal compounds that have generated great interest in the scientific community over recent years [6,7]. Specifically, these two types of natural polysaccharides are valued for their applicability in medicinal and pharmaceutical fields. Although the properties of these value-added compounds vary depending on the chosen extraction methods and subsequent modification procedures, the processes typically result in a residual biomass by-product.

Using macroalgae as feedstock for energy production is not new, with many research groups worldwide studying the potential to make use of this marine resource. The tested technologies are of biochemical or thermochemical nature, where the biomass is converted to energy carriers. The high moisture content and high amounts of low melting point alkali and alkaline earth metals present in the biomass renders it a poor choice for direct combustion. Typically, a particular pathway is chosen based on the desired state of the output fuel. Multiphase, except for solid, fuel precursor production has been demonstrated using seaweed.

Hydrothermal liquefaction (HTL), unlike anaerobic digestion, is capable of fast production of high-quality fuel precursors, typically being carried out in a matter of minutes to tens of minutes [8]. Alkaline homogeneous catalysts are often employed, thus despite the overall high levels of ash, the metals of alkali nature present in the biomass are hypothesized to potentially improve the conversion. Reaction media of elevated pH levels are typically used, as such conditions lead to decreased formation of residual solids, whereas the gases are pushed towards repolymerization. Anastasakis et al. found that in the HTL of macroalgae, experiments in which no external catalyst was added yielded the highest amounts of biocrude [9]. In terms of quality, HTL biocrude is capable of reaching high energy densities, often equivalent to at least 70% of that of fossil crude [9]. The presence of heteroatoms in the biomass is one of the reasons why it is difficult to achieve high energy content in nontreated HTL biocrude. In macroalgae, nitrogen and sulfur are derived from proteins and sulphated carbohydrates, respectively, while all major groups of polysaccharides contain copious amounts of oxygen. When compared to lignocellulose products, algae-derived HTL biocrudes typically are more contaminated due to the high nitrogen and sulfur contents in the initial feedstock [10].

HTL experiments have been carried out with macroalgal feedstock recently [10–13]. However, alginate and fucoidan—two high-value compound extraction residues—have not yet been exposed to such thermochemical conversion conditions. On the other hand, energetic utilization via anaerobic digestion [14], pyrolysis [15], and hydrothermal carbonization [16] has been studied with such algal residues. Since HTL can successfully process sewage sludge and compost, two low-value wet material streams, which suggests that seaweed residues could be susceptible as well [17]. Researchers have recently studied two-stage HTL and co-liquefaction as a means to boost yields of macroalgae-derived biocrude [18,19]. This study focuses on utilizing residual material streams generated at macroalgal factories (i.e., alginate and fucoidan extraction residues). Taking circular economy goals into consideration, such modern bio-refineries could expand and produce both high-value products and HTL biocrudes at high, continuous production capacities. This study was carried out to test whether alginate and fucoidan residues can be effectively liquefied into high-quality biocrudes intended for downstream upgrading and refining of the eventually drop-in quality fuel. Additionally, the study included investigating the effects raw macroalgae demineralization, considering that the post-processed macroalgal biomass is hypothesized to contain copious amounts of inorganics. Finally, extensive neutralization of leaching residues was carried out to test for its necessity, and a shorter reaction time was investigated for potential benefits in supercritical HTL of low-value residual seaweed biomass.

2. Results

2.1. Demineralization

Fucus vesiculosus residues reached stable pH levels of 6 after the water leaching step, a value corresponding to the fresh biomass. This is said to be caused by the macroalgal cell wall polysaccharides that contain acidic functional groups [20]. Four washing steps were necessary to reach pH 7, corresponding to 50 g water/g initial macroalgae. Five washing steps brought the pH level up to 6.9, compared to the initial pH of 5.6 when leaching with HTL water. Citric acid leaching led to the lowest initial pH of 3.8. Eight washing steps only raised the pH to 6.1, corresponding to 100 g water/g initial macroalgae, despite the fact that a diluted acid solution was used.

Ash content was reduced as a result of the dry biomass grinding and subsequent water leaching. A final ash content of 13.71 wt.% was obtained when measured after first reaching pH 7 (i.e., after washing step 4), corresponding to an overall ash reduction of 38.85%. However, the final measurements show that further ash reduction is very limited during neutralization, defined here as 1.14% per four H_2O washes throughout the entire experimental range.

HTL water led to less effective demineralization compared to water leaching. Here, the final ash contents amounted to 16.17 and 14.89 wt.% prior and post neutralization, respectively. The agent's capacity is believed to be limited due to the relatively high amount of inorganics (2.51 wt.%) in the liquid itself. The used catalyst, potassium carbonate, is believed to constitute the majority of the ash and it is hypothesized to add onto the amounts of potassium salts abundant in the seaweed biomass. Neutralization did not offer much in terms of further reductions. The slight reduction amounted to 1.28 wt.%, while an analogous decrease of 1.9 wt.% occurred in the case of water leaching. All in all, diluted HTL water leaching offered inferior ash reduction performance, exhibiting a demineralization potential of 33.59%. The raw data acquired for and used throughout this study are available in the Supplementary Materials.

In the studied fresh seaweed, leaching using a dilute citric acid solution resulted in the highest ash removal efficacy. Final ash contents of 14.53 and 11.85 wt.% were achieved before and after the eight washing steps, respectively. This amounted to ash reductions of 35.19% and 47.15%. The final result is caused by the combined effects of acid and water leaching, as evident from the data. The results of this part of the study are listed in Table 1 and visualized in Figure 1.

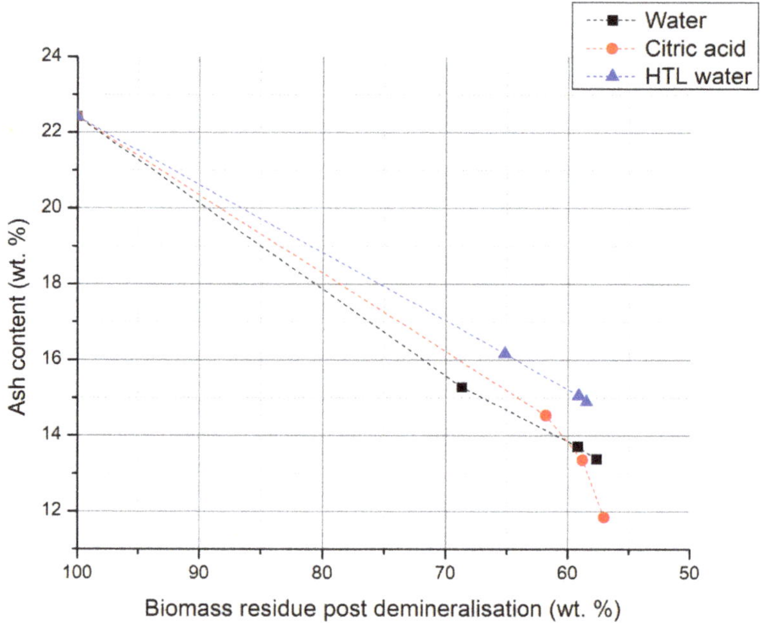

Figure 1. Correlation between macroalgae residue and ash content post demineralization.

Table 1. Demineralization results, highlighting the effects different leaching agents and post-treatment via water washing have on the pH, final ash content and higher heating value of macroalgal biomass.

Leaching	No. of H_2O Washes	pH	Ash [wt.%]	Residue HHV [MJ/kg]
	1	6.4	15.28 ± 0.20	15.18 ± 0.08
Water	4	7	13.71 ± 0.33	15.65 ± 0.05
	8	7	13.38 ± 0.36	15.46 ± 0.04
	1	5.6	16.17 ± 0.91	15.73 ± 0.03
HTL water	4	6.7	15.06 ± 0.96	15.82 ± 0.01
	5	6.9	14.89 ± 0.59	15.80 ± 0.01
	1	4	14.53 ± 0.42	16.16 ± 0.02
Citric acid	4	5.3	13.36 ± 1.36	16.16 ± 0.07
	8	6.1	11.85 ± 0.37	16.03 ± 0.05

Higher heating values (HHVs) of the resultant biomass residues were measured to extend the comparison of the three investigated methods. From an energetic standpoint, higher quality feedstocks were achieved via the use of all three leaching agents. Here, too, the claim that citric acid treatment performs best is valid since the highest HHV recorded was 16.16 MJ/kg compared to the initial 14.95 MJ/kg of the dried *F. vesiculosus* seaweeds.

The final metric here was to measure the amount of solid residues generated post leaching. Differences across the three were observed, with 61.75 wt.%, 65.15 wt.% and 68.66 wt.% of residues generated when leaching with citric acid, HTL water and water, respectively. Thus, the determined differences in ash removal efficacies are confirmed. It is noteworthy that post-treatment water washing equalized the amount of residues by wash number 4 from all three sets of experiments, further suggesting that neutralization and subsequent washing do not depend on the leaching agent used as far as residue generation is concerned

2.2. Value-Added Product Extraction

Despite the high extent of alginate and fucoidan extractions, high quantities of post-extraction residues were produced, amounting to 41.88 and 68.17 wt.% of the initial biomass, respectively.

The residues were exposed to elemental analysis and the effect alginate and fucoidan extractions have on heteroatomic constituents, specifically nitrogen and sulfur, was studied. Fucoidan extraction led to lower amounts of elemental nitrogen, i.e., 1.84 wt.% compared to the initial 3.14 wt.%, but alginate extraction did not affect this parameter of the biomass. Meanwhile, fucoidan extraction resulted in a significantly lower fraction of elemental sulfur. Here, the final content amounted to 0.37 wt.% compared to the initial 1.12 wt.% in the dried seaweeds.

The residues were also studied in terms of how ash content changes after exposure to the two extraction processes. While no significant changes in the final ash content were observed after fucoidan extraction, inorganics were concentrated by the alginate extraction procedure. To confirm, fucoidan and alginate extractives were also ashed. Low amounts of dissolved inorganics were determined in both extracts. Thus, both extraction methods were confirmed to selectively dissolve organics, removing virtually no inorganics. However, the produced ash samples did differ in color (fucoidan extraction residues were light, alginate extraction residues were dark), indicating that of the limited amounts of inorganics removed, the processes do target different compounds. Additionally, as seen in Figure 2, the different residues generated from leaching and extraction did not result in significant differences upon thermal decomposition; divergences are apparent only in the latter stages, when the residue is composed of inorganics to a great extent.

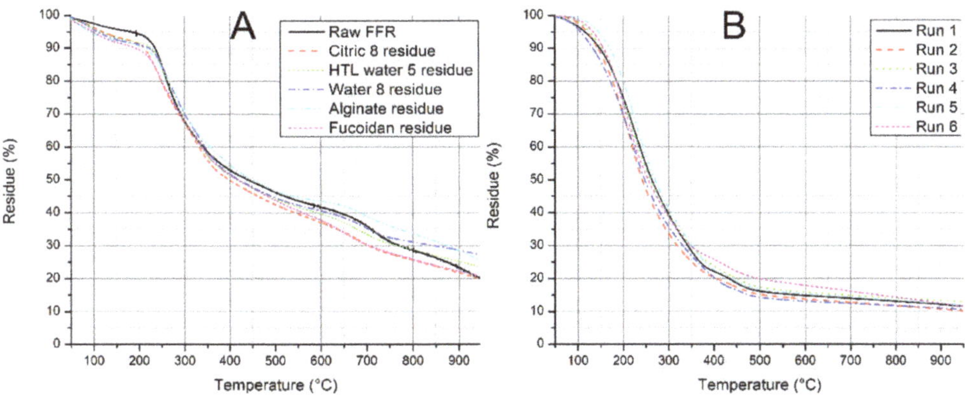

Figure 2. TGA curves representing raw biomass, leaching residues, and extraction residues (**A**), and the six produced biocrudes (**B**).

2.3. Hydrothermal Liquefaction

2.3.1. HTL Yields

The recorded biocrude yields were in the range of 15.23 to 28.21 wt.% on dry, ash-free (DAF) basis. The results, shown in Table 2, highlight that run 6 (short HTL) and run 5 (HTL of neutralized citric acid leaching residues) resulted in the lowest yields. Run 4 (baseline HTL) produced a slightly higher yield of 19.36 wt.%. Finally, as seen by the results of the three remaining biomass treatment runs resulted in improved biocrude yields. Most notably, run 4 (HTL of alginate extraction residue) experiments yielded the highest quantity of products, amounting to 28.21 wt.%.

Table 2. Summary of HTL biocrude yields and key quality parameters, including H/C and O/C ratios and higher heating values and energy recovery rates. DAF basis was used when calculating the biocrude yields.

	Biocrude Yield [wt.%]	H/C	O/C	HHV [MJ/kg]	ER [%]
Run 1	19.36 ± 4.48	1.38	0.35	26.28	30.95
Run 2	21.59 ± 3.73	1.22	0.10	35.35	46.93
Run 3	26.56 ± 5.34	1.29	0.13	34.46	52.83
Run 4	28.21 ± 3.44	1.36	0.13	34.50	45.78
Run 5	17.26 ± 1.1	1.55	0.08	38.05	38.39
Run 6	15.23 ± 3.82	1.54	0.05	39.16	35.29

HTL by-product yields were also quantified in terms of mass yields. Runs 2, 3, and 5 generated the most solids. Run 3, representing the thermochemical conversion of fucoidan extraction, yielded the most solids (0.35 g), whereas run 4 (alginate residues) generated the least (0.18 g). Such a significant difference must be taken into consideration, especially when preparing for continuous operation. In terms of water solubles (WS), runs 2 and 3 yielded the least with 0.15 g and 0.09 g, respectively. Here, the overall average amount of produced WS across all experiments was 0.2 g. Run 1 yielded the most: 0.28 g of WS. Except for run 2, the yields of gaseous by-products were comparable across the experimental range. The conversion of citric acid leaching residues generated the most gases: 0.38 g, compared to the average of 0.35 g. Figure 3 shows the yields of all four products, scaled with respect to each other, and represents the proportional yields at lab scale batch processing. As a general tendency, the data suggest that the slightly worse-performing runs generate higher amounts WS and gas, on average.

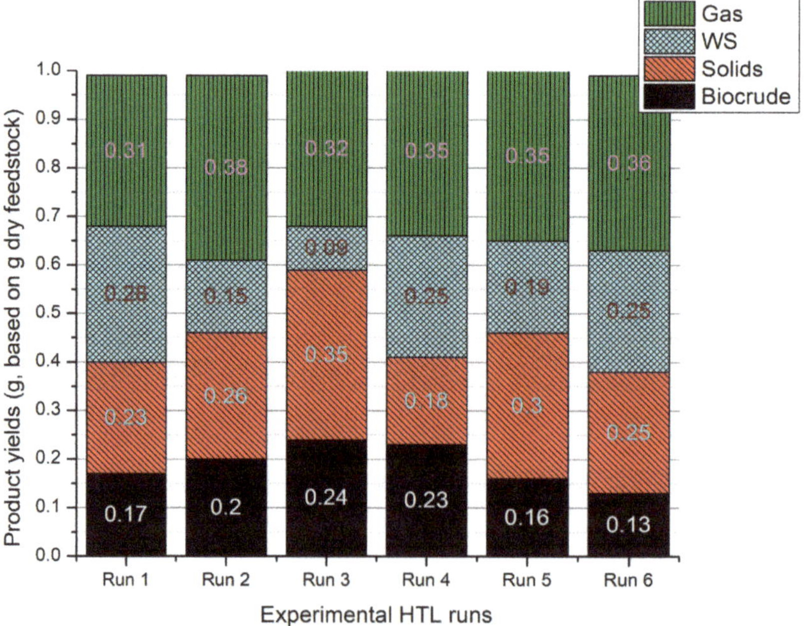

Figure 3. Overview of HTL product yields, including biocrude, solids, water solubles and gases, indiscriminately scaled to a closed mass balance.

The biocrudes were also evaluated in terms of their HHVs and the associated calculated energy recovery rates for each set of experimental runs. The produced biocrudes varied greatly in terms of their HHVs, ranging from as low as 26.28 MJ/kg in run 1, up to 38.05 and 39.16 MJ/kg as estimated for run 5 and 6 biocrudes, respectively. The higher-end values are similar to fossil crudes. As a proxy for the feasibility of converting the different biomass streams via HTL, energy recovery levels in the main fuel product were calculated. As per Table 2, the highest ERs were estimated for runs 2, 3, and 4, with the maximum estimated for fucoidan residue HTL being 52.83%. Conversely, run 1 (HTL of untreated seaweeds) performed the worst, reaching just 30.95% in ER. Finally, the energy contents between 35–38% of the initial feedstocks were estimated for the remaining runs 5 and 6.

2.3.2. Biocrude Quality

Biocrude quality determination of the different conversion runs is no less important than estimating biocrude yield parameters. Elemental H/C and O/C ratios were the first two indicators. While the goal is to have a biofuel precursor with maximal hydrogen and minimal oxygen amounts, the highest H/Cs of ~1.54 with low O/C ratios between 0.05 and 0.08 were determined in the biocrudes of runs 5 and 6. A post-treatment upgrading step including extensive deoxygenation would still be necessary to approach fossil analogues of high enough quality for commercial refining and blending. Adequate quality parameters were recorded in run 2, 3, and 4 biocrudes. Since only baseline run 1 biocrude was a product significantly more contaminated with oxygen, all pre-treatments/conditions, including demineralization, value-added product extraction, and even short retention, can yield superior biocrudes. Table 3 highlights the key elemental constituents in a succinct manner. No significant differences were observed upon proximate analysis of the different products: volatile matter and fixed carbon averaged at 83.32 ± 2.3 and 16.68 ± 2.3 wt.%, respectively. Here, a high fraction of volatiles is an important parameter indicating the potential suitability for use as a fuel precursor for downstream processing into lighter hydrocarbons such as diesel, jet fuel, and gasoline. TGA analysis revealed concerningly high amounts of inorganic residues, averaging at 11.46 ± 0.7%, as shown in Figure 2, further emphasizing the need for biocrude post-treatment.

Table 3. Overview of quantified HTL biocrude sample constituents: carbon, hydrogen, nitrogen, sulfur, and oxygen.

	C [wt.%]	H [wt.%]	N [wt.%]	S [wt.%]	O [wt.%]
Run 1	60.04	6.90	3.07	1.47	28.20
Run 2	77.72	7.91	2.74	0.53	10.78
Run 3	75.28	8.07	2.96	0.49	18.88
Run 4	74.40	8.41	3.01	0.51	13.37
Run 5	77.85	10.03	3.16	n.m.	8.65
Run 6	79.60	10.20	4.32	n.m.	5.57

n.m.—not measured; O calculated by difference, assuming 0.3 wt.% ash.

Differences in biocrude yields and quality between runs 2 and 5 and 1 and 6 can only be done by taking a closer look at the resultant biocrudes; only then is it possible to see whether extensive neutralization or a shorter retention time could hold any advantages. As per Figure 3 and Table 2, biocrude yields of neutralized residue and short retention HTL runs were poor. Only run 5 performed slightly better out of the four. No definitive tendency can be observed from by-product distribution. Out of the two, neutralized residues yielded more gas compared to the leached biomass. Run 1 biocrude had a very high amount of oxygen, as evident in Table 3. Comparing these two, shorter retention seems preferable due to a significantly lower O/C ratio and a slight increase in H/C. When discussing the neutralization extent, however, of a similar O/C ratio, the HTL of neutralized residues yielded a biocrude with a H/C ratio more than 20% higher than that of non-water washed acid leaching residues.

2.3.3. Solids

Organic and inorganic fractions of the generated solid residues were determined and are shown in Figure 4. Averaging at 70.87 wt.%, the organic fractions did not vary significantly across the experimental range. The only exceptions were run 3 (fucoidan residues) solids, of which only 52.54 wt.% were organic in nature. As seen in Figure 3, the amount of gaseous by-product generated during run 3 was average and the yield of WS at the lower end of the spectrum. More inorganics must have been carried through in the solid phase by-product as such a high fraction of the solids was generated. The apparent concentration of organics is, therefore, apparently low. While the ash content of the biomass and HTL solids was measured directly, the missing amount is assigned to the WS fraction, where the corresponding percentage is derived on the basis of the total amount of produced WS. Despite the potential to recycle the water phase, as evidenced in HTL of lignocellulosic biomass, the present study suggests this might not be feasible in HTL of macroalgae nor for demineralization purposes [21]. Upon ICP analysis, the concentration levels of all measured levels except for potassium and sodium were higher in run 1 solids (Figure 5) than those measured in the dried feedstock.

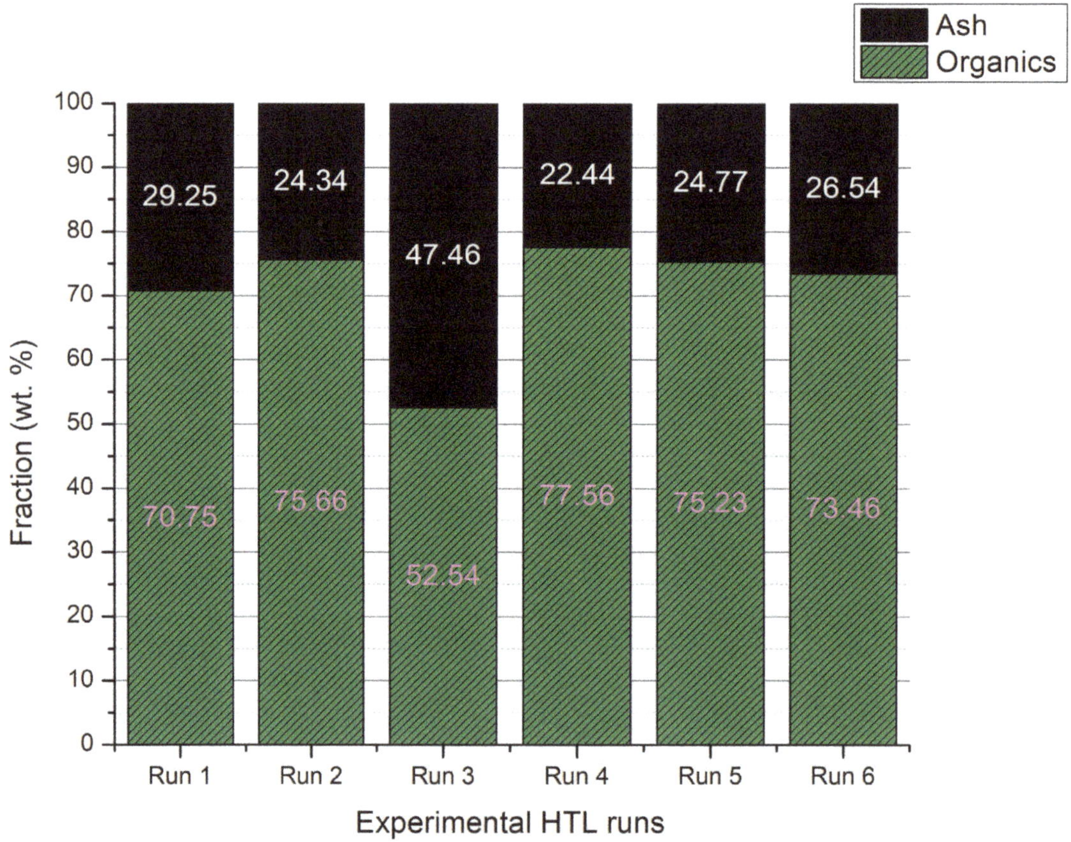

Figure 4. Variation in organic and inorganic constituents of the six produced HTL solids.

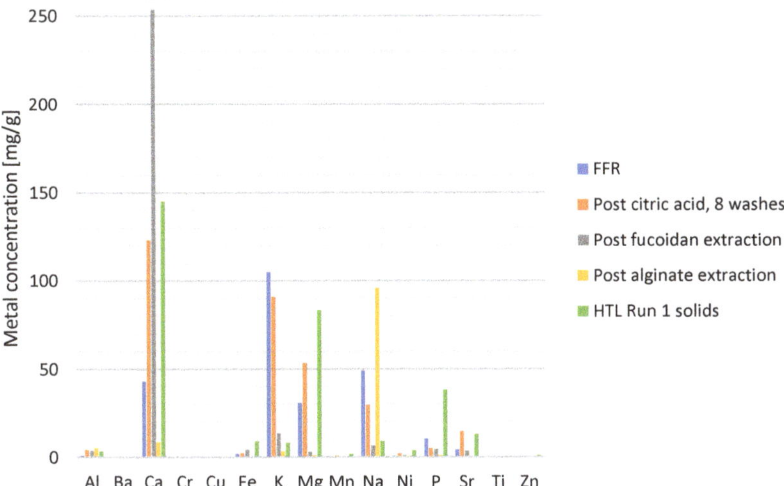

Figure 5. Comparative metal concentrations in raw macroalgae, citric acid leaching, fucoidan, and alginate extraction residues, and the solids produced in HTL run 1, as determined by ICP ash analysis.

2.3.4. Gases

Similar compositions of constituent gases were determined across the gaseous by-products of the six sets of experimental HTL runs (Figure 6). Runs 3 and 4 differentiated from the other samples. While the concentration of CO was significantly higher in run 3 samples, more H_2 was detected in the by-product of run 4. In general, the composition of all product gases was heavily dominated by CO_2, with 84.61 and 91.61 vol.% as the minimal and maximal values, respectively. Typically, only trace amounts of CO, H_2, and CH_4 were detected, confirming that decarboxylation and decarbonylation reactions are behind the removal of oxygen.

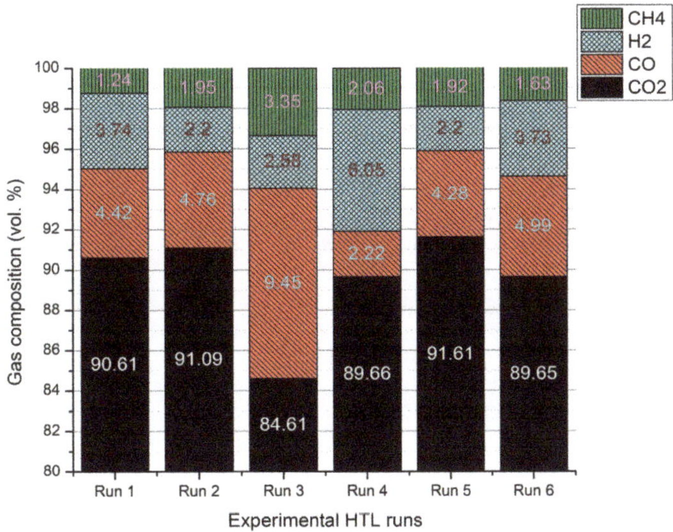

Figure 6. Comparative overview of the compositional profiles of the HTL by-product gases across the experimental range.

3. Discussion

3.1. Demineralization

Should the effect of biomass residue pH be strong and neutrality an important prerequisite for effective HTL of seaweed feedstocks, all demineralization runs would require post-processing neutralization, regardless of the leaching agent that was originally employed. Since it was possible to raise the pH of the HTL water leached residues via water washing, a further synergistic benefit could be acquired by substituting intensive acid pretreatment, given that the leaching agent could be shown to be an effective leaching agent. Here, HTL water usage as a demineralization agent and the subsequent necessity to neutralize the residues deserve further, dedicated studies. There is a potential need to include the use of an external neutralizing agent to render the biomass exposed to citric acid leaching neutral. Alternatively, it is apparent that an alarmingly high water consumption would have to be dealt with downstream. The addition of an alkali catalyst could be synergistically beneficial if both extensive ash removal potential and effective HTL conversion can be shown. However, costs associated with high water consumption and post-treatment are not exclusive to wet acid leaching—each washing step comes at a cost of lost organic matter. This should be taken into account as the neutralization procedure via washing would exacerbate such losses. Reactive solvent citric acid recovery could be a potential way to significantly diminish water demand [22]. While here water washing was studied as a widely available and low-tech method, careful investigations have to be carried out with any potential biomass residue to adequately weigh the pros and cons associated with the presented methods and their extent.

The present study recorded lower HHVs in post-neutralization residues, suggesting that extensive water washing might not be an advantageous method. In fact, the full extent of demineralization is reached before neutralization occurs; thus, any further washes may remove organics more selectively instead. This is confirmed by the elemental analysis data. After citric acid leaching, despite a slightly lower amount of inorganics by 2.68 wt.%, elemental carbon and hydrogen increased only by 1.69% and 0.19%, respectively, when residues after one and eight H_2O washes were compared. This confirms that organics are lost in subsequent washing of macroalgal biomass residues. Finally, the differences in nitrogen were negligible (below 0.1 wt.%), whereas, given the shortage of relevant data, no conclusions can be made about the fate of sulfurous compounds.

It is apparent from the data presented in Table 1 that all three leaching agents lead to higher quality energy feedstocks. Since post-neutralization resulted in decreases in HHVs, it is suggested that extensive water washing might have significant drawbacks. In fact, the highest extent of leaching is seen to take place during the first washing steps, as shown by the average biomass mass losses of 34.82% and 40.98% recorded after the first and fourth wash, respectively. Figure 1 depicts the change in ash content versus biomass residue. As an example, here, a decrease in ash content by 31.85% (i.e., a 7.14% mass loss of the biomass) results in a total biomass mass loss of 31.34%. This indicates that the tested methods are several times more effective at removing organics. Costs associated with feedstock are typically the limiting barrier for biofuel production via HTL and this is no exception for macroalgae [23]. Whether the advantages of demineralization would outweigh the inherent additional costs of processing and leached organics valorization is yet to be shown quantitatively.

Highly selective demineralization of biomass high in inorganics is a topic of high interest in the scientific community. More aggressive acidic de-ashing, such as the use of nitric acid, was shown to be highly effective but still lacked in selectivity [24]. Meanwhile, novel methods, such as treatment with pulsed electric fields, have already been used to reach high demineralization efficiency at a relatively low loss of organics [25]. More importantly, however, it was recently shown that batch aqueous demineralization overestimates water-soluble inorganics and loss of organic matter, indicating the urgent need for large scale semi- or fully continuous pretreatment technique research on the biomass of interest [26]. This seems to be true also for biocrude ash content determination, where even low ash (1–2% db) feedstocks lead to 4–5% ash in continuous supercritical HTL biocrudes [27]. This highlights the possibility that achieving high demineralization extent would not necessarily prevent the need to purify the produced biocrude downstream, as the inorganic homogeneous catalysts commonly used in HTL (e.g., NaOH, K_2CO_3) partially migrate to the product stream. It is the authors view that research is needed in continuous demineralization of biomass high in inorganics and subsequent HTL to further investigate whether it is more economical to develop efficient demineralization treatments or HTL systems capable of processing high ash feedstocks instead.

3.2. Value-Added Product Extraction

The high amounts of residues could be explained by the fact that the *F. vesiculosus* used in the present study had been harvested in late winter, suggesting that the plants had consumed their energy stocks, giving rise to the high amount of inorganics present in the feedstock. Further differences have been recorded when comparing winter-harvested *F. vesiculosus* and autumn-collected *S. latissima*, where the latter contained more than twice more alginate [28,29]. Obviously, genera-specific differences in plant structures could also add to such differences.

The apparent variance in residue coloration served as a solid basis for further ash analysis to identify how this visual difference correlates to compositional differences. Calcium and sodium compounds made up the majority of the total inorganics in fucoidan and alginate extraction residues, respectively. Since $CaCl_2$ and Na_2CO_3 cannot completely be removed in a single post-extraction washing step, significantly increased concentrations of these elements are observed upon ICP analysis. The lightness of the fucoidan residues is a result of the high concentrations of calcium in the biomass. The inorganic constituents of the dried *F. vesiculosus* itself are dominated by alkali and alkali earth metals. Specifically, potassium, sodium, calcium, and magnesium are abundant in the highest amounts, while potassium, magnesium, and manganese are shown to be extracted effectively throughout both of the procedures. However, heavier metals, such as aluminum, copper, nickel, and zinc, remain at nearly unchanged concentrations. Particular to alginate extraction is the enhanced migration of iron, strontium, and phosphorous. This phenomenon is explained by the use of H_2SO_4—this strong acid is known to be a more aggressive demineralization agent [30]. When compared to the initial macroalgae, the HHVs of both extraction residues exhibited improved HHV values.

3.3. Hydrothermal Liquefaction

As reported previously, batch macroalgal biocrude can contain a significant quantity of inorganics, rendering further purification necessary [31]. Several techniques, including filtration, electrocoalescence, washing, and, more recently, washing with carbonated water, utilized in conventional refineries could be employed for this purpose [32,33]. The concentrations of alkali and alkaline earth metals were expected to change due to the high solubility of potassium and sodium salts in water—significant amounts were removed during the product separation procedure, as observed previously in the literature [34]. ICP analysis of the inorganics present in the produced by-product solids indicated that reactor degradation is an area of concern and further studies are necessary to determine specific degradation rates. This is based on the increasing or appearing concentration levels of stainless steel-derived metals, such as chromium, iron, manganese, nickel, and titanium.

Yet another important methodological detail worth noting is that an additional water washing step was included after product filtration to check for significant amounts of water solubles that had precipitated upon rinsing the reactants with acetone. Obviously, using a solvent to empty the reactors and separate the products will impact the results to some extent and the obtained products may not be representative of larger-scale operations. This is made especially clear as gravimetrical separation of biocrude is commonly employed at continuous pilot-scale HTL facilities [21]. Largely varied mass losses between 7.96 to 54.15% were recorded upon the additional water washing step in run 4 and 3 solids, respectively.

The recalcitrance of alginate extraction residue derived HTL solids is suggested to be brought on, once again, by the use of sulfuric acid that had made the residues more stable hydrolytically via structural destruction and effective demineralization. In a previous study crystalline macroalgae structures were shown to be broken down by dilute sulfuric acid in *L. digitata* [31]. This analysis revealed no effect on the amount of acetone WS precipitates by extensive water neutralization or shorter HTL retention time as mass losses of 37 and 22 wt.% were recorded in solids of runs 1 and 6, and 2 and 5, respectively. Nonetheless, given the large variation and possible high misrepresentation of the yields, extra washing makes sense to show quantitively certain data, and reduce the risk of them not being either acetone- and water-insoluble, or representative of large HTL facilities filtered in-line.

4. Materials and Methods

Fresh samples of brown macroalgae *Fucus vesiculosus* (photograph in Figure 7), growing north of the Danish mainland, were acquired for the experimental part of the study. Specifically, these algae were chosen due to their wide distribution in the Baltic Sea. In some western areas, *F. vesiculosus* are the only large, canopy-forming brown macroalgae. They grow along rocky coasts, at low depths [35].

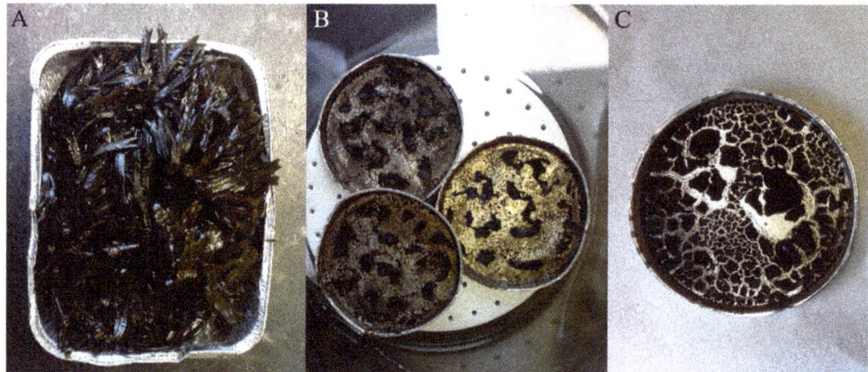

Figure 7. Physical appearance of the samples: fresh biomass (**A**), dried alginate extraction residues (**B**), and dried fucoidan extraction residues (**C**).

Thermogravimetric analysis (TGA) was performed in an inert atmosphere (purged with nitrogen) using a PerkinElmer STA6000 TG/DSC analyzer. Samples of 4–7 mg were heated to 950 °C at a temperature ramp rate of 10 K/min. The nitrogen flow rate was set to 20 mL/min throughout the entire procedure. CHNS analysis was carried out on a Vario Macro Cube simultaneous CHNS analyzer from Elementar. Here, samples of 70–80 mg were analyzed in triplicates. An in-house moisture analysis (KERN MLS) was used to determine the water content in the fresh biomass. Higher heating values (HHVs) of the dried and milled macroalgae samples were measured in triplicates using an IKA C2000 basic bomb calorimeter. Finally, ash content in the biomass samples was determined as the constant mass solid residue post-dry oxidation at 575 ± 25 °C. All compositional information is provided in Table 4.

Table 4. Compositional analysis of freshly harvested winter *F. vesiculosus* macroalgae.

Physical Properties (as Received):	
Water content [%]	77.42 ± 0.5
Proximate Analysis (Dry Basis):	
Volatile matter [%]	54.16
Fixed carbon [%]	20.23
Ash [%]	22.42
Higher heating value [MJ/kg]	14.95 ± 0.01
Elemental Analysis (Dry, Ash-Free Basis):	
Carbon [%]	36.90
Hydrogen [%]	6.06
Nitrogen [%]	3.14
Sulfur [%]	1.12
Oxygen [%] [a]	30.36

[a]—calculated by difference.

4.1. Demineralization

Initial screening tests were carried out on *Laminaria digitata* brown macroalgae as a part of a previous study [30]. The combination of significant ash removal and relatively water-lean neutralization procedure led to evaluating dilute citric acid treatment as optimal.

Demineralization with distilled water was also carried out to establish baseline results. Finally, to investigate an alternative means to utilize one of the by-product streams of continuous HTL, the aqueous phase by-product was used as the third leaching agent. Its acidic nature gives merit to investigate the de-ashing potential, and thus, valorize the otherwise challenging waste stream. The raw aqueous product was a sample previously collected at the local semi-continuous HTL plant and represents a real-world synergistic opportunity. The sample was slightly acidic with a pH level below 5.5 [27].

The raw macroalgae were pre-rinsed with cold water to remove any unbound inorganics as the first step. After the initial rinsing, the biomass was oven-dried and milled (FOSS CyclotecTM 1093, particle size: $\leq$200 µm). In the case of citric acid leaching, the now dry and powdered macroalgae were mixed with a 1 wt.% citric acid solution (12.5 g solution/g macroalgae). The leaching process took place overnight (18 h of continuous stirring at 1000 rpm at room temperature). After leaching, the mixtures were centrifuged (SIGMA 6–16S centrifuge, for 5 min at 4000 rpm) to remove the leachate. Then, the neutralization/rinsing procedure took place. Neutralization is a part of the study to process a non-acidic feedstock. This was done because alkaline processing media were found to suppress char formation from carbohydrates during HTL [21]. Distilled water was added to the residues (12.5 g water/g initial macroalgae) and the mixture was stirred manually. Subsequent centrifugation was utilized for separation. Varying amounts of coupled rinsing-separating steps were enforced to set up for analysis of HTL of post-demineralization macroalgae. The focus here was to determine whether a great neutralization extent is truly necessary for efficient HTL of acid leached macroalgae. The experimental design included drying (at least for 18 h at 105 °C) the residues after one, four, and eight rinsing repetitions (i.e., simulated water consumption ranging from 12.5 to 100 g/g of dry initial macroalgae). Ultimately, we aimed to test the need for water-intensive post-treatment. The pH levels were measured initially, after the leaching period and after each rinsing step (WTW pH 3210 m, accuracy of $\pm$0.2 pH points). All results are reported as average values of triplicate experiments/measurements, unless stated otherwise.

Just 50 mL of HTL aqueous phase was available for the needs of this study. In order to accommodate the required leaching medium, the available 50 mL were diluted with distilled water to reach a total volume of 300 mL. This being said, it is worthwhile to note that the pH of the solution did not change significantly, stabilizing at pH 5.6 prior to mixing. The same acid solution-to-biomass ratio of 12.5 and leaching conditions were kept.

4.2. Value-Added Product Extraction

The experimental flow of the performed alginate removal procedures was adapted from [36]. Three samples (sample size: 5 g) of the winter harvest *F. vesiculosus* were processed. Firstly, the rinsed macroalgae were dried and milled. Then, the powder was mixed with a 0.5 M H_2SO_4 solution (13.58 g solution/g algae) and stored overnight (minimum 21 h) in a dark cabinet. Then, the mixture was centrifuged (5 min at 4000 rpm) and the liquid solution was removed. An intermediary washing step (13.58 g H_2O/g initial algae) with subsequent centrifugation (4000 rpm, 5 min) was performed to remove any residual acid. A 4% Na_2CO_3 solution (19.95 g Na_2CO_3 solution/g initial algae) was added to the residues. The mixture was stirred magnetically (800 rpm) for 2 h. After soaking, the mixture was once again centrifuged to separate the solubles. A washing step (19.95 g water/g initial algae—mix, centrifuge, drain) took place next. All of the above process steps were carried out at room temperature. Finally, the residues were carefully removed from the centrifuge bottles and placed in an induction oven to dry for at least 18 h at 105 °C.

The employed simulative fucoidan extraction procedure was adapted from [15]. Three samples (sample size: 5 g) of winter harvest *F. vesiculosus* were used. The water-rinsed macroalgae were processed mechanically via drying and milling. Subsequently, fucoidan was extracted in a $CaCl_2$ solution. The extraction was finished throughout two steps: samples were exposed two times to 20 min-long magnetic stirring (800 rpm) sessions in 1 wt.% $CaCl_2$ solutions (16.67 g solution/g algae). After each stirring, the mixtures were centrifuged at 4000 rpm for 5 min and the separated liquid was removed. A similar procedure followed the two extraction-separation steps: the solid fucoidan extraction residues were mixed with water (16.67 g water/g initial algae) and centrifuged once more in order to remove any remaining calcium chloride. All steps were carried out at room temperature. Finally, the residues were oven-dried, cooled in a desiccator, weighed, and stored in airtight containers until further processing.

4.3. Hydrothermal Liquefaction

Six separate HTL runs were carried out throughout this study. The experiment list can be seen in Table 5. The main focus of the overall procedure was set on HTL of treated macroalgae, namely de-ashed, post fucoidan extraction and post alginate extraction. Additionally, the effects of post de-ashing neutralization and a shorter retention time were investigated.

Table 5. HTL experimental overview.

Reference	Pre-Treatment	HTL Conditions	Hypothesis/Argument
Run 1	-		Baseline
Run 2	De-ashing		Demineralization improves yield
Run 3	Fucoidan extraction	Normal	Effective HTL with fucoidan extraction residues is possible
Run 4	Alginate extraction		Effective HTL with alginate extraction residues is possible
Run 5	De-ashing and neutralization		Post de-ashing neutralization is not necessary
Run 6	-	Short	High-quality biocrude can be produced at a shorter reaction time

All experiments were carried out in stainless steel (grade 316) 12 mL microreactors. Feedstock dry mass loadings of 20% were used, and all reactions were carried out at 400 °C (±5 °C). Upon feedstock slurry preparation, the specific macroalgae powder was combined with distilled water to form the predefined mixture. A total of 5 g ($\pm$0.1 g) of the slurry was then loaded into the reactors. Nitrogen gas was used to simultaneously leak test (80 bar) and purge the reactors to evacuate atmospheric oxygen. Hereafter, two reactors and thermocouples were mechanically coupled to an agitator, providing mechanical mixing of the reagents inside the reactors while being processed. The two reactors were then submerged into a preheated, fluidized sand bath and held, normally, for 15 min of retention time and 10 min in the short HTL run. The retention time was defined as the time that passes between the moment when the reactors have reached the pre-set temperature of 400 °C ($\pm$5 °C) and the instance of manually submerging the reactors into the cool (~20 °C) water bath. After quenching in water for a minimum of half an hour, the separation procedure begins.

The first step of product separation was the weighing of the gaseous products, gas sampling, and venting the remaining gases via top-mounted valves. The remaining products consisted of solid residues, biocrude, and an aqueous phase. The reactors were washed with acetone to remove all biocrude traces from the reactor. The liquid phase was then separated from the char via vacuum-assisted mechanical filtration (VWR, particle retention: 5–13 μm). The solids present on the filter were then dried overnight at 105 °C and re-filtered with 250 mL of distilled water. The remaining solid residues were dried once again, weighed, and defined as water and acetone insoluble solids. Finally, the produced solids were ashed. This was done to determine how much inorganics are present in the by-product. Acetone was then evaporated from the homogeneous liquid fraction and the biocrude fraction was manually extracted after centrifuge-aided phase separation. The higher density extracts were defined as biocrude, whereas the aqueous by-product was collected, dried, weighed, and denoted as water solubles (WS). Post reaction gases were weighed, adjusted for initial nitrogen addition, and analyzed via GC analysis. The aqueous products were weighed prior and after to show the extent of experimental error due to water losses during acetone evaporation. The produced biocrudes were weighed, their proximate analyses were done via TGA, and their water contents were measured via Karl Fischer titration.

4.4. Calculation Methods

This section describes all calculative methods that were used for determining both product/by-product yields and quality parameters, such as biocrude higher heating value (HHV), hydrogen-to-carbon ratio (H/C), and oxygen-to-carbon ratio (O/C). Both biocrude and gas yields were calculated on a dry and ash-free (DAF) feedstock basis identically as shown in Equation (1). Similarly, the yields of solids were calculated on a dry basis.

$$Yield_{biocrude} = \frac{Mass\ of\ biocrude}{Mass\ of\ dry,\ ash\ free\ feedstock} \cdot 100\% \tag{1}$$

Finally, the yields of WS were determined on a dry feedstock basis, by adding the weighed WS and the amount of solids washed out with water (Equation (2)). This procedure was adopted to better represent the generated amount of WS. Previously utilized methods of presenting the data as process water + WS were shown to be inconsistent (i.e., variations in mass up to 25% among single run triplicate data). Such differences are believed to be caused by the non-automated evaporation step—depending on the duration of this step, more or less process water is lost. However, this does not impair the results of the study as preserving process water was never among the objectives. Furthermore, presenting dry WS data instead is more reliable.

$$Yield_{WS} = \frac{Mass\ of\ WS + mass\ lost\ during\ water\ washing\ of\ solids}{Mass\ of\ dry\ feedstock} \cdot 100\% \tag{2}$$

Due to the inability of measuring HHVs of the produced HTL biocrudes directly (microreactors do not yield sufficient amounts), the study resorted to elemental HHV estimation. In order to represent the biocrude comparably, several HHV estimation formulas were tested against laboratory measurements or raw macroalgae, demineralization, and value-added product extraction residues. The correlation derived by Friedl et al. (Equation (3)) was shown to give the most accurate results—all tested values were within 5% of the experimental measurements [37]. Meanwhile, the correlation proposed by S. Channiwala and P. Parikh (Equation (4)) was used when estimating the HHVs of the produced biocrudes [38]:

$$HHV = 0.00355 \cdot C^2 - 0.232 \cdot C - 2.230 \cdot H + 0.0512 \cdot C \cdot H + 0.131 \cdot N + 20.6 \qquad (3)$$

$$HHV = 0.3491 \cdot C + 1.1783 \cdot H + 0.1005 \cdot S - 0.1034 \cdot O - 0.0151 \cdot N - 0.0211 \cdot Ash \qquad (4)$$

Hydrogen-to-carbon (H/C) and oxygen-to-carbon (O/C) ratios, on an elemental basis, were calculated for each of the produced biocrudes. Here, analyzed sample masses are taken into account. Such quality parameters allow for a direct comparison with biocrudes produced from other biomass sources, different HTL conditions, and even fossil fuels. In the literature, yet another ratio, the effective hydrogen-to-carbon ratio, is often presented to compensate for any water present in the produced biocrude. Contrary to such an approach, the study included measuring the total water content by Karl Fischer titration and subtracting the results both from biocrude yields and elemental composition.

Additionally, to biocrude yield and quality, energy recovered in the form of produced biocrudes was calculated as well to compare the energetics of each HTL run. The recovered ratio is calculated on dry feedstock basis, using Equation (5):

$$ER = \frac{Mass\ of\ biocrude \cdot estimated\ HHV}{Mass\ of\ dry\ feedstock \cdot measured\ HHV} \cdot 100\% \qquad (5)$$

5. Conclusions

All three of the studied demineralization agents led to lower amounts of ash in the macroalgal biomass residues, and correspondingly, superior higher heating values compared to the initial seaweed feedstock. The amounts of generated solid residues leveled out already after the fourth washing step. Citric acid leaching with extensive neutralization was the most effective method for reducing ash, corresponding to a final ash removal efficiency of 47.15%. Ash concentration was observed in biomass residues after alginate extraction. Both alginate and fucoidan extraction residues resulted in improved high heating values.

The recorded hydrothermal liquefaction yields of 26.56 and 28.21 wt.% on a dry ash-free basis were the highest across the experimental range and were obtained when converting fucoidan and alginate extraction residues, respectively. Short retention and neutralized leaching residue experiments exhibited the poorest liquefaction efficiencies. However, these two experiments yielded biocrudes of the highest H/C and lowest O/C ratios, while the baseline dried *F. vesiculosus* product had a higher O/C ratio.

Demineralization led to a slight improvement in biocrude yield. Both fucoidan and alginate extraction residue conversions resulted in relatively high biocrude energy recovery rates of 52.83 and 45.78%. Post macroalgae leaching neutralization is not advised, as it led to significantly poorer biocrude yield and energy recovery rate, and high process water demand. Short retention time hydrothermal liquefaction of *F. vesiculosus* resulted in a low biocrude yield and a high elemental nitrogen content in the product, but promises high quality otherwise, as suggested by advantageous H/C, O/C ratios and a higher heating value of the biocrude.

Supplementary Materials: The following are available online at https://www.mdpi.com/article/10.3390/recycling6030045/s1, MS Excel Workbook: Raw data.

Author Contributions: Conceptualization, L.J. and T.H.P.; methodology, T.H.P. and L.A.R.; data curation, L.J.; writing—original draft preparation, L.J.; writing—review and editing, T.H.P. and

L.A.R.; visualization, L.J.; supervision, T.H.P. and L.A.R. All authors have read and agreed to the published version of the manuscript.

Funding: This research received no external funding.

Institutional Review Board Statement: Not applicable.

Informed Consent Statement: Not applicable.

Data Availability Statement: The data presented in this study are available in the Supplementary Workbook, S1.

Conflicts of Interest: The authors declare no conflict of interest.

References

1. Horn, S.J. *Seaweed Biofuels: Production of Biogas and Bioethanol from Brown Macroalgae*; VDM, Verlag Dr. Müller: Saarbrücken, Germany, 2009.
2. Song, M.; Pham, H.D.; Seon, J.; Woo, H.C. Marine brown algae: A conundrum answer for sustainable biofuels production. *Renew. Sustain. Energy Rev.* **2015**, *50*, 782–792. [CrossRef]
3. Hou, X.; Hansen, J.H.; Bjerre, A.B. Integrated bioethanol and protein production from brown seaweed Laminaria digitata. *Bioresour. Technol.* **2015**, *197*, 310–317. [CrossRef]
4. Adams JM, M.; Ross, A.B.; Anastasakis, K.; Hodgson, E.M.; Gallagher, J.A.; Jones, J.M.; Donnison, I.S. Seasonal variation in the chemical composition of the bioenergy feedstock Laminaria digitata for thermochemical conversion. *Bioresour. Technol.* **2011**, *102*, 226–234. [CrossRef] [PubMed]
5. Ross, A.B.; Jones, J.M.; Kubacki, M.L.; Bridgeman, T. Classification of macroalgae as fuel and its thermo-chemical behaviour. *Bioresour. Technol.* **2008**, *99*, 6494–6504. [CrossRef] [PubMed]
6. Ale, M.T.; Mikkelsen, J.D.; Meyer, A.S. Important determinants for fucoidan bioactivity: A critical review of structure-function relations and extraction methods for fucose-containing sulfated polysaccharides from brown sea-weeds. *Mar. Drugs* **2011**, *9*, 2106–2130. [CrossRef] [PubMed]
7. Smit, A.J. Medicinal and pharmaceutical uses of seaweed natural products: A review. *Environ. Boil. Fishes* **2004**, *16*, 245–262. [CrossRef]
8. Chen, H.; Zhou, D.; Luo, G.; Zhang, S.; Chen, J. Macroalgae for biofuels production: Progress and perspectives. *Renew. Sustain. Energy Rev.* **2015**, *47*, 427–437. [CrossRef]
9. Anastasakis, K.; Ross, A. Hydrothermal liquefaction of the brown macro-alga Laminaria Saccharina: Effect of reaction conditions on product distribution and composition. *Bioresour. Technol.* **2011**, *102*, 4876–4883. [CrossRef]
10. Neveux, N.; Yuen, A.; Jazrawi, C.; He, Y.; Magnusson, M.; Haynes, B.; Masters, A.; Montoya, A.; Paul, N.; Maschmeyer, T.; et al. Pre- and post-harvest treatment of macroalgae to improve the quality of feedstock for hydrothermal liquefaction. *Algal Res.* **2014**, *6*, 22–31. [CrossRef]
11. Jin, B.; Duan, P.; Xu, Y.; Wang, F.; Fan, Y. Co-liquefaction of micro- and macroalgae in subcritical water. *Bioresour. Technol.* **2013**, *149*, 103–110. [CrossRef]
12. Anastasakis, K.; Ross, A. Hydrothermal liquefaction of four brown macro-algae commonly found on the UK coasts: An energetic analysis of the process and comparison with bio-chemical conversion methods. *Fuel* **2015**, *139*, 546–553. [CrossRef]
13. Barreiro, D.L.; Beck, M.; Hornung, U.; Ronsse, F.; Kruse, A.; Prins, W. Suitability of hydrothermal liquefaction as a conversion route to produce biofuels from macroalgae. *Algal Res.* **2015**, *11*, 234–241. [CrossRef]
14. Kerner, K.N.; Hanssen, J.F.; Pedersen, T.A. Anaerobic digestion of waste sludges from the alginate extraction process. *Bioresour. Technol.* **1991**, *37*, 17–24. [CrossRef]
15. Choi, J.; Choi, J.-W.; Suh, D.J.; Ha, J.-M.; Hwang, J.W.; Jung, H.W.; Lee, K.-Y.; Woo, H.-C. Production of brown algae pyrolysis oils for liquid biofuels depending on the chemical pretreatment methods. *Energy Convers. Manag.* **2014**, *86*, 371–378. [CrossRef]
16. Smith, A.; Ross, A.B. Production of bio-coal, bio-methane and fertilizer from seaweed via hydrothermal carbonisation. *Algal Res.* **2016**, *16*, 1–11. [CrossRef]
17. Elliott, D.C.; Biller, P.; Ross, A.B.; Schmidt, A.J.; Jones, S.B. Hydrothermal liquefaction of biomass: Developments from batch to continuous process. *Bioresour. Technol.* **2015**, *178*, 147–156.
18. Wang, S.; Zhao, S.; Cheng, X.; Qian, L.; Barati, B.; Gong, X.; Yuan, C. Study on two-step hydrothermal liquefaction of macroalgae for improving bio-oil. *Bioresour. Technol.* **2020**, *319*, 124176. [CrossRef]
19. Yuan, C.; Wang, S.; Cao, B.; Hu, Y.; Abomohra, A.E.-F.; Wang, Q.; Qian, L.; Liu, L.; Liu, X.; He, Z.; et al. Optimization of hydrothermal co-liquefaction of seaweeds with lignocellulosic biomass: Merging 2nd and 3rd generation feedstocks for enhanced bio-oil production. *Energy* **2019**, *173*, 413–422. [CrossRef]
20. Rey-Castro, C.; Lodeiro, P.; Herrero, R.; de Vicente, M.E.S. Acid− base properties of brown seaweed bi-omass considered as a Donnan gel. A model reflecting electrostatic effects and chemical heterogeneity. *Environ. Sci. Technol.* **2003**, *37*, 5159–5167. [CrossRef] [PubMed]

21. Pedersen, T.H.; Grigoras, I.F.; Hoffmann, J.; Toor, S.S.; Daraban, I.M.; Jensen, C.U.; Rosendahl, L.A. Continuous hydrothermal co-liquefaction of aspen wood and glycerol with water phase recirculation. *Appl. Energy* **2016**, *162*, 1034–1041.

22. Keshav, A.; Norge, P.; Wasewar, K.L. Reactive extraction of citric acid using tri-n-octylamine in nontoxic nat-ural diluents: Part 1—Equilibrium studies from aqueous solutions. *Appl. Biochem. Biotechnol.* **2012**, *167*, 197–213. [CrossRef] [PubMed]

23. Elliott, D.C. Review of recent reports on process technology for thermochemical conversion of whole algae to liquid fuels. *Algal Res.* **2016**, *13*, 255–263. [CrossRef]

24. Díaz-Vázquez, L.M.; Rojas-Pérez, A.; Fuentes-Caraballo, M.; Robles, I.V.; Jena, U.; Das, K.C. Demineralization of Sargassum spp macroalgae biomass: Selective hydrothermal liquefaction process for bio-oil production. *Front. Energy Res.* **2015**, *3*, 6.

25. Robin, A.; Sack, M.; Israel, A.; Frey, W.; Müller, G.; Golberg, A. Deashing macroalgae biomass by pulsed electric field treatment *Bioresour. Technol.* **2018**, *255*, 131–139. [CrossRef] [PubMed]

26. Liaw, S.B.; Wu, H. Leaching characteristics of organic and inorganic matter from biomass by water: Differences between batch and semi-continuous operations. *Ind. Eng. Chem. Res.* **2013**, *52*, 4280–4289. [CrossRef]

27. Sintamarean, I.M.; Grigoras, I.F.; Jensen, C.U.; Toor, S.S.; Pedersen, T.H.; Rosendahl, L.A. Two-stage alka-line hydrothermal liquefaction of wood to biocrude in a continuous bench-scale system. *Biomass Conv. Biorefinery* **2017**, *7*, 425–435. [CrossRef]

28. Obluchinskaya, E.D.; Voskoboinikov, G.M.; Galynkin, V.A. Contents of Alginic Acid and Fuccidan in Fucus Algae of the Barents Sea. *Appl. Biochem. Microbiol.* **2002**, *38*, 186–188. [CrossRef]

29. Schiener, P.; Black, K.D.; Stanley, M.S.; Green, D. The seasonal variation in the chemical composition of the kelp species Laminaria digitata, Laminaria hyperborea, Saccharina latissima and Alaria esculenta. *Environ. Boil. Fishes* **2015**, *27*, 363–373. [CrossRef]

30. Liu, X.; Bi, X.T. Removal of inorganic constituents from pine barks and switchgrass. *Fuel Process. Technol.* **2011**, *92*, 1273–1279. [CrossRef]

31. Toor, S.S.; Jasiunas, L.; Xu, C.C.; Sintamarean, I.M.; Yu, D.; Nielsen, A.H.; Rosendahl, L.A. Reduction of in-organics from macroalgae Laminaria digitata and spent mushroom compost (SMC) by acid leaching and selective hydrothermal liquefaction. *Biomass Conv. Biorefinery* **2018**, *8*, 369–377. [CrossRef]

32. Bateni, H.; Saraeian, A.; Able, C. A comprehensive review on biodiesel purification and upgrading. *Biofuel Res. J.* **2017**, *4*, 668–690. [CrossRef]

33. Blum, S.C.; Sartori, G.; Gorbaty, M.L.; Savage, D.W.; Dalrymple, D.C.; Wales, W.E. Co2 Treatment to Remove Organically Bound Metal Ions from Crude. U.S. Patent 6187175, 13 February 2001.

34. Yang, T.; Wang, W.; Kai, X.; Li, B.; Sun, Y.; Li, R. Studies of Distribution Characteristics of Inorganic Elements during the Liquefaction Process of Cornstalk. *Energy Fuels* **2015**, *30*, 4009–4016. [CrossRef]

35. Graiff, A.; Liesner, D.; Karsten, U.; Bartsch, I. Temperature tolerance of western Baltic Sea *Fucus vesiculosus*—Growth, photosynthesis and survival. *J. Exp. Mar. Biol. Ecol.* **2015**, *471*, 8–16. [CrossRef]

36. Vauchel, P.; Leroux, K.; Kaas, R.; Arhaliass, A.; Baron, R.; Legrand, J. Kinetics modeling of alginate alkaline extraction from Laminaria digitata. *Bioresour. Technol.* **2009**, *100*, 1291–1296. [CrossRef]

37. Friedl, A.; Padouvas, E.; Rotter, H.; Varmuza, K. Prediction of heating values of biomass fuel from elemental composition. *Anal. Chim. Acta* **2005**, *544*, 191–198. [CrossRef]

38. Channiwala, S.; Parikh, P. A unified correlation for estimating HHV of solid, liquid and gaseous fuels. *Fuel* **2002**, *81*, 1051–1063. [CrossRef]

recycling

Article

Fungi and Circular Economy: *Pleurotus ostreatus* Grown on a Substrate with Agricultural Waste of Lavender, and Its Promising Biochemical Profile

Simone Di Piazza [1,*], Mirko Benvenuti [2], Gianluca Damonte [2], Grazia Cecchi [1], Mauro Giorgio Mariotti [1] and Mirca Zotti [1]

1   Department of Life, Earth and Environmental Science (DISTAV), University of Genoa, Corso Europa 26, 16132 Genoa, Italy; grazia.cecchi@edu.unige.it (G.C.); m.mariotti@unige.it (M.G.M.); mirca.zotti@unige.it (M.Z.)
2   Center of Excellence for Biomedical Research (CEBR), Department of Experimental Medicine (DIMES), University of Genoa, Via Leon Battista Alberti 2, 16132 Genoa, Italy; mirko.benvenuti@edu.unige.it (M.B.); gianluca.damonte@unige.it (G.D.)
*   Correspondence: simone.dipiazza@unige.it

Abstract: The increasing production of essential oils has generated a significant amount of vegetal waste that must be discarded, increasing costs for farmers. In this context, fungi, due to their ability to recycle lignocellulosic matter, may be used to turn this waste into new products, thus generating additional income for essential oil producers. The objectives of our work, within the framework of the European ALCOTRA project FINNOVER, were two fold. The first was to cultivate *Pleurotus ostreatus* on solid waste of lavender used for essential oil production. The second was to provide, at the same time, new products that can increase the income of small and medium farms in the Ligurian Italian Riviera. This paper presents two pilot tests in which *P. ostreatus* was grown on substrates with five different concentrations of lavender waste, ranging from 0 to 100% (*w/w*). Basidiomata grown on all the substrates and their biochemical profiles were characterized using high-performance liquid chromatography coupled to mass spectrometry. The biochemical analysis of mushrooms proved the presence of molecules with antioxidant and potential pharmacological properties, in particular in mushrooms grown on lavender-enriched substrates. The results open the possibility of producing mushrooms classified as a novel food. Furthermore, the results encourage further experiments aimed at investigating how different substrates positively affect the metabolomics of mushrooms.

Keywords: essential oil production; agro-waste recycling; mushroom cultivation; closing the loop; HPLC-MS analysis

check for **updates**

Citation: Di Piazza, S.; Benvenuti, M.; Damonte, G.; Cecchi, G.; Mariotti, M.G.; Zotti, M. Fungi and Circular Economy: *Pleurotus ostreatus* Grown on a Substrate with Agricultural Waste of Lavender, and Its Promising Biochemical Profile. *Recycling* **2021**, *6*, 40. https://doi.org/10.3390/recycling6020040

Academic Editor: Leonel Jorge Ribeiro Nunes

Received: 3 May 2021
Accepted: 8 June 2021
Published: 11 June 2021

Publisher's Note: MDPI stays neutral with regard to jurisdictional claims in published maps and institutional affiliations.

Copyright: © 2021 by the authors. Licensee MDPI, Basel, Switzerland. This article is an open access article distributed under the terms and conditions of the Creative Commons Attribution (CC BY) license (https://creativecommons.org/licenses/by/4.0/).

1. Introduction

Due to environmental changes and the high degree of competitiveness of national and international markets, the agri-food industry faces numerous economic challenges; hence, the industry is constantly looking for economically sustainable solutions. These difficulties are manifested in rural areas where small- and medium-sized farms face challenges to market their products and to make them competitive with multinational corporations or foreign products. These difficulties may be caused by the poor efficiency of the production cycles and/or by the lack of competitiveness of the products themselves. A circular economy could be a valid alternative to the habitual "take-make-waste" approach, providing a concrete solution to transform the system into a more efficient, sustainable, and eco-friendly approach. Many studies have highlighted that a circular approach, through closing the production loop, minimizes external inputs and the production of additional waste, making the processes both economically and environmentally sustainable [1]. A circular approach, if properly applied, could allow the survival of many local farm businesses that would otherwise not be economically sustainable. In recent years, the scientific community has worked to optimize production processes from a circular economy perspective.

Concerning bioresources, research has focused on the isolation and selection of particular organisms that are exploitable in circular processes [2]. In this context, fungi—due to their natural roles in ecological cycles—are good candidates to be exploited to recycle and to transform vegetal by-products and waste from agriculture into more valuable products. To date, there are more than 100,000 known species of fungi, but it is estimated that there are over 3,000,000 species that are still unknown [3]. Due to their biological characteristics, fungi are able to colonize almost every environment on Earth, and thus they play a key role in nutrient cycling in trophic chains. For example, lignicolous fungi (that is, wood-decaying fungi) play a crucial role in recycling lignin and cellulose in forest ecosystems [4]. On this basis, it is clear that several species of lignicolous fungi can also be applied fruitfully from a circular economy perspective and can be exploited to turn woody by-products and waste from agriculture into food or other valuable products. Mushrooms, in particular, are a group of fungi characterized as having sporomata (fruiting bodies) visible to the naked eye; they have been collected and cultivated by humans for thousands of years both for food and for medicinal purposes [5]. Almost all traded mushroom species today are saprotrophic or lignicolous fungi cultivated on substrates based on different decaying organic matter, most frequently wood. Recently, the increasing use of mushroom species such as *Pleurotus ostreatus* (Jacq.) P. Kumm., *Lentinula edodes* (Berk.) Pegler, *Ganoderma lucidum* (Curtis) P. Karst. and *Grifola frondosa* (Dicks.) Gray for food [6,7], nutraceutical products, and cosmetics [8,9], has motivated scientists to undertake new research regarding the applications of mushroom-forming fungi. Some examples are enzyme production through solid or liquid fermentation [10,11], bioremediation of polluted environments [12], new composite biomaterial for green building [13], and medical and nutraceutical applications [5,9]. In a recent paper, an international group of researchers [14] further underlined how, due to the research carried out in recent years, fungi can play a central role in our lives and, if properly exploited, they can help address many of the challenges humans are likely to face in the future.

P. ostreatus, also known as the oyster mushroom, is one of the most used and affordable species of cultivated mushrooms globally [6]. Due to its economic, ecological, and medicinal values, *P. ostreatus* is widely cultivated. Several authors have highlighted its usefulness in bioremediation of petroleum and aromatic polycyclic hydrocarbons [15–17]. Furthermore, from a biochemical perspective, *P. ostreatus* is an excellent supplier of different nutrients and well-known molecules with beneficial effects, such as vitamins, amino acids, and essential fatty acids. The nutritional value of these compounds, and their numerous effects, such as anti-inflammatory and antioxidant properties, are well known [18–20]. Within the extensive literature on *P. ostreatus*, research about cultivation techniques and the effects of different substrates on the production of sporomata has shown how different agricultural waste can be used for sporomata production [21–26]. These findings confirmed the versatility of this mushroom and suggest the possibility of cultivating it on other types of agricultural waste from a circular perspective.

True lavender (*Lavandula angustifolia* Miller) and its cultivars are widely utilized in the essential oil industry. The natural range of *L. angustifolia* lies in the south of France, the Pyrenees (Spain) and, partially, the north-western Italian Riviera [27]. Today, in addition to France and Spain, Bulgaria, the United Kingdom, China, Ukraine, and Morocco are among the biggest global producers. Bulgaria, in particular, has recently become the world's leading producer, producing 100 tons of lavender oil per year. During the extraction process, the relatively low quantity of essential oil in fresh lavender (0.8–1.3%) results in enormous quantities of solid residues (tens of thousands of tons globally) that still have a high content of useful substances. These wastes are usually discarded directly in nearby locations or disposed of as special waste, leading to potential environmental issues.

In this work, within the framework of the European project ALCOTRA 1198 FINNOVER (http://www.interreg-finnover.com accessed on 3 May 2021), we hypothesized that *P. os-*

treatus can be cultivated on substrates enriched with waste derived from the extraction of lavender essential oils. The pilot tests were established at the Stalla Company, a small agricultural business in Liguria (north-western Italy). The company was founded in 1900 and in the past 20 years has cultivated and hybridized several species of flowers, including the *L. angustifolia* cultivar *imperia* used in this work. The activities were carried out in two pilot tests with the main goal of confirming the feasibility of cultivating *P. ostreatus* mushrooms on waste of the *L. angustifolia* cultivar *imperia* in a small local and rural business. The biochemical profile of sporomata grown on lavender-based substrates was analyzed through high-performance liquid chromatography coupled with mass spectrometry (HPLC-MS) to evaluate the variation due to the lavender enrichment. The biochemical data showed that the mushrooms grown on the enriched substrates have a high content of useful substances that can add value to the final product. The results of this work highlight the possibility of using lavender residues according to the circular economy principle for the production of mushrooms.

2. Materials and Methods

2.1. Pilot Tests

P. ostreatus was used in the tests because it has several interesting characteristics: it grows quickly, is relatively easy to manage, has good organoleptic characteristics, and has interesting, well-known medicinal properties. Moreover, it also grows spontaneously in many areas of Liguria and northern Italy.

In 2017, two strains of *P. ostreatus* called POA (autochthonous wild strain, isolated from a Ligurian locality in the province of Savona) and POC (domesticated strain) were isolated and tested for the pilot plant cultivation. The two strains were isolated under a laminar flow on Malt Extract Agar (MEA) and Potato Dextrose Agar (PDA) at 24 °C for 15 days. Once purified, the two strains were preserved on PDA slants at 6 °C in the ColD collection of the Laboratory of Mycology DiSTAV (University of Genoa). The two tests described started in December (2018 and 2019, respectively) with spawn production and incubation, and the cultivation phases were carried out in spring at the Stalla Company.

Substrates were prepared according to Yang et al. [23] and modified as follows. In December, plant residues of the *L. angustifolia* cultivar *imperia* derived from the extraction of lavender essential oil were properly shredded and mixed with barley straw for the preparation of growth substrates with different lavender/straw ratios (Table 1). The water content of the final mixed substrates was adjusted to 70%.

Table 1. Details of the different percentages of lavender present in the substrates.

	% Barley Straw	% Lavender
1	100	0
2	0	100
3	50	50
4	60	40
5	70	30

The prepared substrates were placed in 25 × 45 cm thermoresistant polypropylene bags (600–700 g per bag), sealed, and autoclaved at 125 °C for 30 min. Once cooled, the substrates were inoculated under a laminar flow hood using 30 g of 7 day old cultures of the POA and POC strains grown on PDA. The spawn was incubated in the dark at 24 °C and 65–70% relative humidity (RH) for 50–60 days to allow the mycelium to colonize the entire substrate.

The pilot cultivation test was carried out in April at the Stalla Company. The pilot plant was set up in a company greenhouse adapted for cultivation. The 25 m^2 pilot area was set up using special sterilized plastic tarps. Cultivation benches (90 cm high) were set up in the pilot area, adjacent to a fog irrigation system to manage the RH during the cultivation phase. In addition, to prevent contamination, a small anteroom was set up to provide double protection against contaminants to the pilot area. Before starting cultivation, all surfaces were sanitized using a 20% hypochlorite solution. The spawn was placed in the pilot area: all bags were placed on the benches, spaced 20 cm apart from each other. To enable the formation of primordia, six holes of 2 cm in diameter were made using a sterile scalpel on each spawn bag. During the cultivation period, the fog irrigation system was used to keep the RH at about 80%; moreover, the irrigation contributed to control the temperature at around $22 \pm 3\,°C$ during the entire period of cultivation.

The basidiomata were harvested when the pileus margins were flat to slightly rolled upwards. They were counted, weighed, and preserved at $-20\,°C$ for the subsequent biochemical analysis.

The weight data of the collected basidiomata were processed using Microsoft Excel to calculate the mean, perform one-way analysis of variance (ANOVA), and create charts of production on the different substrates. In addition, the percentage yield on the different substrates (Y) was calculated using the following equation:

$$Y = \frac{\text{basidiomata fresh weight}}{\text{weight of substrate}} \times 100$$

2.2. Sample Preparation and Biochemical Analysis

The basidiomata were divided into aliquots of 2 g fresh weight each, taking care to eliminate growth substrate usually attached to hyphae. Each aliquot was placed within an Eppendorf tube. The samples obtained were dried using a Speed-Vac freeze dryer slightly heated to $35\,°C$ to facilitate the evaporation of the water in the vegetation. The samples were finely pulverized in a mortar to obtain a greater contact surface with the solvent, placed inside heat-resistant glass vials with 2 mL of anhydrous ethanol, methanol, or acetonitrile next to a magnetic anchor, and the vials were hermetically closed. Water was not used as an extraction solvent because in samples rich in polymeric sugars such as those examined, there is a rehydration of the sample itself, which is not desirable because it prevents effective filtration and centrifugation. The sample was left for 2 h on a heating plate at $55\,°C$ and then filtered through filter paper and transferred to settle in a test tube in a refrigerator for 30 min. This allowed agglutination of the residual excess carbohydrates, causing it to precipitate. A final centrifugation at 13,000 rpm for 10 min was used to obtain clear extract without solid residue.

The HPLC-ESI-MS analysis was performed using an Agilent 1100 chromatograph directly coupled with an MSD ion trap mass spectrometer. Chromatographic separation was conducted using a C-18 Symmetry column (Waters Corporations). The column was chosen considering the complexity of the matrix to be analyzed and the reproducibility of the method.

For the mobile phase, HPLC-grade acetonitrile (Merck, Darmstadt, Germany) and Milli-Q water (Millipore Corp., Bedford, Italy) were used, and were both filtered, degassed, and added, respectively, to 0.5 and 1% formic acid (Carlo Erba, Sabadell, Spain) to facilitate and improve ionization. SIAD (Bergamo, Italy) supplied the research-grade nitrogen (>99.995%). Absolute ethanol (Carlo Erba) was used for the extraction. The characterization of a given signal can be undertaken by analyzing the "full scan" and tandem spectra with Massbank EU, an exhaustive "open access" tool available on the Internet. The m/z ratio of the parent ion was entered and, to focus the search and exclude some substances, the m/z ratios of one or more fragment ions were also entered. For each search, a relative intensity normalized to 100 and an adequate tolerance regarding the accuracy and resolution of the instrument used was set. In our case, although the instrument used has an accuracy of 0.05 UMA when considering the m/z ratios, the tolerance was set at 0.3 UMA.

The main issue with this kind of sample is the dominant presence of long-chained polysaccharides, such as chitin and related polymers, whose behavior as "chemical sponges" can interfere with the optimal separation in HPLC and ionization in the mass spectrometer. To overcome these problems, great care was applied in the filtration and centrifugation processes; these steps are essential to ensure the quality of the samples that are injected in the HPLC-MS system.

3. Results

3.1. Pilot Tests

The entire mushroom production process on substrates enriched with lavender waste was established to be cost effective and well integrated, with the lowest impact on the usual activities of the company. As shown in Figure 1, the mushroom production process began with the waste derived from essential oil extraction (center of Figure 1). After the mushroom production process, the exhausted substrate consisting of vegetable material partially degraded by the biological activity of the mushrooms was used within the company as a soil conditioner or to produce biocompost.

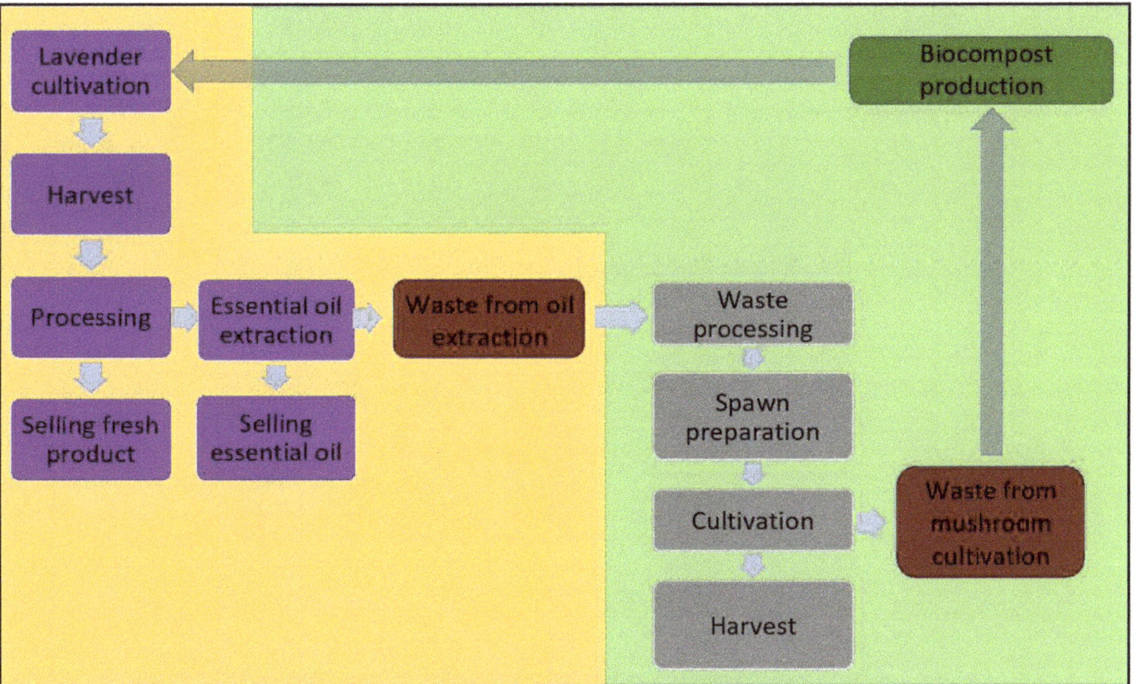

Figure 1. The entire circular process for mushroom cultivation. The yellow area concerns the cultivation of lavender and the essential oil production process. The green area represents the mushroom production process and the valorization of the spent substrates for biocompost production.

In 2018 (the first pilot test), during the incubation phase, 97% of the inoculated bags completed the incubation phase, whereas 3% were contaminated by a common microfungus of the genus *Trichoderma*. In 2019 (the second pilot test), all of the incubated bags reached the optimal colonization. The incubation time required to achieve complete colonization of the substrates was 40 ± 5 days in 2018 and 47 ± 7 in 2019. The substrates consisting of 100% lavender residues needed 12 ± 8 days longer in 2018 and 15 ± 7 days longer in 2019 to complete colonization compared with the control substrate. There were no significant differences between the two strains used regarding growth. During the cultivation phase, 90 and 95% of the bags in 2018 and 2019, respectively, reached the production of primordia and developed fruit bodies.

The weight of the basidiomata grown on the different substrates in 2018 and 2019 is shown in Table 2. The one-way ANOVA confirmed that the substrate composition affected the weight ($p < 0.01$), but there was no difference in the growth between the two strains tested.

Table 2. Mushroom production (expressed in grams) on the different substrates tested.

	2018		2019	
Substrate	POA	POC	POA	POC
1	493	537	851	826
2	287	294	456	490
3	370	407	623	538
4	424	435	594	545
5	452	439	793	753

As shown in Figure 2, the percentage yield (Y) for both pilot tests showed a lower production efficiency for the substrate containing high lavender waste concentrations compared with the control (100% straw), which had a Y between 27.4 and 29.8. The substrate with 100% lavender had a Y between 15.2 and 16.9, the substrate with 50% lavender had a Y between 17.9 and 22.6, and the substrate with 40% lavender had a Y between 18.2 and 24.2. The Y for the substrate containing 30% lavender—between 24.4 and 26.4—was closest to the control substrate.

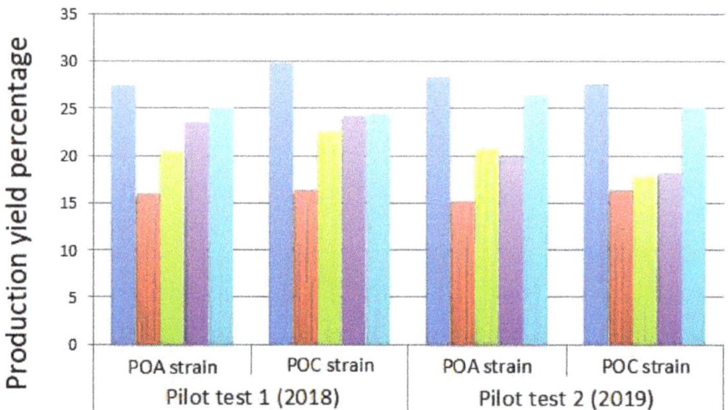

Figure 2. Production yield expressed as the percentage of the two strains grown on different substrates (each color refers to different straw/lavender ratios: blue = 100/0; red = 0/100; green = 50/50; violet = 60/40; light blue = 70/30) containing different concentrations of lavender waste in 2018 and 2019.

3.2. Biochemical Analysis

The analysis of sporomata extracts grown on the substrates with lavender showed that an average mass of extract between 1.6- and 2.1-fold higher in weight was produced compared with the sporomata grown on the control substrate. In addition, domesticated sporomata grown on the control substrate tended to produce fewer metabolites than the wild strain. The analysis was performed using multiple samples of mushrooms grown under the same conditions. This examination demonstrated the repeatability of the extraction and analytical methods. As shown in Figures 3 and 4, the chromatograms of the extracted samples overlap perfectly.

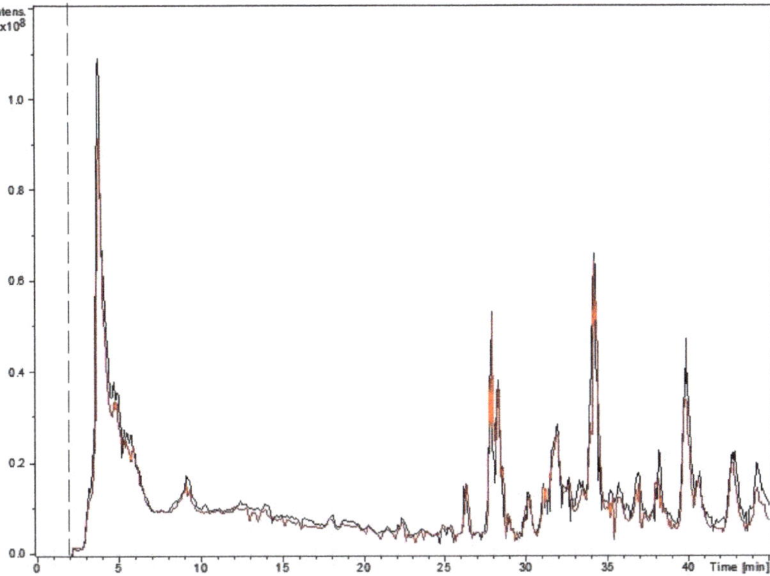

Figure 3. A comparison between samples of the same strain grown under the same conditions and extracted in ethanol. The perfect overlap of the different extracts confirms the reliability and reproducibility of the extraction and analytical methods.

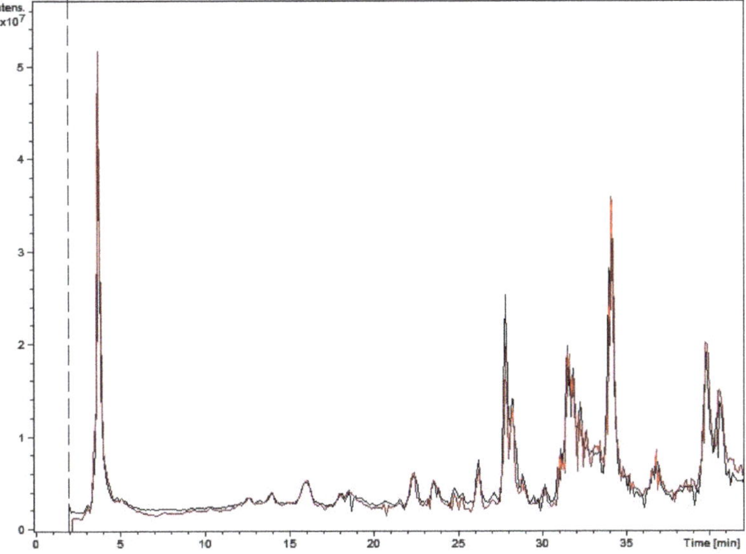

Figure 4. A comparison between samples of the same strain grown under the same conditions and extracted in acetonitrile.

This method isolated and identified >50 molecules belonging to numerous chemical families, such as di-tripeptides, fatty acids, and their epoxides. There were no major differences between the domesticated and wild strains regarding production of biomass and metabolites, although there were differences in the metabolite profile of *P. ostreatus* harvested on only straw compared with the enriched substrates. As shown in Table 3, the

HPLC-MS method allowed dividing the various classes of compounds into five groups relative to their retention time within the chromatographic analysis. The obtained separation could be very useful because it is predictive of the biological activities that each fraction, or specific compounds, from *P. ostreatus* extracts may have for different applications, including for human health.

Table 3. Classification of the isolated and identified compounds relative to the percentage of eluent B and their retention time in high-performance liquid chromatography.

Group	% of Acetonitrile	Retention Time (min)
Group 1 carboxylic acids, alcohols, and nucleosides	0–11	3–5
Group 2 carbohydrates, amino acids, dipeptides, tripeptides, and their derivatives (Amadori products)	15–32	7–15
Group 3 nucleotides, polyphenols, and lactones	35–48	16.5–20
Group 4 fatty acids and their derivatives	50–69	22.5–30
Group 5 cholanic acid and apolar compounds	70–100	35–40

4. Discussion

The two pilot tests carried out demonstrated the application of mushroom cultivation on a lavender farm following the circular economy principle. This process, as already proposed by other authors, enables farmers to reuse the agricultural waste produced during the main production cycle to obtain a new product [11,20,26]. The particular substrate leads to distinct metabolomic profiles, a phenomenon that improves the commercial value compared with mushrooms grown on traditional substrates. This difference compensates for the lower yield of mushrooms. These data are consistent with the feasibility study carried out as part of the FINNOVER project (www.interreg-finnover.com accessed on 3 May 2021). The vegetal waste partially inhibited the mycelial development, increasing the incubation time when grown on the substate with lavender, and, later, the development of basidiomata (lowering the production yield compared with the control). Based on the average yields reported in Table 1, there were significant differences in the weight of basidiomata produced on the substrates tested ($p < 0.001$), but there were no differences in the biomass of the produced sporomata between the two strains tested. This result was further confirmed by calculating the production efficiency index Y, as shown in Figure 2. These calculations revealed that the yields on a substrate composed of 100% of lavender residue were lower than those on the control substrate composed exclusively of straw. These differences in biomass production suggest the use of lavender residues to grow mushrooms may be unsuitable because, as shown in Figure 2, the Y value is inversely proportional to the lavender waste content in the substrate. However, as discussed below, the metabolomic characteristics and added nutraceutical value provided by the lavender to the cultivated basidiomata could compensate for a small drop in production. Because the substrate containing 30% lavender residue had slightly lower yields (ranging from 1 to −5.4%) than those observed in samples without lavender residue, this substrate may provide an appropriate balance between biomass production and improved metabolomic profile.

P. ostreatus is a well-known food that is rich in essential dietary elements [28,29]. Specifically, we found that the wild strain is a slightly better producer of metabolites in response to the presence of an environmental substrate than the domesticated strain, if dried extracts and just secondary metabolites are considered. This variable production probably depends on the fact that the domesticated strain, which has been cultivated for generations in a protected and controlled environment through cloning of the mycelium—a type of asexual reproduction that does not allow genetic mixing—depresses or silences in successive generations various secondary metabolic pathways that are no longer useful for maintaining physiological activity. These metabolic pathways are not silenced in the wild strain, because in nature it is in direct contact with environmental stressors, and direct competitors and pathogens. The substrate enriched with lavender, a plant particularly rich in terpenes and other molecules it produces as antibacterial and antifungal agents, stimulates the fungus to activate its secondary metabolism, as evidenced by the presence of compounds that are not part of the fungus's usual primary metabolomic pattern. Although weight data from the different growth phases of the fungus showed that the fungus grown on straw generally produced greater biomass, HPLC-MS and tandem mass analysis showed that the levels of *n*-acetylglucosamine and its precursors remained constant in each sample, independently of the type of substrate used for growth.

Of the many molecules extracted and characterized, significant interest exists in the amino acids, glucosides, and small dipeptides, that are particularly known in the fungi kingdom. There are myriad roles for these peptides and their derivatives; it has been shown that these peptides have antifungal, antimicrobial, immunostimulant, and growth-promoting properties [30]. The evolution of these peptides can be seen in the large cyclized peptide molecules produced by some fungi, such as amatoxins of the *Amanita* genus, a clear evolutionary trend that leads from simple molecules to complex cycled peptides with defensive functions. Amanitin toxins, in general, are among the most effective toxins produced by superior fungi, with a clear defensive role. Therefore, the condition by which peptides and dipeptides are produced in *P. ostreatus* appears to be archaic or, at least, less evolved [30]. Of significant interest is the notable amount of glutathione in the samples. Its action is highly relevant regarding free radicals and peroxide ions, and largely justifies the powerful antioxidant action of *P. ostreatus* extracts analyzed on cell cultures treated with oxidizing agents (personal data).

Of note is the presence of fatty acid epoxides, particularly those of myristoleic, linoleic, and linolenic acids produced by the fungus grown on lavender-enriched substrates. Fungi are an excellent source of polyunsaturated fatty acids, whose role has long been recognized in the prevention of inflammatory heart and other diseases [31]. Moreover, it is evident that a balanced intake of polyunsaturated fatty acids is fundamental for human health. We also found a notable share of palmitic acid, a saturated fatty acid. The samples of *P. ostreatus* grown on substrate enriched with lavender also showed the presence of numerous molecules derived from fatty acids, in particular a series of fatty acid epoxides. These molecules are produced by the fungus in response to environmental stresses through the activation of the monooxygenase domain of cytochrome P450, which is common to all fungi and has been highly preserved during evolution [32]. In particular for linoleic and linolenic acid, in the samples cultivated on substrates enriched with lavender, we noticed a partial decrease in the content of polyunsaturated fatty acids in favor of the presence of fatty acid epoxides. This trend was also confirmed by the presence of cholanic acid in the samples of fungi cultivated on substrates enriched with lavender. These molecules of steroid origin are similar to those found in human bile salts and mammals in general. These molecules are probably involved in the fatty acid mobilization inside fungal cells, consistent with the presence of lipid drops in the fungal cells. The presence of fatty acid epoxides in fungi is interesting. It is known that fatty acids often have cytotoxic activity or otherwise are harmful to health [33]. It should be noted that epoxides of fatty acids were only present in fungi cultivated on enriched substrates, and were absent or present in trace amounts in fungi grown on only straw, confirming that lavender acts as a stressor, and suggesting the role of fatty acids epoxides as biomarkers of stress in Basidiomycota. In contrast to plants, in which the role of peroxygenases in the production of fatty acid epoxides is known, in fungi it appears that only the monooxygenase domain of cytochrome P450 is involved in their synthesis [34]. Nevertheless, the role of fatty acid epoxides in the regulation and suppression of inflammatory processes has been recognized [35]. Several toxic and tumor growth-promoting activities are known [36], so further investigation to verify the activity of fatty acid epoxides produced by *P. ostreatus* would be useful.

5. Practical Implications of This Study

The pilot tests conducted allowed us to evaluate the technical feasibility of exploiting the cultivation of mushrooms to recycle agricultural residues in small rural farms. The results confirmed the feasibility of this approach, but showed a lower production yield compared with the standard substrate in optimal conditions. The reasons for this reduced efficiency are due to the different substrate and to the fact that the Stalla Company, which produces lavender essential oil, is not specialized in the production of mushrooms. Although the yields were low, a positive aspect that emerged from these tests is the presence of interesting substances within the basidiomata. This positive effect, due to the different substrates, should be investigated because it could add value to the product and make it extremely profitable for the farmers. Future tests will contribute to optimizing and improving the efficiency of the process, and to better understand how different residues can influence the biochemical composition of the fungi produced.

6. Conclusions

The pilot tests carried out in this work showed that mushrooms can be fruitfully exploited in the agricultural circular economy. In particular, we found that a lavender based waste product could be recycled, resulting in interesting characteristics from a food that is currently commercially available, making it a niche product and a potential novel food. The same model could be tested using different residues and fungi. By evaluating different combinations, other interesting products could emerge that may be exploited in different rural contexts. The results obtained in this work should spur further research on this topic.

Author Contributions: Conceptualization, S.D.P., M.B., G.D., G.C., M.G.M., M.Z.; methodology S.D.P., M.B., G.D., M.Z.; validation M.Z. and G.D.; formal analysis; investigation S.D.P., M.B.; data curation, S.D.P., M.B., G.C.; writing—original draft preparation, S.D.P., M.B., writing—review and editing, S.D.P., M.B., G.D., G.C., M.G.M., M.Z.; supervision, G.D. and M.Z.; project administration, M.Z.; funding acquisition, M.Z. All authors have read and agreed to the published version of the manuscript.

Funding: This research was funded by a Interreg-Alcotra European Union project called FINNOVER (n° 1198), http://www.interreg-finnover.com/ accessed on 3 May 2021.

Institutional Review Board Statement: Not applicable.

Informed Consent Statement: Not applicable.

Acknowledgments: The authors acknowledge Franco Stalla for the technical work during the pilot tests.

Conflicts of Interest: The authors declare no conflict of interest.

References

1. Zabaniotou, A.; Rovas, D.; Libutti, A.; Monteleone, M. Boosting circular economy and closing the loop in agriculture: Case study of a small-scale pyrolysis–biochar based system integrated in an olive farm in symbiosis with an olive mill. *Environ. Dev.* **2015**, *14*, 22–36. [CrossRef]
2. Branduardi, P. Closing the loop: The power of microbial biotransformations from traditional bioprocesses to biorefineries, and beyond. *Microb. Biotechnol.* **2021**, *14*, 68–73. [CrossRef]
3. Blackwell, M. The fungi: 1, 2, 3 . . . 5.1 million species? *Am. J. Bot.* **2011**, *98*, 426–438. [CrossRef] [PubMed]
4. López-Mondéjar, R.; Brabcová, V.; Štursová, M.; Davidová, A.; Jansa, J.; Cajthaml, T.; Baldrian, P. Decomposer food web in a deciduous forest shows high share of generalist microorganisms and importance of microbial biomass recycling. *ISME J.* **2018**, *12*, 1768–1778. [CrossRef] [PubMed]
5. Wasser, S.P. Medicinal mushroom science: Current perspectives, advances, evidences, and challenges. *Biomed. J.* **2014**, *37*, 345–356. [CrossRef] [PubMed]

5. Royse, D.J.; Baars, J.; Tan, Q. Current Overview of Mushroom Production in the World. In *Edible and Medicinal Mushrooms: Technology and Applications*; Zied, D.C., Pardo-Giménez, A., Eds.; Wiley-Blackwell: Hoboken, NJ, USA, 2017; pp. 5–13.

7. Ferraro, V.; Venturella, G.; Pecoraro, L.; Gao, W.; Gargano, M.L. Cultivated mushrooms: Importance of a multipurpose crop, with special focus on Italian fungiculture. *Plant Biosyst.* **2020**, *11*, 1–19. [CrossRef]

8. Taofiq, O.; González-Paramás, A.M.; Martins, A.; Barreiro, M.F.; Ferreira, I.C.F.R. Mushrooms extracts and compounds in cosmetics, cosmeceuticals and nutricosmetics-A review. *Ind. Crops Prod.* **2016**, *90*, 38–48. [CrossRef]

9. Ma, G.; Yang, W.; Zhao, L.; Pei, F.; Fang, D.; Hu, Q. A critical review on the health promoting effects of mushrooms nutraceuticals. *Food Sci. Hum. Wellness* **2018**, *7*, 125–133. [CrossRef]

10. Fazenda, M.L.; Seviour, R.; McNeil, B.; Harvey, L.M. Submerged Culture Fermentation of "Higher Fungi": The Macrofungi. *Adv. Appl. Microbiol.* **2008**, *63*, 33–103.

11. Hyde, K.D.; Xu, J.; Rapior, S.; Jeewon, R.; Lumyong, S.; Niego, A.G.T.; Abeywickrama, P.D.; Aluthmuhandiram, J.V.S.; Brahamanage, R.S.; Brooks, S.; et al. The amazing potential of fungi: 50 ways we can exploit fungi industrially. *Fungal Divers.* **2019**, *97*, 1–136. [CrossRef]

12. Vaverková, M.D.; Adamcová, D.; Radziemska, M.; Voběrková, S.; Mazur, Z.; Zloch, J. Assessment and Evaluation of Heavy Metals Removal from Landfill Leachate by Pleurotus ostreatus. *Waste Biomass Valorization* **2018**, *9*, 503–511. [CrossRef]

13. Appels, F.V.W.; Camere, S.; Montalti, M.; Karana, E.; Jansen, K.M.B.; Dijksterhuis, J.; Krijgsheld, P.; Wösten, H.A.B. Fabrication factors influencing mechanical, moisture- and water-related properties of mycelium-based composites. *Mater. Des.* **2019**, *161*, 64–71. [CrossRef]

14. Meyer, V.; Basenko, E.Y.; Benz, P.J.; Braus, G.H.; Caddick, M.X.; Csukai, M.; de Vries, R.P.; Endy, D.; Frisvad, J.C.; Gunde-Cimerman, N.; et al. Growing a circular economy with fungal biotechnology: A white paper. *Fungal Biol. Biotechnol.* **2020**, *7*, 5. [CrossRef]

15. Galli, E.; Brancaleoni, E.; Di Mario, F.; Donati, E.; Frattoni, M.; Polcaro, C.M.; Rapanà, P. Mycelium growth and degradation of creosote-treated wood by basydiomycetes. *Chemosphere* **2008**, *72*, 1069–1072. [CrossRef] [PubMed]

16. Zitte, L.F.; Awi-Waadu, G.D.B.; John, A.U. Effect of oyster mushroom (*Pleurotus ostreatus*) mycelia on petroleum hydrocarbon contaminated substrate. *J. Agric. Soc. Res.* **2012**, *12*, 115–121.

17. Mohammadi-Sichani, M.; Assadi, M.M.; Farazmand, A.; Kianirad, M.; Ahadi, A.M.; Hadian-Ghahderijani, H. Ability of *Agaricus bisporus, Pleurotus ostreatus* and *Ganoderma lucidum* compost in biodegradation of petroleum hydrocarbon-contaminated soil. *Int. J. Environ. Sci. Technol.* **2019**, *16*, 2313–2320. [CrossRef]

18. Cardwell, G.; Bornman, J.F.; James, A.P.; Black, L.J. A Review of Mushrooms as a Potential Source of Dietary Vitamin D. *Nutrients* **2018**, *10*, 1498. [CrossRef] [PubMed]

19. Scalbert, A.; Johnson, I.T.; Saltmarsh, M. Polyphenols: Antioxidants and beyond. *Am. J. Clin. Nutr.* **2005**, *81*, 215S–217S. [CrossRef]

20. Venturella, G.; Ferraro, V.; Cirlincione, F.; Gargano, M.L. Medicinal Mushrooms: Bioactive Compounds, Use, and Clinical Trials. *Int. J. Mol. Sci.* **2021**, *22*, 634. [CrossRef]

21. Hernández, D.; Sánchez, J.E.; Yamasakia, K. A simple procedure for preparing substrate for *Pleurotus ostreatus* cultivation. *Bioresour. Technol.* **2003**, *90*, 145–150. [CrossRef]

22. Nirmalendu, D.; Mukherjeeb, M. Cultivation of *Pleurotus ostreatus* on weed plants. *Bioresour. Technol.* **2007**, *98*, 2723–2726.

23. Yang, D.; Liang, J.; Wang, Y.; Sun, F.; Tao, H.; Xu, Q.; Zhang, L.; Zhang, Z.; Ho, C.T.; Wan, X. Tea waste: An effective and economic substrate for oyster mushroom cultivation. *J. Sci. Food. Agric.* **2016**, *96*, 680–684. [CrossRef]

24. Tagkouli, D.; Kaliora, A.; Bekiaris, G.; Koutrotsios, G.; Christea, M.; Zervakis, G.I.; Kalogeropoulos, N. Free Amino Acids in Three Pleurotus Species Cultivated on Agricultural and Agro-Industrial By-Products. *Molecules* **2020**, *25*, 4015. [CrossRef] [PubMed]

25. Marlina, L.; Sukotjo, S.; Marsudi, S. Potential of oil palm Empty Fruit Bunch (EFB) as media for oyster mushroom, Pleurotus ostreatus cultivation. *Procedia Chem.* **2015**, *16*, 427–431. [CrossRef]

26. Obodai, M.; Cleland-Okine, J.; Vowotor, K.A. Comparative study on the growth and yield of *Pleurotus ostreatus* mushroom on different lignocellulosic by-products. *J. Ind. Microbiol. Biotechnol.* **2003**, *30*, 146–149. [CrossRef] [PubMed]

27. Despinasse, Y.; Moja, S.; Soler, C.; Jullien, F.; Pasquier, B.; Bessière, J.-M.; Noûs, C.; Baudino, S.; Nicolè, F. Structure of the Chemical and Genetic Diversity of the True Lavender over Its Natural Range. *Plants* **2020**, *9*, 1640. [CrossRef]

28. Alam, N.; Amin, R.; Khan, A.; Ara, I.; Shim, M.J.; Lee, M.W.; Lee, T.S. Nutritional Analysis of Cultivated Mushrooms in Bangladesh—*Pleurotus ostreatus, Pleurotus sajor-caju, Pleurotus florida* and *Calocybe indica*. *Mycobiology* **2007**, *36*, 228–232. [CrossRef]

29. Cateni, F.; Gargano, M.L.; Procida, G.; Venturella, G.; Cirlincione, F.; Ferraro, V. Mycochemicals in wild and cultivated mushrooms: Nutrition and health. *Phytochem. Rev.* **2021**, *3*, 1–45.

30. Ng, T.B. Peptides and proteins from fungi. *Peptides* **2004**, *25*, 1055–1073. [CrossRef]

31. Node, K.; Huo, Y.; Ruan, X.; Yang, B.; Spiecker, M.; Ley, K.; Zeldin, D.C.; Liao, J.K. Anti-inflammatory properties of cytochrome P450 epoxygenase-derived eicosanoids. *Science* **1999**, *285*, 1276–1279. [CrossRef]

32. Pednault, K.; Angers, P.; Avis, T.J.; Gosselin, A.; Tweddel, R.J. Fatty acid profiles of polar and non-polar lipids of *Pleurotus ostreatus* and *P. cornucopiae* var. 'citrino-pileatus' grown at different temperatures. *Mycol. Res.* **2007**, *111*, 1228–1234. [CrossRef] [PubMed]

33. Kuksis, A.; Pruzanski, W. Chapter 3—Epoxy Fatty Acids: Chemistry and Biological Effects. In *Fatty Acids: Chemistry, Synthesis, and Applications*; Ahmad, M.U., Ed.; Academic Press and AOCS Press: Cambridge, MA, USA, 2017; pp. 83–119.

34. Kennedy, M.J.; Reader, S.L.; Davies, R.J. Fatty acid production characteristics of fungi with particular emphasis on gamma linolenic acid production. *Biotechnol. Bioeng.* **1993**, *42*, 625–634. [CrossRef] [PubMed]

35. Corrêa, R.C.G.; Brugnari, T.; Bracht, A.; Peralta, R.M.; Ferreira, I.C.F.R. Biotechnological, nutritional and therapeutic uses of *Pleurotus* spp. (Oyster mushroom) related with its chemical composition: A review on the past decade findings. *Trends Food Sci Technol.* **2016**, *50*, 103–117. [CrossRef]
36. Greene, J.F.; Newman, J.W.; Williamson, K.C.; Hammock, B.D. Toxicity of Epoxy Fatty Acids and Related Compounds to Cells Expressing Human Soluble Epoxide Hydrolase. *Chem. Res. Toxicol.* **2000**, *13*, 217–226. [CrossRef]

 *recycling*

Review

Biodegradation of Hemicellulose-Cellulose-Starch-Based Bioplastics and Microbial Polyesters

Mateus Manabu Abe [1], Marcia Cristina Branciforti [2] and Michel Brienzo [1,*]

1   Institute for Research in Bioenergy (IPBEN), University of São Paulo State (UNESP), Rio Claro 13500-230, Brazil; mateusabe40@gmail.com
2   Department of Materials Engineering, São Carlos School of Engineering (EESC), University of São Paulo (USP), São Carlos 13566-590, Brazil; marciacb@sc.usp.br
*   Correspondence: michel.brienzo@unesp.br

Abstract: The volume of discarded solid wastes, especially plastic, which accumulates in large quantities in different environments, has substantially increased. Population growth and the consumption pattern of societies associated with unsustainable production routes have caused the pollution level to increase. Therefore, the development of materials that help mitigate the impacts of plastics is fundamental. However, bioplastics can result in a misunderstanding about their properties and environmental impacts, as well as incorrect management of their final disposition, from misidentifications and classifications. This chapter addresses the aspects and factors surrounding the biodegradation of bioplastics from natural (plant biomass (starch, lignin, cellulose, hemicellulose, and starch) and bacterial polyester polymers. Therefore, the biodegradation of bioplastics is a factor that must be studied, because due to the increase in the production of different bioplastics, they may present differences in the decomposition rates.

Keywords: biodegradation; bioplastics; lignocellulosic fibers; microbial polyesters

 check for updates

Citation: Abe, M.M.; Branciforti, M.C.; Brienzo, M. Biodegradation of Hemicellulose-Cellulose-Starch-Based Bioplastics and Microbial Polyesters. *Recycling* **2021**, *6*, 22. https://doi.org/10.3390/recycling6010022

Received: 28 January 2021
Accepted: 11 March 2021
Published: 22 March 2021

Publisher's Note: MDPI stays neutral with regard to jurisdictional claims in published maps and institutional affiliations.

Copyright: © 2021 by the authors. Licensee MDPI, Basel, Switzerland. This article is an open access article distributed under the terms and conditions of the Creative Commons Attribution (CC BY) license (https://creativecommons.org/licenses/by/4.0/).

1. Introduction

Consumption demands for industrialized materials such as plastics in their various applications have increased over the past years. This consumption is generating residues, which require alternatives for their proper disposal and recycling. Disposal, recycling, and plastic substitution are potential research areas towards urgent and necessary solutions. Most commercial plastics come from the petrochemical industry, which uses natural gas and fossil hydrocarbons as feedstock. Such synthetic plastics are biodegradable and degradable only for a long period. Therefore, they are considered neither biodegradable nor renewable [1]. Synthetic polymers, such as polypropylene (PP), polyethylene (PE), polytetrafluoroethylene (PTFE), nylon, polyester (PS), and epoxy are examples of plastic components of high resistivity, chemical and biological inertness, resistance, flexibility, and other interesting properties [2–5].

At the beginning of the large-scale production of synthetic plastic materials, their properties seemed adequate for good quality development. However, such materials are non-biodegradable, thus generating large accumulations of residues in different landscapes. Thus, they have been a cause of growing concerns due to environmental problems. New materials based on biological sources have been developed towards solving or reducing the above-mentioned problems. However, in addition to be renewable and biodegradable, bioplastics must have vapors barrier properties and mechanical properties that meet the different applications of this material, and the attention has now evolved towards the possible ecotoxic effects of bioplastics and active properties for a cover of food.

The names of biodegradable and/or bioplastic products given by companies and reported in the literature, when drawn up wrongly, can lead to misunderstandings by the general public due to incorrect classifications of the polymeric materials [6–9]. A bioplastic can be biodegradable or not. However, a biodegradable material does not necessarily come from a biological source. Towards the avoidance of errors, the following definitions reported in this article, must be clarified:

- Plastics are polymeric matrices comprised of organic polymers of high molecular weight and other substances, such as fillers, colors, and additives [6]. In general, the synthetic route is predominant in the synthesis of the material.
- Bioplastic refers to materials that are biodegradable, bio-based, or both. Although the term bioplastic is generally used to distinguish polymers derived from fossil resources, it is worth mentioning that bioplastics may come from petroleum [6]. The prefix "bio" of bioplastic does not necessarily mean this material is environmentally friendly [6].
- Biomass is a source of natural organic carbon that may originate from animals or vegetables raised/cultivated by humans or that spontaneously emerge in terrestrial and marine environments [10].

Different biotic and abiotic factors contribute to the different degradation processes [11]. Thermal, mechanical, and chemical degradation, as well as photodegradation, are examples of abiotic degradation. A degradation process is related to the fragmentation of material into small elements or molecules, or just physical and chemical changes in a polymer. Due to high temperatures, polymers can be thermally degraded. The chemical bonds in their chains are broken by a thermo-degradation effect [12].

Mechanical degradation is an abiotic degradation mechanism that occurs through shear forces (due to aging, turbulence in water and air, snow pressure, and other factors), tension and/or compression. Under environmental conditions, it acts synergistically with different abiotic factors [13].

Abiotic chemical degradation occurs by the degradative effect of chemicals substances and represent one of the most important mechanisms of abiotic degradation, since the polymer matrix is affected by atmospheric or agrochemical pollutants, such as oxygen (i.e. O_2 or O_3), which produce free radicals through oxidation, attacking covalent bonds [13]. Abiotic chemical degradation differs from biotic chemical degradation, mainly regarding the origin of the chemical with a degrading effect.

Photodegradation is the process of degradation of polymers by the action of light, resulting in the oxidation of the material. UV rays interact with chromophores groups of polymers (carbonyl, hydroxyls, and aldehydes), which are degraded by chain fission, photoionization, crosslinking, and oxidation reaction [11,13–15].

Microbial biodegradation is a degradation process of polymers and other materials through the action of microorganisms [11] resulting in CO_2 and/or methane, water, cell biomass, and energy. However, in the natural environment and even in the process of controlled biodegradation, abiotic effects help or even occur synergistically with biodegradation. This consideration of synergism is important for the elaboration of biodegradation procedures.

With environmental concern, this review evaluated the biodegradation process (considering the synergistic action of biotic and abiotic agents) of bioplastics elaborated with polysaccharides from plant biomass and microbial polyesters. Moreover, addressing the definitions, biodegradation mechanism, and factors that affect the biodegradative process of bioplastics. The scope of this review does not address the biodegradation of bioplastics produced from polymers of animal origin (natural polymer), and bioplastics derived from petroleum (PBAT, PBS, PVA, PCL, and PGA [16]. However, the definitions presented in this review do not exclude these types of bioplastics.

2. Problems Related to Plastics

The current geological era, the so-called Anthropocene, is exposed to the influence of human actions in different environments. Indicators from such anthropic actions are biodiversity reduction, deforestation, climate, and other environmental changes [17]. However, materials produced by human society, like plastics, are also indicators of the Anthropocene

Plastics directly (i.e., environmental impacts from the plastic production chain) affect different environments (e.g., terrestrial and marine).

An environment in which plastic waste currently generates several problems, is the oceans, due to the large accumulation of these materials. The plastic that reaches the oceans mostly is generated in coastal population regions, where the disposal and management of this waste is destined for uncontrolled landfills [18]. Due to urban runoff and inland waterways, such plastics reach oceans and are transported via tide and winds. It is estimated that between 4.8 and 12.7 million metric tons of plastics produced in the continent (distribution varies according to the analyzed location) reached the marine environment in 2010 [18].

In recent years, environmental concerns (e.g., harmful effects of plastics on the environment, since they are not biodegradable [19], or slowly degraded) have been intensified. Large accumulations of floating plastics in the oceans have been reported- approximately 1.8 trillion pieces of plastic have been quantified in the Great Pacific Garbage Patch (GPGP) [20]. The ingestion of plastic fragments by the marine fauna is a major concern due to their small size [18,21,22], the so-called microplastics, which are smaller than 5 mm [21]. Besides, since plastic fragments are present on the surface and floor of oceans, as well as in several maritime regions (coastal areas) and the Arctic sea ice, strategies, as a reduction in inputs [18] and the elaboration/utilization of biodegradable materials, would be adequate measures to reduce the impacts of plastics.

Even with the area of studies on the impacts of plastics on fauna and for the various organisms still under development, some studies point to the occurrence of toxicological effects of this synthetic waste [23,24]. A plastic intake and entanglement can lead to the lower life quality of organisms, loss of mobility, external and internal injuries, blockage of digestion, and other harms [25]. Goldstein and Goodwin [22] identified the presence of microplastics (mainly PE, PS, and PP) in the digestive tract of 33.5% of *Gooseneck crustaceans* (*Lepas* spp.).

The development of innovative technologies represents a means for both sustainable development and the growth of emerging countries, such as Brazil, whose sustainable energy has been highlighted by innovation technologies. Thus, even with bioplastic not representing a material for total replacement of non-biodegradable plastic, researches and production of bioplastics are a technological alternative for the development of a more sustainable and balanced society. Therefore, aligning with the current trend (socio-political and environmental), in which concerns with the environment is growing.

3. Biodegradation Process

The microbial biodegradation of materials occurs by the action of microorganisms, such as fungi and bacteria [26], and is classified as physical, chemical, and enzymatic according to modifications in the materials. Biodegradation is a natural process of vital importance for nutrients and energy recycling [27]. Microorganisms use organic material as a source of nutrition for their metabolism; except for the substances used in metabolic incorporation, the rest is oxidized by cellular respiration, thus leading to the formation of simple and small submetabolites, released in the environment [28,29].

Biodegradation due to physical degradation occurs from the adhesion of microorganism species to the surface of organic materials through the secretion of a gum [30] produced by microorganisms. This gum represents a complex matrix made of natural polymers (e.g., polysaccharides and proteins). Such a thick complex, together with microorganisms, infiltrates the material and changes its volume, size, pores distribution, moisture content, and thermal transfers. A few microorganisms (e.g., filamentous fungi) lead to cracks in the materials due to mycelial growth, i.e., both their durability and resistance properties are reduced [13,31]. Microorganism biofilms are a matrix that protects microorganisms from different environmental conditions and results in a major change in materials [13,32].

Biodegradation by chemical degradation refers to the production of chemical substances by living organisms, which facilitate and increase the speed of the process. Emulsifying substances produced by microorganisms help the exchange between hydrophobic and hydrophilic phases, which are important interactions for the penetration of microorganisms in the polymeric material [33]. Such a lime formation (polymers secreted by microorganisms mixed with different microbial species) improves the material deterioration. It represents a point of accumulation of polluting and chemical substances (abiotic chemical degradation), thus benefitting microbial proliferation [34].

Examples of chemical substances released into the environment by microorganisms, which play an important role in chemical biodegradation, are nitrous acid, nitric acid, and sulfuric acid. All of these compounds are produced by chemolithotrophic bacteria, such as *Nitrosomonas* spp., *Nitrobacter* spp., and *Thiobacillus* spp., respectively [29,33,35]. Apart from the action of chemical substances generated by those organisms, chemoorganotrophic microorganisms generate organic acids with potential for chemical degradation (e.g., oxalic, citric, gluconic, glutaric, glyoxalic, oxaloacetic, and fumaric acids).

The action mechanisms of such acids (organic or inorganic) are diverse and include an increase in surface erosion when adhering to the material surface [36]. The use of those acids as nutrients benefits the growth of filamentous fungi and bacteria [13]. Another action mechanism of biotic chemical degradation is the oxidation of organic material. Certain fungi and bacteria have specific proteins in their membrane that capture iron-chelating compounds (siderophores) [37]. With this mechanism microorganisms capture cations from a matrix.

Biodegradation by enzymatic degradation occurs due to the depolymerization of polymeric chains of a matrix through the action of hydrolase enzymes that catalyze the reactions of chemical bonds breakage adding a water molecule. These bonds are ether, peptide-like, and ester, present in biodegradable bioplastics. The main enzymes are amylases and cellulases, which cleave starch and cellulose polymers, respectively. However, other enzymes (breakage of ester bonds), such as esterases and lipases, can degrade co-polyesters.

A mechanism that explains the action of hydrolases (e.g., depolymerase) in polyesters hydrolysis (synthetic and natural) through biodegradation is related to three amino acids, namely serine, histidine, and aspartate. A hydrogen bond is formed when a component reacts with the histidine ring, thus guiding interaction between histidine and serine, and forming an alcohol group of high nucleophilic character (-O). Histidine plays a deprotonating role for serine, i.e., as a base. The alkoxide group includes an ester bond and generates an acyl-enzyme and an alcohol group. Finally, a free enzyme and a terminal carboxyl group are generated by the action of the water molecule under an acyl-enzyme. This entire enzymatic degradation process is termed catalytic triad [30,38,39], and the products generated are metabolized or not by microorganisms that have depolymerizing enzymes. Therefore, a consortium of microorganisms is important for complete biodegradation [13]. Figure 1 depicts the mechanism of action of depolymerizes and the catalytic triad.

Apart from the biodegradation of cellulose, starch, and polyesters, hemicellulose is another polymer that can be degraded by microbial enzymes. A catalytic action of hemicellulases (enzymatic pool) on different types of hemicellulose polysaccharides produces monomeric sugars, acetic acids [40]. For example, enzymes that degrade xylan (hemicellulose from grasses) are endo-1,4-β-xylanase (cleavage results in oligosaccharides), xylan 1,4-β-xylosidase (cleavage of oligosaccharides generate by xylan, which forms xylose monomers), and accessory enzymes, such as xylan-esterases, ferulic and p-coumaric-esterases, α-L-arabinofuranosidases, and α-4-O-methyl glucuronidase [41]. Both enzymes act synergistically so that xylans and hemicellulose mannans of some types of plant cell walls are depolymerized [42]. Nevertheless, some polymers are not biodegraded by common enzymatic hydrolysis, i.e., polymers can be oxidized by enzymes such as laccase, dioxygenase, peroxides, monooxygenase, and oxidases [43]. Thus, such enzymes are not hydrolases and influence the cleavage process of polymers differently from hydrolases (oxygen insertion, hydroxylation, oxidation, and free radical formation lead to polymer cleavage) [43]. Figure 2 depicts the enzymatic biodegradation process.

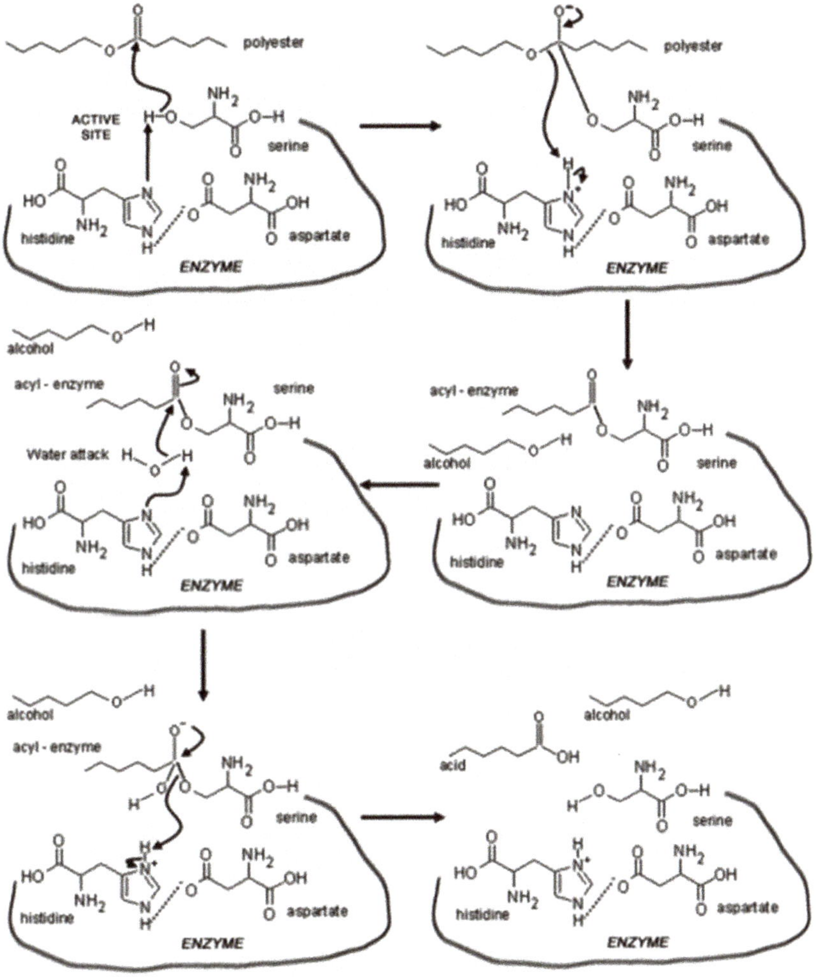

Figure 1. Enzymatic hydrolysis of polymers and catalytic site of depolymerase enzymes [13].

The result of biodegradation, for example of bioplastic from natural polymers (e.g., polysaccharides) is the generation of small molecules from a polymer. Microorganisms cannot employ large substances insoluble in water for obtaining organic or inorganic nutrients for their metabolism. They produce enzymes and chemicals used in extracellular environments and, therefore, depolymerize the materials. After hydrolysis and/or oxidative action of microorganism enzymes on different polymers, which results in monomers, metabolism oxidation occurs. In this system, organic compounds lead to a loss of electrons and the consequent production of ATP molecule (adenosine triphosphate). This is the last biodegradation stage, in which organic matter is mineralized. The microorganisms use smaller and simple organic molecules, such as oligomers and monomers, for their metabolic activities. However, byproducts are generated from microbial metabolism (e.g., carbon dioxide—aerobic degradation), water, biomass, methane, and hydrogen sulfide (anaerobic degradation) [45,46]. Figure 3 displays the biotic and abiotic degradation of plastic.

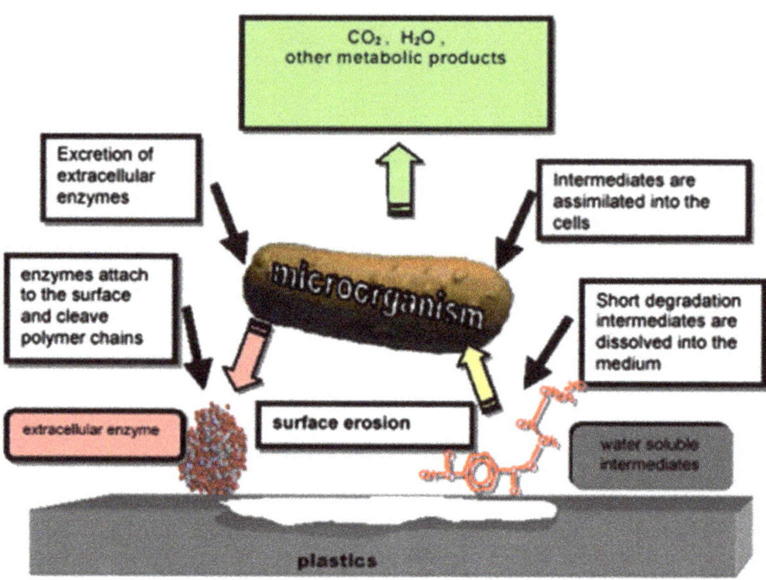

Figure 2. Enzymatic biodegradation process [44].

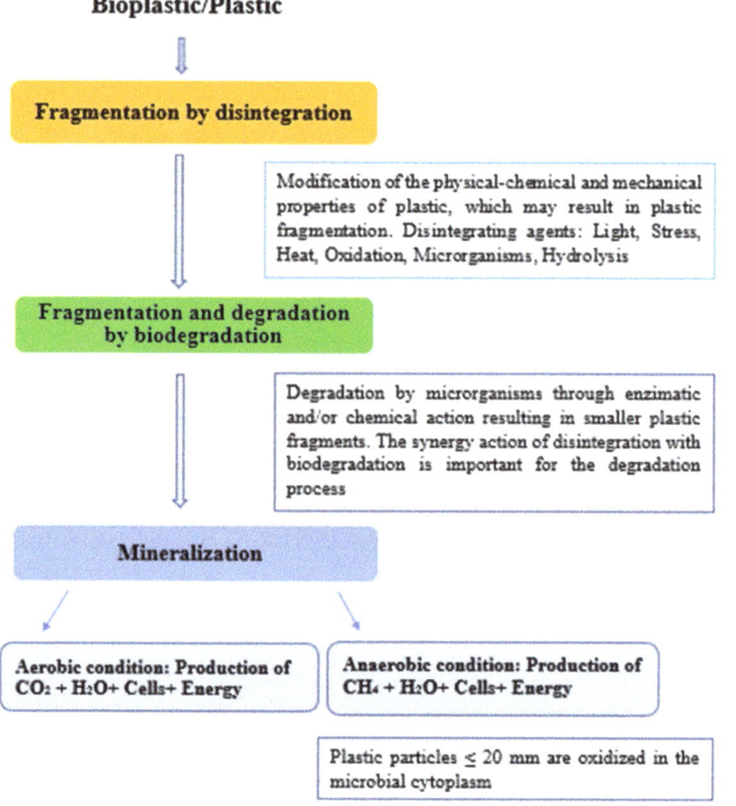

Figure 3. Disintegration, biodegradation, and mineralization process of plastic polymeric materials Adapted [47].

3.1. Factors That Influence Biodegradation

The microbial population available is a key factor for biodegradation in an environment (soil, air, and water), and several properties (e.g., the chemical constitution of materials) affects the efficiency of the biodegradation process. Chemical composition influences the biodegradation of plastics through different patterns of crystallinity, hydrophilic and hydrophobic character, conformational flexibility, polymer accessibility, surface area, molecular weight, melting temperature, hydrolyzable and oxidizable bonds in polymer chains, morphology, and stereoconfiguration [27,48,49].

Crystallinity influences biodegradability because it affects the accessibility of the enzyme to the material polymer. More organized regions of polymers (crystalline) tend to hinder enzymatic hydrolysis since catalytic proteins diffuse with greater difficulty. On the other hand, water molecules diffuse more easily between amorphous (less organized) regions, and enzymes can easily access the material polymers in such regions [28].

The polarity of bioplastics directly influences biodegradation, since materials developed with hydrophobic polymers are less susceptible to enzymatic attack. Degrading microorganisms depend on a hydrophilic surface to adhere to and catalyze the depolymerization reaction by means of hydrolytic enzymes. However, this enzymatic accessibility to the material is reduced on hydrophobic polymeric surfaces. This impediment occurs not only because the microorganisms and enzymes are more hydrophilic, but also due to the aqueous medium (usual water), in which the enzyme is contained, to have their contact with the material (bioplastic) reduced. For example, glycolic polyacids (PGA) are more easily biodegraded than poly (lactic acids) (PLA), since PLA is more hydrophobic) [50].

Blends in a polymeric bioplastic matrix are common when it is desired to obtain materials with certain characteristics, and also interfere with biodegradation (increase or reduce biodegradation), since the different components of biocomposites can influence the accessibility of the enzyme to the polymeric material in different ways.

The molecular weight of polymers affects the biodegradability of plastics, since the heavier the molecular weight, the greater the difficulty for microorganisms to break it down and assimilate. Therefore, the lower the molecular weight of the polymer, the easier the biodegradation, since the need for extracorporeal digestion is reduced. Aliphatic polyester is one of the few biodegradable polymers of high molecular weight [13]. However, it is worth mentioning that in addition to the molecular weight, the types of bonds in the polymeric chain (considering that bioplastic, like plastic, is formed by a polymeric matrix), and different chemical groups in polymers influence the biodegradation process.

Although the term "bio" degradation is directly correlated with the fragmentation of a polymer by the action of microorganisms, these microorganisms do not act in isolation on the polymeric material, since abiotic agents influence the fragmentation efficiency. The abiotic degradation of organic matter such as thermal, mechanical, chemical, and by the action of light are examples of degradative processes. These processes work synergistically with biodegradation, reducing the material to dimensions that allow microbial assimilation [13,51].

3.2. Assessment and Biodegradation Quantification

Biodegradation can be measured through metabolic products, physical and chemical properties of plastics/bioplastic, acidification of the medium, and other ways. CO_2 is a product of biodegradation, more specifically, of the oxidation of organic matter, and can be used for direct or indirect measurement of material biodegradation over a period of time. Its content released in a degradation process is quantified by the respirometry technique, which can use a closed CO_2 production and a capture system. International methods, such as ASTM D5338-15 [52] and ISO 14855-2: 2018 [53] are applied for the quantification of the CO_2 produced in a microbial degradation process.

The measurement of consumed oxygen (ISO 17556: 2003) [54] is another method of quantifying biodegradation by respirometry. Respirometry involves techniques that measure parameters indicative of cellular respiration. The higher the consumption of oxygen and the release of CO_2 by microorganisms, the better the biodegradation indicator. For details and examples of other standard methods of respirometry analysis (ASTM, EN and ISO), see specialized literature [55].

Methane molecules can also be used for measurements of materials biodegradation. However, unlike the above-mentioned respirometry techniques, CH_4, CO_2, and other gases quantification is generally conducted under anaerobic conditions. Analysis methods such as ASTM D5511-02 [56], is used for this purpose.

Apart from microbial proliferation in plastic/bioplastics materials, analyses of color change, surface roughness, cracks, and holes are also alternatives for checking the deterioration of materials [13,36]. Analysis parameters can be used especially for materials of difficult biodegradation and low CO_2 release. However, the results of such analyses (e.g., microbial growth in the polymeric matrix) are not recommended for the conclusion of biodegradation or abiotic degradation directly [13]. Additional techniques, such as electron microscopy, photon microscopy, microscopy of polarization, and atomic force microscopy reinforce the results [13,57,58].

The physical properties of plastics/bioplastics (e.g., tensile strength, elongation at break, modulus of elasticity, crystallinity, cold crystallization temperature, and glass transition temperature) can be measured as biodegradation indicators. The weight loss of a sample determined by the burial method can be used in plastic/bioplastic biodegradation analyses, although it may result from the solubility and volatility of certain substances [13]. The analysis of weight loss of bioplastics by burying in soil, or composting systems, may result in conclusion errors, since in addition to the mass of the soil or compost account for the variation in the bioplastic mass, in bioplastics washing processes (a step which precedes weighing procedures), can cause fragmentation and loss of material derived from bioplastic. Thus, even though the method of analyzing mass loss is frequently reported in the literature, as is usual in determining the biodegradation of bioplastics, this technique ends up being difficult to perform [59]. Recent articles evaluating the biodegradation of bioplastics by burying in soil and compost has used image evaluation as a tool for analysis, that is, the reduction of the area of bioplastics, detected by image registration (from the insertion of the bioplastic in a mold/grid with known dimensions) [60,61].

The indication of biodegradation through products generated by microorganisms is another way of measuring the process. For example, the biodegradation of polymeric cellulose materials can be measured according to the release of glucose [62], or the quantification of 1,4-butanediol as an indicator of the biodegradation of PBA and PBS polymers [63].

The increase in microbial biomass (weight or number of cells) is indicative of a biodegradation process since a single source of carbon (plastic or bioplastic material) in a closed environment can point out the occurrence of biodegradation and/or surface changes and molecular rearrangements. However, conclusive statements about the amount of mineralized material cannot be directly made.

The evaluation and quantification of bioplastic and/or plastic biodegradation by the above-mentioned methods can be conducted in an aqueous medium and soil. However, each condition of analysis imposes different requirements, which leads to different responses from different methods.

3.3. Biodegradation of Bio-Based Polymers Bioplastics

In this topic, biodegradation of bioplastics developed with polysaccharides from plant biomass/lignocellulose and microbial polyesters was followed as the scope of this review. It was exemplified the biodegradation of a category of bioplastics, those developed with natural polymers (vegetable and microbial). Therefore, this review does not intend to address issues related to the development of bioplastics of vegetable and microbial origin, advantages and disadvantages in addition to the viability of this material (related to the economic aspects and properties of bioplastics). To obtain this information, it is recommended reading of the specialized literature [6,64–66].

3.3.1. Biodegradation of Plant-Based Polymers Bioplastics

The mass loss of bioplastics from rice straw showed complete degradation after 105 days [67]. Rice straw bioplastics were composed mainly of cellulose and trifluoroacetic acid. On the first day of contact with the soil, the bioplastic showed an increase in mass, due to the phenomenon of water absorption by the material. According to the authors, its mechanical properties are similar to those of polystyrene (bioplastic in the dry state).

The mass loss of bioplastics consisting of acetylated starch and acetylated sugarcane fibers (lignin, hemicellulose, and cellulose) resulted in 24.2 to 39.3% degradation after 5 weeks [68]. The acetyl group may have created stable biodegradable sites; however, an increasing effect on the crystallinity of the bioplastic with the addition of cellulose may have contributed to the low biodegradation rate due to the restriction effect of the microbial enzyme's activity. In addition to the crystallinity and chemical structure of cellulose, microbial diversity, carbon availability and the period of biodegradation considered can influence its depolymerization.

Bioplastics (glycerol, acetylated starch, and acetylated nanocellulose composition) subjected to biodegradation in a petri dish with *Trametes versicolor* were completely degraded in 60 days, and after 40 days with starch and non-acetylated reinforcement. The starch bioplastics were completely biodegraded after 30 days, and the addition of cellulose to the formulation of bio-based plastics resulted in a longer biodegradation time [69]. Water and moisture absorption is important in the biodegradation process of bioplastics [70]. The starch-based bioplastics investigated in this study were composed of different concentrations of oxidation starch (20, 40, and 60%). Oxidation decreases biodegradation due to reduced swelling and water absorption from the soil by bioplastic.

Hemicellulose is another natural plant-derived polymer of potential application for the development of bioplastics. However, in addition to the elaboration that biomaterial, the study of the biodegradation of these carbohydrates in bioplastics have not received attention, as the area of use of hemicellulose for bioplastic focus on physicochemical properties and modifications of this macromolecule. The bioplastic based on xylan (of the hemicellulose type of grasses) and blended with gelatine was completely biodegraded after 15 days of conditioning (determined by the burial procedure) [71]. This bioplastic was considered 100% biodegradable since the sample could not be recovered for weighing. A bioplastic made with 50% xylan (from beechwood) and PVA (polyvinyl alcohol) was 56% biodegraded after 30 days by burial in soil [72]. PVA reduced the biodegradation of the bioplastic produced by the PVA/xylan mixture. The sample with 25% xylan was 42.2% biodegraded after 30 days of burial in soil.

Xylan was grafted with poly-(ε-caprolactone) (PCL), and biodegradation was evaluated by BOD (biological oxygen demand). The biodegradation (aerobic and activated sludge) kinetics of bioplastics with high concentrations of PCL was delayed in comparison to materials made with pure hemicellulose or with lower graft concentrations [66]. Despite changes in the kinetics, the biodegradation property of the bioplastic was not altered and ranged between 95.3 and 99.7%.

Recalcitrant substances also influence the biodegradation of natural polymers. Lignin is a constituent of lignocellulosic fibers, shows the highest degree of recalcitrance in the plant cell wall [73,74]. This polymeric complex of phenylpropane units hinders the biodegradation of the material or products that contain it, such as bioplastics, and reduces the contact surface of lignocellulosic fibers with degrading enzymes [74]. Lignin requires different enzymes to degrade due to the different units that comprise its polymeric complex [75]. In anaerobic environments lignin may persist biodegradation for a longer time, with this process is primarily more efficient in aerobic environments [76], due to the catalytic action involved in oxygen.

Starch and lignin (lignosulfonate) bioplastics were completely biodegraded after 4-month burial [77]. Biodegradation was measured through the analysis of CO_2 and morphological characteristics. The samples with lignin analyzed after 5 weeks of biodegradation tests were fragmented, however, small residues of the bioplastic were identified.

After the 2-month burial, the samples with lignin showed a significant biodegradation effect, with small fragments of the material still observed. After 4 months of testing, residues of bioplastic fragments were no longer detected. A bioplastic made from the addition of lignin (1.2% w/t) to the bio-PTT matrix (Bio-poly (trimethylene terephthalate)) increased its weight loss through biodegradation in soil [78]. In 140-day burial, bio-PTT/lignin bioplastic showed more than 50% mass weight loss.

A higher CO_2 emission was reported from films with lignin in comparison to the bioplastic composed only of starch, due to the greater amount of carbon atoms in its formulation [77]. However, such CO_2 may have originated from the metabolism of soil organic compounds, i.e., the bioplastic may have stimulated the microbial degradation of stable organic compounds in the soil through the priming effect. A strategy for the biodegradation of bioplastics composed of lignin, due to the recalcitrance of this phenolic complex, is the application of UV radiation prior to chemical, microbiological and/or enzymatic treatments. Lignin is susceptible to photodegradation due to the UV effect [79]. After photodegradation, other treatment combinations can be applied for the degradation or biodegradation of lignocellulosic fibers, such as enzymatic or oxidative treatments. One of the advantages of using lignin in the development of thermoplastic formulations is its processing at high temperatures [80]. However, studies using lignin in the bioplastic formulation, have not received much attention.

3.3.2. Biodegradation/Enzymatic Degradation of Plant-Based Polymers Bioplastics in Relation to Derivatization

The assessment of biodegradation, disintegration, and enzymatic degradation of bioplastics made with natural polymers (such as proteins, starch, cellulose, and hemicellulose) is not recurrent in the literature. Biodegradation has received lower attention when compared to the objective of most studies, which is to evaluate the physicochemical and mechanical properties of the materials. However, this limitation in the studies is even greater when compared to the biodegradation of bioplastics made with modified polymers.

The comparison between bioplastics developed from unmodified and modified hemicellulose presents few studies intending to analyze the enzymatic degradation [81], and biodegradation. This low number of studies with hemicellulose could be related to the difficulties in obtaining a plastic polymer matrix from this heterogeneous vegetable polysaccharide. However, in addition to the analysis of the physicochemical properties of modified bioplastics, the effects of chemical, physical, and biological (and enzymatic) modifications of polymers on biodegradation must be considered. The enzymes involved in the enzymatic degradation of unmodified and modified polysaccharides may be different. Moreover, a more complex enzymatic pool will be required for modified polysaccharides.

As pending groups are attached to the polysaccharides chain, new enzymes will be required for further hydrolysis. According to a recent review article, physical modifications of polysaccharides hardly result in a change in the biodegradation process [82]. However, chemical changes result in different degradation mechanisms. Considering a chemical similarity, the enzymatic degradation of cellulose acetate can be catalyzed by acetyl esterases, an enzyme common for xylan deacetylation. The modification or functionalization of polysaccharides may result in a reduction in biodegradation since modified bioplastics (acetylated cellulose, acetylated xylan, acetylated starch, starch propionate, starch butyrate, starch valerate, and starch hexanoate) showed a reduction in anaerobic biodegradation [83]. For example, the degree substitution (DS) > 1.5, 1.5, 1.2 for starch, cellulose, and modified xylan (acetylated) respectively, represented the minimum modification necessary to delay the biodegradation of bioplastics.

The chemical modifications of the polysaccharides that make up bioplastics, such as acetylation, increase the degree of hydrophobicity of the polymers and the plastic matrix. This has the advantage of reducing the solubilization of the polymers in polar solutions. However, resulted in a decrease the enzymatic degradation. It was observed a reduction in two mannases of *Cellvibrio japonicus* (CjMan5A and CjMan26A), with reduced catalytic activities on galactoglucomannan substrates (hemicellulose) due to the decrease of the solubility of the polymers [81]. Other studies in the literature showed the influence of chemical modification of hemicellulose in relation to solubility, thermal resistance, crystallinity [84], and biodegradation rate [85]. Therefore, the diffusion of water by the composite and biodegradation is a parameter affected by chemical derivatization.

Modified xylans with an increase in the DS reduced enzymatic degradation by xylanolitic enzyme [86]. However, a rapid biodegradation rate (80%) on the first day of the evaluation was achieved for (hydroxypropyl)xylan. Substitutions above 1.5 reduced enzymatic degradability by 10%. However, the modification of cellulose with hydroxypropyl led to a reduction in biodegradation (20% in 18 days). Regarding the DS and the enzymatic activity, the article justifies the limitation of the recognition of the xylanolitic enzyme to the substrate due to chemical modification. In addition to the sterile impediment, when it changes the polysaccharide polarity through modification, it may be another explanation for the degradability reduction [87].

Modifications of polysaccharides may result in a less hydrophobic bioplastic, favoring the process of biological and abiotic degradation. Xylan carboxymethylation for bioplastic production showed an increase in water absorption at high relative humidity, demonstrating, therefore, the hydrophilic character of the carboxymethyl groups [88]. Carboxymethylation is a procedure for the production of hemicellulose-based bioplastics with increases in hydrophilic characteristics [89]. This procedure results in the development of environmentally favorable materials considering biodegradation.

A modification of hemicellulose by subtraction of chemical constituents may result in a different biodegradation process. An enzymatic modification of arabinoxylan resulted in an increase in the bioplastic crystallinity as the arabinose content was reduced [90,91]. In both of these studies, the effects of enzymatic modification of hemicellulose in relation to biodegradation were not evaluated. However, the increase in crystallinity may be a retarding factor in the bioplastic biodegradation due to the degree of organization of the molecules limiting enzymatic action, probably reducing the water absorption effect and reducing microbial growth.

The different modifications in natural bio-based polymers (for example, polysaccharides) may result in a difficulty in biodegradation or enzymatic degradation. The rate of degradation of these materials can reduce in a given period. However, the material can still be metabolized or degraded using enzymes. For example, acetylated xylan is the form found in natural lignocellulosic materials, therefore, although acetyl groups result in a delay in biodegradation, these polysaccharides are biodegradable by microbial enzymes, such as xylanases and esterases, whereas the acetylated xylan form is predominant in the environment.

3.3.3. Biodegradation of Microbe-Based Polymers Bioplastics

Under the nutritional abundance of carbon and nitrogen, some bacteria can synthesize energy reserve polymer (inclusions). Polymers like polyhydroxyalkanoate (PHA) (intracellular granules), can be produced via microbial fermentation of biomass (animal or vegetable). Regarding applications, these natural polymers are an important alternative for the manufacture of bioplastic materials since they are biodegradable and biocompatible, and used in the medical field [92]. With the 41% increase in world production of PHAs between 2010–2017, this polyester has become a polymer of significant interest in the development of bioplastics. The properties of this microbial polyester can contribute to a reduction of environmental impacts due to the closed carbon cycle generated by biodegradation [93].

There is a growing interest in the development of materials formulated with PHAs, the study of the biodegradability of these materials. However, factors that influence the degradation of composites and bioplastics are necessary. Some of the marine microorganisms that are known to degrade PHAs [94] are *Aestuariibacter halophilus* S23; *Alcanivorax* sp. 24; *Alcanivorax dieselolei* B-5; *Pseudoalteromonas haloplanktis*; *Alteromonas* sp. MH53; *Bacillus* sp.; *Bacillus* sp. strain NRRL B-14911; *Bacillus* sp. MH10; *Comamonas testosteroni* YM1004;

Enterobacter sp.; *Aliiglaciecola lipolytica*; *Gracilibacillus* sp.; *Marinobacter* sp. NK-1; *Nocardiopsis aegyptia*; *Pseudoalteromonas* sp. NRRL B-30083; *Pseudoalteromonas gelatinilytica* NH153; *Pseudoalteromonas shioyasakiensis* S35; *Pseudomonas stutzeri* YM1006; Psychrobacillus sp. PL87; *Rheinheimera* sp. PL100; *Shewanella* sp. JKCM-AJ-6,1α; *Streptomyces* sp. SNG9. Terrestrial microbial representatives degraders PHAs [95] are *Alcaligenes faecalis*; *Pseudomonas lemoignei*; *Acientobacter* sp.; *Acientobacter schindleri*; *Bacillus* sp.; *Pseudomonas* sp.; *Stenotrophomonas maltophilia*; *Variovorax paradoxus*; *Stenotrophomonas rhizophilia*; *Penicillium* sp.; *Purpureocillium lilacinum*; *Verticillium lateritium*; *Burkholderia* sp.; *Nocardiopsis* sp.; *Streptomyces* sp.; *Bacillus cereus*; *Burkholderia* sp.; *Cupriavidus* sp.; *Gongronella butleri*; *Penicillium oxalicum*.

As in polysaccharide-based bioplastics, crystallinity in polyester bioplastics from microbial synthesis plays an important role in the biodegradation process. In bioplastics with higher proportions of amorphous regions, depolymerization occurs more quickly through abiotic or biotic action. For example, higher biodegradation was obtained with hydroxybutyrate (PHB), hydroxybutyrate-co-hydroxyvalerate (PHBV-40), PHBV-20, and P (3HB, 4HB) (10% mol of 4HB) and PHBV-3 [96]. According to the quantification of CO_2 in a composting vessel, PHBV-40 and P (3HB, 4HB) (10% mol 4HB) showed the highest degrees of biodegradation, due to a reduction in crystallinity with the addition of higher percentages of HV (valerate hydroxide—indicated by the numbering in front of the acronym) and 4HB. Biodegradation was 90.5%, 89.3%, 80.2%, 90.3% and 79.7% in 110 days of analysis for bioplastics formulated by PHBV-40, PHBV-20, PHBV-3, P (3HB, 4HB) and PHB, respectively.

The advantage of using PHBV in comparison to PHB is the ease of processing and good toughness. Certain PHBV disadvantages such as low thermal stability and a high degree of crystallinity must be overcome [96]. Improvements, mediated by chemical changes, must be performed together with the preservation of the material's biodegradation property, which depending on the HV percentage, maybe rigidity or flexibility, similarly to commercial synthetic plastics (polyethylene, polypropylene, and polyvinylchloride), and assurance of biodegradation of the formulated bioplastic [96].

A commercial Ecoflex bioplastic (commercial product of BASF) was compared to PHB and poly (3-hydroxybutyrate-co-3-hydroxyhexanoate) (PHBHHx) in activated sludge for 18 days. Bioplastics composed of PHBHHx showed a higher degree of biodegradability than Ecoflex and PHB, with weight losses of 40, 20, and 5%, respectively [97]. The low crystallinity and morphology of the surface of the bioplastic proved a determining factor in the biodegradation process, observed mainly in bioplastics with 12% HHx (hydroxyhexanoate), which displayed a rough and porous surface before and after undergoing activated exposure to sludge and lipases (Figure 4).

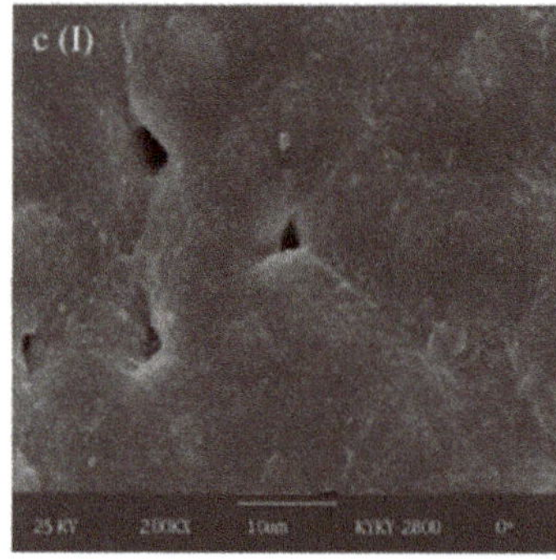
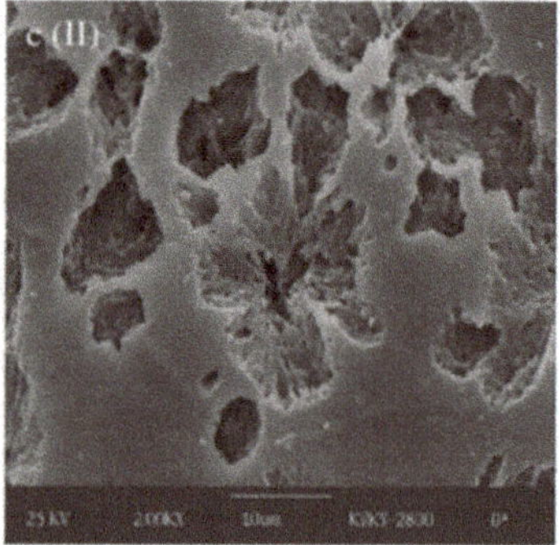

Figure 4. Surface morphology of bioplastic made with PHB (12% HHx) before (**left**) and after degradation (**right**) [97].

Besides surface morphology and crystallinity of the bioplastics, other factors, such as mixing components, depth of burial (due to environmental and/or microbial differences), and time of exposure to the soil also determine the biodegradability degree. In the study performed by Weng et al. [98], evaluating through appearance and fragmentation, the following results were achieved for the biodegradation of polymeric blends (poly (3-hydroxybutyrate-co-4-hydroxybutyrate and poly (lactic acid)—(P (3HB, 4HB)/PLA)): In the first month of testing, blends composed of 100% P (3HB, 4HB) and those with 25% PLA showed loss of integrity (appearance), whereas in the second month, both bioplastics had been almost completely biodegraded. This behavior was similar for the different depths of burial used (20 and 40 cm); however, at 20 cm and 2 months of testing, a greater difficulty was observed in the collection of fragments of blends with 75% of P (3HB, 4HB). For both depths of burial, the higher the concentration of PLA in the blends, the longer the biodegradation time. However, the biodegradation behavior was the opposite for higher concentrations of P (3HB, 4HB).

Polymer blends with 100% and 75% P (3HB, 4HB) were degraded more easily at 20 cm depth, although the presence of PLA in the bioplastics represented a delay in biodegradability at both depths tested. At 40 cm, PLA suffered greater disintegration, due to the anaerobic conditions, providing better conditions for degradation of the PLA, as reported by the authors [98].

In addition to temperature, bioplastic composition, crystallinity, degree of hydrophilicity and environmental conditions in relation to oxygen concentration, another factor that must be taken into account is the abundance of microbial biomass and the efficiency of fungi and bacteria biodegradation in different environments. The biodegradation of microbial polyesters by fungi was reported as dominant in soil [95]. However, the biodegradation of polymers in aquatic (marine) medium was faster with the use of bacteria [99,100].

PHA bioplastics as well as other bioplastics have limitations in their applications and achievements due to the high cost, low mechanical resistance, and impairment of biodegradation in functionalization processes and mixtures with other polymers [93]. In addition to the low ductility property, one of the main disadvantages of using PHAs in the production of bioplastics is the formation of brittle bioplastics, properties that can be improved by mixing biodegradable polymers from oil. However, the sustainability of bioplastic manufacturing is affected since the use of oil in the extraction and refinement stage generates the carbon dioxide production [101]. Another impact that should be considered in the production of PHA bioplastics with synthetic polymers such as the use of PCL is the reduction in the rate of biodegradation [102]. However, the development of polyester bioplastics with bio-based fibers origin (blends development), such as lignocellulosic fibers, can assist in overcoming the low ductility property of bioplastics [93]. This blend reduces costs, ensuring a biodegradable and renewable product.

The application of polyhydroxyalkanoate as a bioplastic has major limitations (e.g., its production costs for replacing conventional plastics [103]). A potential alternative for the optimization of PHA production technologies is the use of organic residues, such as lignocellulosics [104]. As an example, hemicellulose [105–107], cellulose [108] and a mixture of hemicellulose and cellulose hydrolyzate [109] have been used for the production of microbial polyesters, such as PHA and P3HB (PHB). However, the use of lignocellulosic fibers poses limitations mainly related to the production yield and generation of inhibitory substances for PHA-producing microorganisms [104]. Table 1 shows some properties and biodegradation times of different biodegradable bioplastics produced from biopolymers.

Poly(lactic acid) (PLA) is another polyester that may be partially derived from the microbial fermentation of biopolymers. The lactic acid produced by the bacteria is polymerized by a chemical route, thus forming PLA, which offers several advantages, such as rigidity and miscibility with other biodegradable plastics. However, in several application areas (e.g., manufacture of 3D printers), its fibers have been used by Brazilian companies due to the PLA lower heat loss in comparison to oil-derived plastics [110]. Nevertheless, bioplastics from bacterial polyesters should be considered, since PHAs like PBH present several advantages in comparison to PLA [111]. Table 1 shows some properties and biodegradation times of different biodegradable bioplastics produced from biopolymers.

Table 1. Properties and biodegradation time of different biodegradable bioplastics produced from biopolymers.

Bioplastic Type	Polymer Type	CONB	BPR	PAB	TS (MPa)	E (%)	Reference
PHB nanofiber	Polyester	Soil, 30 °C and 80% humidity	100% in 21 days	Weight loss	…	…	[112]
PHB/Starch	Polyester	Sludge, 35 °C, anaerobic	93.8% in 190 days	Biogas quantification	…	…	[113]
PHB	Polyester	Compost, 55 °C and 70% humidity	Approx. 80% in 28 days	CO_2 quantification	1015	…	[114]
PHA	Polyester	Seawater	100% in 1.5 to 3.5 years	Literature review	…	…	[115]
PHA	Polyester	Soil, 12–15 cm depth and 35% humidity	30% in 60 days	Weight loss	16.2	600.3	[116]
PHA/Rice husk	Polyester/Fibers	Soil, of 12–15 cm depth and 35% humidity	>90% in 60 days	Weight loss	7.5	<400	[116]
PHA	Polyester	Soil, 20 °C and 60% humidity	48.5% in 280 days	CO_2 quantification	…	…	[117]
PHBV/Starch	Polyester/Polysaccharide	Liquid medium	100% in 31 days	CO_2 quantification	21.01	10.85	[118]
PHBV/NPK	Polyester/Fertilizer	Soil, 25–30 °C and 65% humidity	68.66% in 112 days	Weight loss	…	…	[119]
PHBV/Starch	Polyester/ Polysaccharide	Soil, 25 °C and 20% humidity	>60% in 150 days	Weight loss	~7	~3.2	[120]
Starch	Polysaccharide	Soil, 3.5 cm depth and 65% humidity	30% in 5 days	Weight loss	1.88	0.45	[70]
Starch	Polysaccharide	Marine water with sediment	Approx. 69% in 236 days	BOD	4.7 *	211	[121]
Starch	Polysaccharide	Microorganisms in a plate, 25 °C and 75% humidity	100% in 30 days	Weight loss	8.6	52	[69]
Starch/Cellulose	Polysaccharide/Modified	Microorganisms in plate, 25 °C and 75% humidity	100% in 60 days	Weight loss	14.7	50	[69]
Cellulose	Polysaccharide	Soil	100% in 105 days	Weight loss	45	6.1	[68]
Cellulose/Starch	Polysaccharide	Soil/ humus, 25 °C and 75% humidity	24.2 to 39.3% in 35 days	Weight loss	5.6 to 35.0	13.1 to 21.7	[69]
Hemicellulose/ Gelatine	Polysaccharide/Partially hydrolyzed protein	Soil/manure	100% in less than 15 days	Weight loss	…	…	[71]
PLA/Starch	Poly(lactic acid)/ Polysaccharide	Compost and 58 °C	79.7% in 90 days	Weight loss	…	…	[122]

CONB = Biodegradation conditions; BPR = Biodegradation period and rate; PAB = Biodegradation analysis procedure; TS = Tensile strength; E = elongation; … = not reported; * = Newton (N).

Bioplastics are renewable and/or biodegradable and display good mechanical properties such as tensile strength similar to certain synthetic plastics in common use (Table 1). Polypropylene and polystyrene, fossil-based synthetic plastics, show TS between 25–40 MPa and 30–55 MPa, respectively [68], whereas it ranges between 55–124 and 9–17 for Cellophane^{TM} and low-density polyethylene (LDPE), respectively [123,124]. This similarity of mechanical properties of bioplastics and plastics was also reported by Hansen et al. [125]. Therefore, even if at present and in the future, the total replacement of non-biodegradable plastic from petroleum, is something unlikely, for some applications, such as bioplastics for use in agriculture (mulch) and packaging (short lifetime), this technology can represent one of the alternatives (along with other actions and technologies) to mitigate the environmental impacts related to plastics.

3.4. Effect of the Bio-Based Polymer Addition on the Biodegradation Rate of PHAs Bioplastics

The addition of bio-based polymers in polyester microbial biocomposites is vast in the literature [93,126,127]. The bio-based polymers application, mainly lignocellulosic fibers, is due to the improvement in the biodegradation rate of the formulated bioplastic [93]. This improvement in the biodegradation of PHAs biocomposite is related to the increase in hydrophilicity and water absorption by bioplastics. A mixture of 30% Sisal fibers (wt) and PHBV resulted in increased water absorption of 14% compared to pure PHBV (0.8%) [128]. The use of Kenaf fibers (main cellulose) in the blend with poly(3-hydroxybutyrate-co-3-hydroxyhexanoate) [P(3HB-co-3HHx)], resulted in a greater loss of mass in biodegradation in soil due to greater water absorption and microbial binding sites in the microbial polyester from binding with Kenaf fiber [129]. This study suggested that the accelerated deterioration of the blend (reduction of mechanical properties), after 6 burial weeks (soil) was due to the weakening of the adhesion between the fiber/[P(3HB-co-3HHx)], with the access of water to the internal hydrophobic regions of the polymer.

The use of hemicellulose with PHAs is also an alternative to increase the blends biodegradation rate in relation to pure microbial polyesters [93]. Bioplastics from PHBV/ Peach Palm Particles (lignocellulosic fiber with considerable hemicellulose content) were biodegraded faster than pure PHBV in soil [130]. The authors reported cracks, corrosion, and discoloration after 2 months of biodegradation. The poor adhesion between the fiber/PHBV interface, which resulted in greater water absorption and accessibility of soil microorganisms, was suggested as contributed to the deterioration.

Starch, another polysaccharide from vegetable biomass, can also be used in the production of bioplastics with reduced biodegradation time. The mixture of starch and PHBV (50/50% wt) was fully biodegraded in the soil after 33 days, i.e., there was a 50% reduction in biodegradation time compared to pure PHBV [131]. The addition of starch reduced the crystallinity of the blend, facilitating the absorption of water by the matrix and increased the enzymatic activity on the surface and in the inner region of the blend. There was an increase in the biodegradation of PHA/starch blends as the starch content increased in the formulation, with biodegradation of PHA/30% starch (wt) corresponding to 44% in 6 months of burial in soil [132].

Chemical modifications of polysaccharides, such as cellulose acetylation (cellulose acetate) can result in partial or total inhibition of blends biodegradation, due to reduced solubilization and hydrophobicity of the fibers. The acetyl and butyryl group in cellulose reduced the rate of biodegradation of the PHB blend/modified cellulose due to the impediment of the substituents and reduced the blends/water interactions [133]. However, in the same study, the mechanical properties were improved by increasing the concentration of cellulose acetate butyrate. Related to cellulose, the degree of substitution above 2.5 results in the inhibition of biodegradation [93]. However, some chemical modifications can positively influence the biodegradation of microbial and bio-based polymer blends. This improvement in biodegradation is due to the increase in the contact area surface of the fibers, resulting from the surface treatments of the fibers, such as, an increase in the fiber rugosities with the application of NaOH, which removes the hemicellulose and lignin fibers [93].

Lignin is the most recalcitrant constituent of lignocellulosic fibers due to the complexity of the composition of this phenolic macromolecule [82,134]. The lignin enzymatic catalytic degradation needs different enzymes [75,82], or even the synergy of an enzyme complex. The inclusion of lignin in the blends of PHAs and PLA results in steric impediment of the enzyme and reduction of the degree of hydrophilicity, which is shown in the literature as a factor in reducing the biodegradation of polyester blends [93,132]. For example, the biodegradation in soil of the PHA/lignin blend was 4% after 24 weeks which was lower than the rate of biodegradation of the PHA/10% starch, PHA/cellulose (11.1 and 100% respectively) [132]. An alternative for obtaining bioplastics from microbial polyesters, with guaranteed polymer biodegradability, is the use of enzymes and microorganisms capable of catalyzing the breakdown of lignin. The main enzymes involved in lignin oxidoreduction are laccases (Lac), lignin peroxidase (LiP), and manganese peroxidase (MnP) [82], and the recently discovered enzymes dye-decolorizing peroxidases and unspecific peroxygenase [134,135].

The application of specific microorganisms that degrade lignin and/or PHAs can be an alternative to improve the biodegradation of the blends of these polymers. In nature, these enzymes act synergistically, and some microorganisms can produce the three enzymes, while others only produce a few of the necessary enzymes [82]. LiP has a key role in the degradation of lignin, due to the distinct characteristics of the active site of the enzyme. However, the catalytic action of LiP is mediated by H_2O_2, which is generated by Lac [135]. The effectiveness of the microorganisms is essential for the biodegradation of PHA/lignin blends, as some microorganisms such as *Phanerochaete chrysosporium* are considered excellent for the degradation of lignin [134,136]. However, Brown-rot fungi due to the degradation mechanisms of lignin not being oxidative, presents a reduced degradation process of lignin [82]. Another example is the case of the bacteria *Streptomyces viridosporus*, which can result in a reduced degradation process of lignin since this bacterium acts in the non-phenolic regions of lignin [82,134].

4. Conclusions

Advances in the development of materials and technologies with fewer environmental impacts are highly expected, mainly due to the progress in the area of biopolymers over the past two decades. However, biopolymers application and use in the various sectors of society is limited, i.e., the annual production of bioplastics compared to plastics is still low. In this way, the use of plant biomass and microbial polyesters can help the development of bioplastic feasible, due to the availability of resources, biocompatibility, biodegradability and generally does not result in ecotoxicity. However, the physicochemical and biodegradation properties must be considered for the study of the optimization of bioplastic from natural polymers. Several actions must be taken so that bioplastic can become a reality on a large scale. The state of São Paulo (Brazil) has established a law that prohibits the supply of disposable plastic products to commercial establishments, which may increase the production scale of some bioplastics, thus reducing costs. The approval of a law by the Chinese government that prohibits the import of international plastic wastes for recycling can also encourage the production of bioplastics. An increase in the production and distribution of bioplastics is not sufficient for the development of a more conscious and sustainable society, i.e., care must be taken for the identification of a bioplastic and/or biodegradable material towards no final consumers' mistakes and no unsuitable actions or disposal habits.

Funding: The authors acknowledge the São Paulo Research Foundation (FAPESP) for the financial support of this research project (process number 2019/16853-9; 2019/12997-6; and 2017/22401-8).

Conflicts of Interest: The authors declare no conflict of interest.

References

1. Singh, A.A.; Afrin, S.; Karim, Z. Green Composites: Versatile Material for Future. *Green Energy Technol.* **2017**, 29–44. [CrossRef]
2. Nagalakshmaiah, M.; El Kissi, N.; Dufresne, A. Ionic Compatibilization of Cellulose Nanocrystals with Quaternary Ammonium Salt and Their Melt Extrusion with Polypropylene. *ACS Appl. Mater. Interfaces* **2016**, *8*, 8755–8764. [CrossRef]
3. Mariano, M.; Pilate, F.; De Oliveira, F.B.; Khelifa, F.; Dubois, P.; Raquez, J.M.; Dufresne, A. Preparation of Cellulose Nanocrystal-Reinforced Poly(lactic acid) Nanocomposites through Noncovalent Modification with PLLA-Based Surfactants. *ACS Omega* **2017**, *2*, 2678–2688. [CrossRef]
4. Nagalakshmaiah, M.; Nechyporchuk, O.; El Kissi, N.; Dufresne, A. Melt extrusion of polystyrene reinforced with cellulose nanocrystals modified using poly[(styrene)-co-(2-ethylhexyl acrylate)] latex particles. *Eur. Polym. J.* **2017**, *91*, 297–306. [CrossRef]
5. Nagalakshmaiah, M.; Afrin, S.; Malladi, R.P.; Elkoun, S.; Robert, M.; Ansari, M.A.; Svedberg, A.; Karim, Z. Biocomposites: Present trends and challenges for the future. *Green Compos. Automot. Appl.* **2019**, 197–215. [CrossRef]
6. Lackner, M. Bioplastics-biobased plastics as renewable and/or biodegradable alternatives to petroplasticsle. *Kirk-Othmer Encycl. Chem. Technol.* **2015**, *6*, 1–41. [CrossRef]
7. Nazareth, M.; Marques, M.R.C.; Leite, M.C.A.; Castro, Í.B. Commercial plastics claiming biodegradable status: Is this also accurate for marine environments? *J. Hazard. Mater.* **2019**, *366*, 714–722. [CrossRef]
8. Harding, K.G.; Gounden, T.; Pretorius, S. "Biodegradable" Plastics: A Myth of Marketing? *Procedia Manuf.* **2017**, *7*, 106–110. [CrossRef]
9. Iwata, T. Biodegradable and bio-based polymers: Future prospects of eco-friendly plastics. *Angew. Chemie Int. Ed.* **2015**, *54*, 3210–3215. [CrossRef] [PubMed]
10. Bonechi, C.; Consumi, M.; Donati, A.; Leone, G.; Magnani, A.; Tamasi, G.; Rossi, C. *Biomass: An Overview, Bioenergy Systems for the Future: Prospects for Biofuels and Biohydrogen*; Woodhead Publishing: Cambridge, UK, 2017; pp. 3–42. [CrossRef]
11. Kabir, E.; Kaur, R.; Lee, J.; Kim, K.H.; Kwon, E.E. Prospects of biopolymer technology as an alternative option for non-degradable plastics and sustainable management of plastic wastes. *J. Clean. Prod.* **2020**, *258*, 120536. [CrossRef]
12. Crawford, C.B.; Quinn, B. Physiochemical properties and degradation. *Microplastic Pollut.* **2017**, 57–100. [CrossRef]
13. Lucas, N.; Bienaime, C.; Belloy, C.; Queneudec, M.; Silvestre, F.; Nava-Saucedo, J.E. Polymer biodegradation: Mechanisms and estimation techniques—A review. *Chemosphere* **2008**, *73*, 429–442. [CrossRef]
14. Fairbrother, A.; Hsueh, H.-C.; Kim, J.H.; Jacobs, D.; Perry, L.; Goodwin, D.; White, C.; Watson, S.; Sung, L.-P. Temperature and light intensity effects on photodegradation of high-density polyethylene. *Polym. Degrad. Stab.* **2019**, *165*, 153–160. [CrossRef]
15. Niaounakis, M. Properties. In *Biopolymers: Applications and Trends*; Elsevier Science: Amsterdam, The Netherlands, 2015; pp. 91–138.
16. Bátori, V.; Åkesson, D.; Zamani, A.; Taherzadeh, M.J.; Sárvári Horváth, I. Anaerobic degradation of bioplastics: A review. *Waste Manag.* **2018**, *80*, 406–413. [CrossRef] [PubMed]
17. Crutzen, P.J. The anthropocene. *Earth Syst. Sci. Anthr.* **2006**, 13–18. [CrossRef]
18. Jambeck, J.R.; Ji, Q.; Zhang, Y.-G.; Liu, D.; Grossnickle, D.M.; Luo, Z.-X. Plastic waste inputs from land into the ocean. *Science.* **2015**, *347*, 768–771. [CrossRef] [PubMed]
19. Geyer, R.; Jambeck, J.R.; Law, K.L. Production, use, and fate of all plastics ever made. *Sci. Adv.* **2017**, *3*. [CrossRef]
20. Lebreton, L.; Slat, B.; Ferrari, F.; Sainte-Rose, B.; Aitken, J.; Marthouse, R.; Hajbane, S.; Cunsolo, S.; Schwarz, A.; Levivier, A.; et al. Evidence that the Great Pacific Garbage Patch is rapidly accumulating plastic. *Sci. Rep.* **2018**, *8*, 1–15. [CrossRef]
21. Wright, S.L.; Thompson, R.C.; Galloway, T.S. The physical impacts of microplastics on marine organisms: A review. *Environ. Pollut.* **2013**, *178*, 483–492. [CrossRef]
22. Goldstein, M.; Goodwin, D. Gooseneck barnacles (*Lepas* spp.) ingest microplastic debris in the North Pacific Subtropical Gyre. *PeerJ.* **2013**, *1*, 184. [CrossRef] [PubMed]
23. De Sá, L.C.; Luís, L.G.; Guilhermino, L. Effects of microplastics on juveniles of the common goby (*Pomatoschistus microps*): Confusion with prey, reduction of the predatory performance and efficiency, and possible influence of developmental conditions. *Environ. Pollut.* **2015**, *196*, 359–362. [CrossRef]
24. Canesi, L.; Ciacci, C.; Bergami, E.; Monopoli, M.P.; Dawson, K.A.; Papa, S.; Canonico, B.; Corsi, I. Evidence for immunomodulation and apoptotic processes induced by cationic polystyrene nanoparticles in the hemocytes of the marine bivalve Mytilus. *Mar. Environ. Res.* **2015**, *111*, 34–40. [CrossRef]
25. Gregory, M.R. Environmental implications of plastic debris in marine settings—Entanglement, ingestion, smothering, hangers-on, hitch-hiking and alien invasions. *Philos. Trans. R. Soc. B Biol. Sci.* **2009**, *364*, 2013–2025. [CrossRef]
26. Pathak, V.M.; Navneet. Review on the current status of polymer degradation: A microbial approach. *Bioresour. Bioprocess.* **2017**, *4*, 15. [CrossRef]
27. Palmisano, A.C.; Pettigrew, C.A. Biodegradability of Plastics. *Bioscience* **1992**, *42*, 680–685. [CrossRef]
28. De Paoli, M.A. Degradação e Estabilização de Polímeros. *Artliber* **2008**, *1*, 286.
29. Crispim, C.A.; Gaylarde, C.C. Cyanobacteria and biodeterioration of cultural heritage: A review. *Microb. Ecol.* **2005**, *49*, 1–9. [CrossRef]
30. Capitelli, F.; Principi, P.; Sorlini, C. Biodeterioration of modern materials in contemporary collections: Can biotechnology help? *Trends Biotechnol.* **2006**, *24*, 350–354. [CrossRef]
31. Bonhomme, S.; Cuer, A.; Delort, A.M.; Lemaire, J.; Sancelme, M.; Scott, G. Environmental biodegradation of polyethylene. *Polym. Degrad. Stab.* **2003**, *81*, 441–452. [CrossRef]

32. Flemming, H.C. Relevance of biofilms for the biodeterioration of surfaces of polymeric materials. *Polym. Degrad. Stab.* **1998**, *59*, 309–315. [CrossRef]
33. Warscheid, T.; Braams, J. Biodeterioration of stone: A review. *Int. Biodeter. Biodegr.* **2000**, *46*, 343–368. [CrossRef]
34. Zanardini, E.; Abbruscato, P.; Ghedini, N.; Realini, M.; Sorlini, C. Influence of atmospheric pollutants on the biodeterioration of stone. *Int. Biodeterior. Biodegrad.* **2000**, *45*, 35–42. [CrossRef]
35. Rubio, C.; Ott, C.; Amiel, C.; Dupont-Moral, I.; Travert, J.; Mariey, L. Sulfato/thiosulfato reducing bacteria characterization by FT-IR spectroscopy: A new approach to biocorrosion control. *J. Microbiol. Methods* **2006**, *64*, 287–296. [CrossRef]
36. Lugauskas, A.; Levinskaite, L.; Peĉiulyte, D. Micromycetes as deterioration agents of polymeric materials. *Int. Biodeterior. Biodegrad.* **2003**, *52*, 233–242. [CrossRef]
37. Pelmont, J. Biodégradations et métabolismes. Les bactéries pour les technologies de l'environnement. EDP. *Sciences* **2005**, *1*, 798.
38. Abou Zeid, D.-M. Anaerobic Biodegradation of Natural and Synthetic Polyesters. Ph.D. Dissertation, Technischen Universität Carolo-Wilhelmina zu Braunschweig, Braunschweig, Germany, 2001.
39. Belal, E.S. Investigations on Biodegradation of Polyesters by Isolated Mesophilic Microbes. Ph.D. Dissertation, Technischen Universität Carolo-Wilhelmina zu Braunschweig, Braunschweig, Germany, 2003.
40. Freitas, C.; Carmona, E.; Brienzo, M. Xylooligosaccharides production process from lignocellulosic biomass and bioactive effects. *Bioact. Carbohydr. Diet. Fibre.* **2019**, *18*, 100–184. [CrossRef]
41. Terrone, C.C.; Nascimento, J.M.F.; Terrasan, C.R.F.; Brienzo, M.; Carmona, E.C. Salt-tolerant α-arabinofuranosidase from a new specie Aspergillus hortai CRM1919: Production in acid conditions, purification, characterization and application on xylan hydrolysis. *Biocatal. Agric. Biotechnol.* **2020**, *23*, 101460. [CrossRef]
42. Pérez, J.; Muñoz-Dorado, J.; De La Rubia, T.; Martínez, J. Biodegradation and biological treatments of cellulose, hemicellulose and lignin: An overview. *Int. Microbiol.* **2002**, *5*, 53–63. [CrossRef]
43. Costa, C.Z.; Albuquerque, M.C.C.; Brum, M.; Castro, A. Microbial and enzymatic degradation of polymers: A review. *Quim Nova.* **2015**, *38*, 259–267. [CrossRef]
44. Mueller, R.J. Biological degradation of synthetic polyesters-Enzymes as potential catalysts for polyester recycling. *Process Biochem.* **2006**, *41*, 2124–2128. [CrossRef]
45. Premraj, R.; Doble, M. Biodegradation of Polymers. *Indian J. Biotechnol.* **2005**, *4*, 186–193.
46. Kumar, S.; Maiti, P. Controlled biodegradation of polymers using nanoparticles and its application. *RSC Adv.* **2016**, *6*, 67449–67480. [CrossRef]
47. Krzan, A.; Hemjinda, S.; Miertus, S.; Corti, A.; Chiellini, E. Standardization and certification in the area of environmentally degradable plastics. *Polym. Degrad. Stab.* **2006**, *91*, 2819–2833. [CrossRef]
48. Rani-Borges, B.; Faria, A.U.; De Campos, A.; Gonçalves, S.P.C.; Martins-Franchetti, S.M. Biodegradation of additive PHBV/PP-co-PE films buried in soil. *Polimeros* **2016**, *26*, 161–167. [CrossRef]
49. Priyanka, N.; Archana, T. Biodegradability of Polythene and Plastic by the Help of Microorganism: A Way for Brighter Future. *J Environ. Anal. Toxicol* **2011**, *1*, 111. [CrossRef]
50. Gunatillake, P.A.; Adhikari, R. Biodegradable synthetic polymers for tissue engineering. *Eur. Cells Mater.* **2003**, *5*, 1–16. [CrossRef] [PubMed]
51. Proikakis, C.S.; Mamouzelos, N.J.; Tarantili, P.A.; Andreopoulos, A.G. Swelling and hydrolytic degradation of poly(d,l-lactic acid) in aqueous solutions. *Polym. Degrad. Stab.* **2006**, *91*, 614–619. [CrossRef]
52. ASTM D5338-15. *Standard Test Method for Determining Aerobic Biodegradation of Plastic Materials Under Controlled Composting Conditions, Incorporating Thermophilic Temperatures*; ASTM Int.: West Conshohocken, PA, USA, 2015.
53. ISO 14855-2. *Determination of the Ultimate Aerobic Biodegradability of Plastic Materials under Controlled Composting Conditions—Method by Analysis of Evolved Carbon Dioxide—Part 2: Gravimetric Measurement of Carbon Dioxide Evolved in a Laboratory-Scale Test*; ISO: Geneva, Switzerland, 2018.
54. ISO 17556:2003. *Determination of the Ultimate Aerobic Biodegradability in Soil by Measuring the Oxygen Demand in a Respirometer or the Amount of Carbono Dioxide Evolved*; ISO: Geneva, Switzerland, 2003.
55. Pavel, S.; Khatun, A.; Haque, M.M. Nexus Among Bio-plastic, Circularity, Circular Value Chain & Circular Economy. *SSRN Electron. J* **2020**, *47*. [CrossRef]
56. ASTM D5511-02. *Standard Test Method for Determining Anaerobic Biodegradation of Plastic Materials under High-Solids Anaerobic-digestion Conditions*; ASTM Int.: West Conshohocken, PA, USA, 2002.
57. Zhao, J.H.; Wang, X.Q.; Zeng, J.; Yang, G.; Shi, F.H.; Yan, Q. Biodegradation of poly(butylene succinate-co-butylene adipate) by Aspergillus versicolor. *Polym. Degrad. Stab.* **2005**, *90*, 173–179. [CrossRef]
58. Tsuji, H.; Echizen, Y.; Nishimura, Y. Photodegradation of biodegradable polyesters: A comprehensive study on poly(l-lactide) and poly(ε-caprolactone). *Polym. Degrad. Stab.* **2006**, *91*, 1128–1137. [CrossRef]
59. Medina Jaramillo, C.; Gutiérrez, T.J.; Goyanes, S.; Bernal, C.; Famá, L. Biodegradability and plasticizing effect of yerba mate extract on cassava starch edible films. *Carbohydr. Polymers.* **2016**, *151*, 150–159. [CrossRef] [PubMed]
60. Balaguer, M.P.; Villanova, J.; Cesar, G.; Gavara, R.; Hernandez-Munoz, P. Compostable properties of antimicrobial bioplastics based on cinnamaldehyde cross-linked gliadins. *Chem. Eng. J.* **2015**, *262*, 447. [CrossRef]
61. Piñeros-Hernandez, D.; Medina-Jaramillo, C.; López-Córdoba, A.; Goyanes, S. Edible cassava starch films carrying rosemary antioxidant extracts for potential use as active food packaging. *Food Hydrocoll.* **2017**, *63*, 488–495. [CrossRef]

62. Aburto, J.; Alric, I.; Thiebaud, S.; Borredon, E.; Bikiaris, D.; Prinos, J.; Panayiotou, C. Synthesis, characterization, and biodegradability of fatty-acid esters of amylose and starch. *J. Appl. Polym. Sci.* **1999**, *74*, 1440–1451. [CrossRef]

63. Lindström, A.; Albertsson, A.C.; Hakkarainen, M. Quantitative determination of degradation products an effective means to study early stages of degradation in linear and branched poly(butylene adipate) and poly(butylene succinate). *Polym. Degrad. Stab.* **2004**, *83*, 487–493. [CrossRef]

64. Yang, J.; Ching, Y.C.; Chuah, C.H. Applications of Lignocellulosic Fibers and Lignin in Bioplastics: A Review. *Polymers* **2019**, *11*, 751. [CrossRef]

65. Farhat, W.; Venditti, R.; Ayoub, A.; Prochazka, F.; Fernández-de-Alba, C.; Mignard, N.; Taha, M.; Becquart, F. Towards thermoplastic hemicellulose: Chemistry and characteristics of poly-(ε-caprolactone) grafting onto hemicellulose backbones. *Mater. Des.* **2018**, *153*, 298–307. [CrossRef]

66. Ochi, S. Development of high strength biodegradable composites using Manila hemp fiber and starch-based biodegradable resin. *Compos. Part A Appl. Sci. Manuf.* **2006**, *37*, 1879–1883. [CrossRef]

67. Bilo, F.; Pandini, S.; Sartore, L.; Depero, L.E.; Gargiulo, G.; Bonassi, A.; Federici, S.; Bontempi, E. A sustainable bioplastic obtained from rice straw. *J. Clean. Prod.* **2018**, *200*, 357–368. [CrossRef]

68. Fitch-Vargas, P.R.; Camacho-Hernández, I.L.; Martínez-Bustos, F.; Islas-Rubio, A.R.; Carrillo-Cañedo, K.I.; Calderón-Castro, A.; Jacobo-Valenzuela, N.; Carrillo-López, A.; Delgado-Nieblas, C.I.; Aguilar-Palazuelos, E. Mechanical, physical and microstructural properties of acetylated starch-based biocomposites reinforced with acetylated sugarcane fiber. *Carbohydr. Polym.* **2019**, *219*, 378–386. [CrossRef]

69. Babaee, M.; Jonoobi, M.; Hamzeh, Y.; Ashori, A. Biodegradability and mechanical properties of reinforced starch nanocomposites using cellulose nanofibers. *Carbohydr. Polym.* **2015**, *132*, 1–8. [CrossRef]

70. Oluwasina, O.O.; Olaleye, F.K.; Olusegun, S.J.; Oluwasina Olayinka, O.; Mohallem, N.D.S. Influence of oxidized starch on physicomechanical, thermal properties, and atomic force micrographs of cassava starch bioplastic film. *Int. J. Biol. Macromol.* **2019**, *135*, 282–293. [CrossRef]

71. Lucena, C.A.A.; Da Costa, S.C.; Eleamen, G.R.D.A.; Mendonça, E.A.D.M.; Oliveira, E.E. Desenvolvimento de biofilmes à base de xilana e xilana/gelatina para produção de embalagens biodegradáveis. *Polimeros* **2017**, *27*, 35–41. [CrossRef]

72. Wang, S.; Ren, J.; Li, W.; Sun, R.; Liu, S. Properties of polyvinyl alcohol/xylan composite films with citric acid. *Carbohydr. Polym.* **2014**, *103*, 94–99. [CrossRef]

73. Schmatz, A.A.; Tyhoda, L.; Brienzo, M. Sugarcane biomass influenced by lignin. *Biofuels Bioprod. Bioref.* **2020**, *14*, 469–480. [CrossRef]

74. Melati, R.B.; Shimizu, F.L.; Oliveira, G.; Pagnocca, F.C.; de Souza, W.; Sant'Anna, C.; Brienzo, M. Key Factors Affecting the Recalcitrance and Conversion Process of Biomass. *Bioenergy Res.* **2019**, 12. [CrossRef]

75. Figueiredo, P.; Lintinen, K.; Hirvonen, J.T.; Kostiainen, M.A.; Santos, H.A. Properties and chemical modifications of lignin: Towards lignin-based nanomaterials for biomedical applications. *Prog. Mater. Sci.* **2018**, *93*, 233–269. [CrossRef]

76. Van Soest, P.J. *Nutritional Ecology of the Ruminant*; Cornell University Press: Ithaca, NY, USA, 1994; p. 2.

77. Campagner, M.R.; Da Silva Moris, V.A.; Pitombo, L.M.; Do Carmo, J.B.; De Paiva, J.M.F. Polymeric films based on starch and lignosulfonates: Preparation, properties and evaluation of biodegradation. *Polimeros* **2014**, *24*, 740–751. [CrossRef]

78. Gupta, A.K.; Mohanty, S.; Nayak, S.K. Influence of addition of vapor grown carbon fibers on mechanical, thermal and biodegradation properties of lignin nanoparticle filled bio-poly(trimethylene terephthalate) hybrid nanocomposites. *RSC Adv.* **2015**, *5*, 56028–56036. [CrossRef]

79. Thakur, V.K.; Thakur, M.K.; Gupta, R.K. Review: Raw Natural Fiber-Based Polymer Composites. *Int. J. Polym. Anal. Charact.* **2014**, *19*, 256–271. [CrossRef]

80. Agrawal, A.; Kaushik, N.; Biswas, S. Derivatives and Applications of Lignin–An Insight. *Sci. Tech. J.* **2014**, *1*, 30–36.

81. Arnling Bååth, J.; Martínez-Abad, A.; Berglund, J.; Larsbrink, J.; Vilaplana, F.; Olsson, L. Mannanase hydrolysis of spruce galactoglucomannan focusing on the influence of acetylation on enzymatic mannan degradation. *Biotechnol. Biofuels.* **2018**, *11*, 1. [CrossRef]

82. Polman, E.M.N.; Gruter, G.J.M.; Parsons, J.R.; Tietema, A. Comparison of the aerobic biodegradation of biopolymers and the corresponding bioplastics: A review. *Sci. Total Environ.* **2021**, *753*, 141953. [CrossRef] [PubMed]

83. Rivard, C.; Moens, L.; Roberts, K.; Brigham, J.; Kelley, S. Starch esters as biodegradable plastics: Effects of ester group chain length and degree of substitution on anaerobic biodegradation. *Enzyme Microb. Technol.* **1995**, *17*, 848–852. [CrossRef]

84. Tserki, V.; Matzinos, P.; Kokkou, S.; Panayiotou, C. Novel biodegradable composites based on treated lignocellulosic waste flour as filler. Part I. Surface chemical modification and characterization of waste flour. *Compos. Part A Appl. Sci. Manuf.* **2005**, *36*, 965–974. [CrossRef]

85. Tserki, V.; Matzinos, P.; Panayiotou, C. Novel biodegradable composites based on treated lignocellulosic waste flour as filler. Part II. Development of biodegradable composites using treated and compatibilized waste flour. *Compos. Part A Appl. Sci. Manuf.* **2006**, *37*, 1231–1238. [CrossRef]

86. Glasser, W.G.; Ravindran, G.; Jain, R.K.; Samaranayake, G.; Todd, J. Comparative Enzyme Biodegradability of Xylan, Cellulose, and Starch Derivatives†. *Biotechnol. Prog.* **1995**, *11*, 552–557. [CrossRef]

87. Mitchell, D.; Grohmann, K.; Himmel, M.; Dale, B.; Schroeder, H. Effect of the degree of acetylation on the enzymatic digestion of acetylated xylans. *J. Wood Chem. Technol.* **1990**, *10*, 111–121. [CrossRef]

88. Alekhina, M.; Mikkonen, K.S.; Alén, R.; Tenkanen, M.; Sixta, H. Carboxymethylation of alkali extracted xylan for preparation of bio-based packaging films. *Carbohydr. Polym.* **2014**, *100*, 89–96. [CrossRef]

89. Geng, W.; Venditti, R.A.; Pawlak, J.J.; Pal, L.; Ford, E. Carboxymethylation of hemicellulose isolated from poplar (Populus grandidentata) and its potential in water-soluble oxygen barrier films. *Cellulose* **2020**, *27*, 3359–3377. [CrossRef]

90. Heikkinen, S.L.; Mikkonen, K.S.; Pirkkalainen, K.; Serimaa, R.; Joly, C.; Tenkanen, M. Specific enzymatic tailoring of wheat arabinoxylan reveals the role of substitution on xylan film properties. *Carbohydr. Polym.* **2013**, *92*, 733–740. [CrossRef]

91. Höije, A.; Stememalm, E.; Heikkinen, S.; Tenkanen, M.; Gatenholm, P. Material properties of films from enzymatically tailored arabinoxylans. *Biomacromolecules* **2008**, *9*, 2042–2047. [CrossRef]

92. Kjeldsen, A.; Price, M.; Lilley, C.; Guzniczak, E.; Archer, I. A Review of Standards for Biodegradable Plastics. *Ind. Biotechnol. Innov. Cent. IBioIC* **2018**, 33.

93. Meereboer, K.W.; Misra, M.; Mohanty, A.K. Review of recent advances in the biodegradability of polyhydroxyalkanaoate (PHA) bioplastics and their composites. *Green Chem.* **2020**, *22*, 5519–5558. [CrossRef]

94. Suzuki, M.; Tachibana, Y.; Kasuya, K. Biodegradability of poly(3-hydroxyalkanoate) and poly(ε-caprolactone) via biological carbon cycles in marine environments. *Polym. J.* **2020**, *53*, 47–66. [CrossRef]

95. Volovaa, T.G.; Boyandina, A.N.; Prudnikovab, S.V. Biodegradation of polyhydroxyalkanoates in natural soils. *J. Sib. Fed. Univ. Biol.* **2015**, *2*, 152–167. [CrossRef]

96. Weng, Y.X.; Wang, X.L.; Wang, Y.Z. Biodegradation behavior of PHAs with different chemical structures under controlled composting conditions. *Polym. Test.* **2011**, *30*, 372–380. [CrossRef]

97. Wang, Y.W.; Mo, W.; Yao, H.; Wu, Q.; Chen, J.; Chen, G.Q. Biodegradation studies of poly(3-hydroxybutyrate-co-3-hydroxyhexanoate). *Polym. Degrad. Stab.* **2004**, *85*, 815–821. [CrossRef]

98. Weng, Y.X.; Wang, L.; Zhang, M.; Wang, X.L.; Wang, Y.Z. Biodegradation behavior of P(3HB,4HB)/PLA blends in real soil environments. *Polym. Test.* **2013**, *32*, 60–70. [CrossRef]

99. Morohoshi, T.; Ogata, K.; Okura, T.; Sato, S. Molecular characterization of the bacterial community in biofilms for degradation of poly (3-Hydroxybutyrate-co-3-Hydroxyhexanoate) films in seawater. *Microbes Environ.* **2018**, 33. [CrossRef]

100. Shruti, V.C.; Kutralam-Muniasamy, G. Bioplastics missing link in the era of microplastics. *Sci. Total Environ.* **2019**, *697*, 134139. [CrossRef]

101. Gironi, F.; Piemonte, V. Bioplastics and petroleum-based plastics: Strengths and weaknesses. *Energ. Sources Part A* **2011**, *33*, 1949–1959. [CrossRef]

102. Narancic, T.; Verstichel, S.; Reddy Chaganti, S.; Morales-Gamez, L.; Kenny, S.T.; De Wilde, B.; Babu Padamati, R.; O'Connor, K.E. Biodegradable Plastic Blends Create New Possibilities for End-of-Life Management of Plastics but They Are Not a Panacea for Plastic Pollution. *Environ. Sci. Technol.* **2018**, *52*, 10441–10452. [CrossRef] [PubMed]

103. Kourmentza, C.; Plácido, J.; Venetsaneas, N.; Burniol-Figols, A.; Varrone, C.; Gavala, H.N.; Reis, M.A.M. Recent advances and challenges towards sustainable polyhydroxyalkanoate (PHA) production. *Bioengineering* **2017**, *4*, 55. [CrossRef] [PubMed]

104. Jiang, G.; Hill, D.J.; Kowalczuk, M.; Johnston, B.; Adamus, G.; Irorere, V.; Radecka, I. Carbon sources for polyhydroxyalkanoates and an integrated biorefinery. *Int. J. Mol. Sci.* **2016**, *17*, 1157. [CrossRef]

105. Bertrand, J.I.; Ramsay, B.A.; Ramsay, J.A.; Chavarie, C. Biosynthesis of poly-βhydroxyalkanoates from pentoses by Pseudomonas pseudoflava. *Appl. Environ. Microbiol.* **1990**, *6*, 3133–3138. [CrossRef] [PubMed]

106. Lopes, M.S.G.; Rocha, R.C.S.; Zanotto, S.P.; Gomez, J.G.C.; Silva, L.F. Screening of bacteria to produce polyhydroxyalkanoates from xylose. *World J. Microbiol. Biotechnol.* **2009**, *25*, 1751–1756. [CrossRef]

107. Lee, H.S.; Cho, D.; Han, S.O. Effect of natural fiber surface treatments on the interfacial and mechanical properties of henequen/polypropylene biocomposites. *Macromol. Res.* **2008**, *16*, 411–417. [CrossRef]

108. Nduko, J.M.; Suzuki, W.; Matsumoto, K.; Kobayashi, H.; Ooi, T.; Fukuoka, A.; Taguchi, S. Polyhydroxyalkanoates production from cellulose hydrolysate in Escherichia coli LS5218 with superior resistance to 5-hydroxymethylfurfural. *J. Biosci. Bioeng.* **2012**, *113*, 70–72. [CrossRef]

109. Cesário, M.T.; Raposo, R.S.; de Almeida, M.C.M.D.; van Keulen, F.; Ferreira, B.S.; da Fonseca, M.M.R. Enhanced bioproduction of poly-3-hydroxybutyrate from wheat straw lignocellulosic hydrolysates. *New Biotechnol.* **2014**, *31*, 104–113. [CrossRef]

110. Jones, F. A Promessa dos Bioplásticos. FAPESP. Available online: https://revistapesquisa.fapesp.br/a-promessa-dos-bioplasticos/ (accessed on 20 June 2020).

111. Chen, G. A microbial polyhydroxyalkanoates (PHA) based bio- and materials industry. *Chem. Soc. Rev.* **2009**, *38*, 2434–2446. [CrossRef]

112. Altaee, N.; El-Hiti, G.A.; Fahdil, A.; Sudesh, K.; Yousif, E. Biodegradation of different formulations of polyhydroxybutyrate films in soil. *Springerplus* **2016**, *5*, 762. [CrossRef] [PubMed]

113. Gutierrez-Wing, M.T.; Stevens, B.E.; Theegala, C.S.; Negulescu, I.I.; Rusch, K.A. Anaerobic biodegradation of polyhydroxybutyrate in municipal sewage sludge. *J. Environ. Eng.* **2010**, *136*, 709–718. [CrossRef]

114. Tabasi, R.Y.; Ajji, A. Selective degradation of biodegradable blends in simulated laboratory composting. *Polym. Degrad. Stab.* **2015**, *120*, 435–442. [CrossRef]

115. Dilkes-Hoffman, L.S.; Lant, P.A.; Laycock, B.; Pratt, S. The rate of biodegradation of PHA bioplastics in the marine environment: A meta-study. *Mar. Pollut. Bull.* **2019**, *142*, 15–24. [CrossRef] [PubMed]

116. Wu, C.S. Preparation and Characterization of Polyhydroxyalkanoate Bioplastic-Based Green Renewable Composites from Rice Husk. *J. Polym. Environ.* **2014**, *22*, 384–392. [CrossRef]

17. Gómez, E.F.; Michel, F.C. Biodegradability of conventional and bio-based plastics and natural fiber composites during composting, anaerobic digestion and long-term soil incubation. *Polym. Degrad. Stab.* **2013**, *98*, 2583–2591. [CrossRef]
18. Coelho, N.S.; Almeida, Y.M.B.; Vinhas, G.M. A biodegradabilidade da blenda de poli(β-Hidroxibutirato-co-Valerato)/amido anfótero na presença de microrganismos. *Polímeros* **2008**, *18*, 270–276. [CrossRef]
19. Harmaen, A.S.; Khalina, A.; Ali, H.M.; Azowa, I.N. Thermal, Morphological, and Biodegradability Properties of Bioplastic Fertilizer Composites Made of Oil Palm Biomass, Fertilizer, and Poly(hydroxybutyrate-co-valerate). *Int. J. Polym. Sci.* **2016**, 1–8. [CrossRef]
20. Magalhães, N.F.; Andrade, C.T. Properties of melt-processed poly (hydroxybutyrate-co-hydroxyvalerate)/starch 1:1 blend nanocomposites. *Polimeros* **2013**, *23*, 366–372. [CrossRef]
21. Tosin, M.; Weber, M.; Siotto, M.; Lott, C.; Innocenti, F.D. Laboratory test methods to determine the degradation of plastics in marine environmental conditions. *Front. Microbiol.* **2012**, *3*, 1–9. [CrossRef]
22. Sarasa, J.; Gracia, J.M.; Javierre, C. Study of the biodisintegration of a bioplastic material waste. *Bioresour. Technol.* **2009**, *100*, 3764–3768. [CrossRef]
23. Allen, W.R. Structural applications for flexible packaging: Innovations in pouch forms and uses. *Polym. Plast. Technol. Eng.* **1986**, *25*, 295–320. [CrossRef]
24. Briston, J.H. *Plastic Films*, 3rd ed.; Wiley: New York, NY, USA, 1988.
25. Hansen, N.M.L.; Blomfeldt, T.O.J.; Hedenqvist, M.S.; Plackett, D.V. Properties of plasticized composite films prepared from nanofibrillated cellulose and birch wood xylan. *Cellulose* **2012**, *19*, 2015–2031. [CrossRef]
26. Sánchez-Safont, E.; Arrillaga, A.; Anakabe, J.; Cabedo, L.; Gamez-Perez, J. Toughness Enhancement of PHBV/TPU/Cellulose Compounds with Reactive Additives for Compostable Injected Parts in Industrial Applications. *Int. J. Mol. Sci.* **2018**, *19*, 2102. [CrossRef] [PubMed]
27. Albuquerque, R.M.; Meira, H.M.; Silva, I.D.L.; Galdino, C.J.S.; Almeida, F.C.G.; Amorim, J.D.P.; Vinhas, G.M.; Costa, A.F.S.; Sarubbo, L.A. Production of a bacterial cellulose/poly(3-hydroxybutyrate) blend activated with clove essential oil for food packaging. *Polym. Polym. Compos.* **2020**. [CrossRef]
28. Dangtungee, R.; Tengsuthiwat, J.; Boonyasopon, P.; Siengchin, S. Sisal natural fiber/clay-reinforced poly(hydroxybutyrate co hydroxyvalerate) hybrid composites. *J. Thermoplast. Compos. Mater.* **2014**, *28*, 6–879. [CrossRef]
29. Joyyi, L.; Ahmad Thirmizir, M.Z.; Salim, M.S.; Han, L.; Murugan, P.; Kasuya Ki Maurer, F.H.J.; Zainal Arifin, M.I.; Sudesh, K. Composite properties and biodegradation of biologically recovered poly(3HB-co-3HHx) reinforced with short kenaf fibers. *Polym. Degrad. Stab.* **2017**, *137*, 100–108. [CrossRef]
30. Batista, K.C.; Silva, D.A.K.; Coelho, L.A.F.; Pezzin, S.H.; Pezzin, A.P.T. Soil Biodegradation of PHBV/Peach Palm Particles Biocomposites. *J. Polym. Environ.* **2010**, *18*, 346–354. [CrossRef]
31. Rosa, D.S.; Rodrigues, T.C.; Graças Fassina Guedes, C.; Calil, M.R. Effect of thermal aging on the biodegradation of PCL, PHB-V, and their blends with starch in soil compost. *J. Appl. Polym. Sci.* **2003**, *89*, 3539–3546. [CrossRef]
32. Kratsch, H.A.; Schrader, J.A.; McCabe, K.G.; Srinivasan, G.; Grewell, D.; Graves, W.R. Performance and biodegradation in soil of novel horticulture containers made from bioplastics and biocomposites. *Hort. Technol.* **2015**, *25*, 119–131. [CrossRef]
33. Wang, T.; Cheng, G.; Ma, S.; Cai, Z.; Zhang, L. Crystallization behavior, mechanical properties, and environmental biodegradability of poly(?-hydroxybutyrate)/cellulose acetate butyrate blends. *J. Appl. Polym. Sci.* **2003**, *89*, 2116–2122. [CrossRef]
34. Datta, R.; Kelkar, A.; Baraniya, D.; Molaei, A.; Moulick, A.; Meena, R.S.; Formanek, P. Enzymatic degradation of lignin in soil: A review. *Sustainability* **2017**, *9*, 1163. [CrossRef]
35. Santacruz-Juarez, E.; Buendia-Corona, R.E.; Ramírez, R.E.; Sanchez, S. Fungal enzymes for the degradation of polyethylene: Molecular docking simulation and biodegradation pathway proposal. *J. Hazard. Mater.* **2021**, *411*, 125118. [CrossRef]
36. Chio, C.; Sain, M.; Qin, W. Lignin utilization: A review of lignin depolymerization from various aspects. *Renew. Sust. Energ. Rev.* **2019**, *107*, 232–249. [CrossRef]

recycling

Article

Characterization of Used Lubricant Oil in a Latin-American Medium-Size City and Analysis of Options for Its Regeneration

Carlos Sánchez-Alvarracín *, Jessica Criollo-Bravo, Daniela Albuja-Arias, Fernando García-Ávila and M. Raúl Pelaez-Samaniego

Faculty of Chemical Sciences, University of Cuenca, Av. 12 de Abril and Av. Loja, Cuenca 10202, Ecuador; jessica.criollo@ucuenca.edu.ec (J.C.-B.); daniela.albuja@ucuenca.edu.ec (D.A.-A.); fernando.garcia@ucuenca.edu.ec (F.G.-Á.); manuel.pelaez@ucuenca.edu.ec (M.R.P.-S.)
* Correspondence: carlos.sancheza@ucuenca.edu.ec; Tel.: +593-991-708-800

Abstract: Petroleum-derived products, such as lubricant oils, are non-renewable resources that, after use, must be collected and processed properly to avoid negative environmental impacts. A circular economy of used oils requires the re-refining and reuse of the same. Similar to most countries in Latin America, the management of used oils in Ecuador is still incipient and few cities collect and treat this material properly. In Cuenca, the ETAPA company collects ~1344 t/year of used oils, which are subjected to pretreatment operations prior to their use as fuel in a cement factory. However, combustion generates polluting gases and disallows the adding of value to the used oils. The lack of studies on the characterization and methods utilized for recovering used oils under the conditions found in medium-size Latin-American cities (e.g., Cuenca), alongside a lack of government policies, have hindered the adoption of re-refining operations. The objective of this work is to characterize the used lubricant oils in Cuenca, to compare them with the properties of used oils from other countries, and to suggest some re-refining technologies for oils with similar properties. Used oil samples were collected from mechanic shops and car-lubricating shops for characterization. Its physicochemical properties and metal contents are comparable to the used oils in other countries globally. Specifically, the flash point, kinematic viscosity, TBN, and concentrations of Zn, Cd, and Mg are similar to the properties of used oils in Iraq, Egypt, and the United Arab Emirates. Based on these results, the best re-refining option for used oils in Cuenca is extraction with solvents in which sedimentation and dehydration (already conducted in Cuenca) is followed by a solvent reaction process, a vacuum distillation process, a finishing process with bentonite, and a final filtration step.

Keywords: waste lubricating oil; characterization; used oil management; circular economy

Citation: Sánchez-Alvarracín, C.; Criollo-Bravo, J.; Albuja-Arias, D.; García-Ávila, F.; Pelaez-Samaniego, M.R. Characterization of Used Lubricant Oil in a Latin-American Medium-Size City and Analysis of Options for Its Regeneration. *Recycling* **2021**, *6*, 10. https://doi.org/10.3390/recycling6010010

Received: 15 October 2020
Accepted: 12 January 2021
Published: 2 February 2021

Publisher's Note: MDPI stays neutral with regard to jurisdictional claims in published maps and institutional affiliations.

Copyright: © 2021 by the authors. Licensee MDPI, Basel, Switzerland. This article is an open access article distributed under the terms and conditions of the Creative Commons Attribution (CC BY) license (https://creativecommons.org/licenses/by/4.0/).

1. Introduction

Lubricating oil (or engine lubricant oil), a product derived from petroleum refining, is a mixture of essential oils (either virgin or processed, mineral or synthetic) and additives. Lubricant oils are used in equipment with moving parts to reduce friction and surface wear off [1], which makes these oils widely employed for the operation of internal combustion engines (ICEs; both gasoline- and diesel-type engines). During the operation of ICEs, the additives in the oil are partially consumed and the oil quality decreases over time due to degradation by oxidation, decomposition of oil and its additives, and contamination with water, gasoline (or diesel), dirt, metals, and carbon particles [2]. The degradation of lubricant oil reduces its life, which severely limits its reusability. Thus, it needs to be replaced with new oil [3]. The used lubricant oil must be collected and stored adequately to avoid environmental pollution, such as from spillage. Often, used lubricant oils are burnt without any treatment, which emits harmful gases into the environment. Sometimes these materials are even spilled in rivers [4].

Used lubricating oils contain heavy metals (e.g., Cr, Cd, As, and Pb) and harmful chemical compounds, such as polynuclear aromatic hydrocarbons, benzene, chlorinated solvents, polychlorinated biphenyl (PCBs), and polycyclic aromatic hydrocarbons (PAHs). Therefore, proper management of used oils is necessary to avoid the negative effects inflicted on human health and nature [5]. From an environmental point of view, combustion of used oils is not recommended [5], since the improper incineration of 5 L of used oil could pollute an amount of air equivalent to that needed by a person to live over three years [6]. The adsorption of Cr and its compounds released by burning used oil can cause some types of cancer [5]. The negative impacts of the improper management of used oil on the environment and on human health [7] require the exploration of options for regenerating these oils with the intent of producing new lubricants and other petroleum-derived products. The yield of lubricant oils through the re-refining of used oils is higher than the yield from virgin crude petroleum refining [2], promoting a reduction of about 90% of the environmental impacts that otherwise could result from the production of petroleum-derived lubricant oils [7]. Some companies in the world, particularly large industries that consume lubricants, regenerate used oil in their own facilities by physicochemical processes to remove the contaminants. Then, the regenerated oil is added to new oils to operate ICEs or is used in other industrial applications, such as gear lubricants, cutting oil, and metal-rolling lubricants [8]. This option has the advantage of reducing losses in each processing step, although the regenerated oil cannot be used again in the engines or gearboxes in vehicles due to degradation. Thus, lubricating oils must be re-refined for further use as lubricants in ICEs [9].

Our literature review suggests that there is a limited amount of studies reporting on the collection methods, characterization, and reuse techniques for used lubricating oils in Latin-American countries. Mexico and Brazil are the leading countries in the region on the reuse of used lubricant oils. In Mexico, Bravo Energy recovers around 100 million L of used oil in the form of fuels from Mexico and other countries (the US, Argentina, and Chile) [8]. Brazil consumes around 1 billion L/year of lubricating oil, 60% of which corresponds to automobile ICEs. More than 36% of the collected used oil in the country is derived from cars' ICEs [10]. The Sindicato Nacional da Indústria do Rerrefino de Óleos Minerais, "Sindirrefino" [11], ensures that the oil-refining activities in Brazil follow the latest technology, obtaining products duly specified by the National Petroleum Agency, according to Portaria ANP-130/1999. The industrial park has three differentiated technologies: (a) an acid/clay system with a "cracking term", obtaining a neutral heavy base oil; (b) a flash distillation or skin evaporation system, which allows obtaining a light and medium neutral base oil; and (c) an extraction system selective of propane solvents to obtain a medium neutral base oil. Currently, there are several Latin-American cities that possess private re-refining plants. Some examples of re-refining plants include CILCA (Lima, Peru) [12], Biofactor (Duran, Ecuador) [13], Nueva Energía S.A. (Buenos Aires, Argentina) [14], and other plants in Medellin and Bogota (Colombia) [15]. In September 2019, the IDB (Inter-American Development Bank) approved a loan to co-finance an oil recovery plant in Central America in partnership with the Costa Rican company Metalub, which currently collects and ships used oils to the US for re-refinement [16]. However, in the region, re-refining is mostly solely conducted at the city scale; thus, countries as a whole are not involved in the efficient collection and re-refining systems of used oils.

In Ecuador, the State, through the Agency of Regulation and Control of Hydrocarbon Fuels (ARCH) [1], regulates and controls all activities related to the production, distribution, and use of petroleum and petroleum-derived fuels and lubricants. According to the Central Bank of Ecuador, from January to July 2018, the country imported 70,081 t of lubricants, corresponding to an FOB value above US$ 100 million [17]. In May 2019, the Ministry of the Environment of Ecuador (Ministerio del Ambiente) promulgated Law 042 to regulate the management of used lubricant oils and the containers of these materials; it establishes general guidelines on the storage, collection, and final disposal of used lubricating oils, and the minimum collection goal corresponds to 20% calculated on the total tons of lubricating oil imported or manufactured annually [18]. However, the law does not make specific reuse operations or the re-refining of used oils mandatory. According to data from Arc

& Pieper S.A. (from Quito), around 84 million L/year of used oils are not recovered in the country. From it, 30% correspond to automotive and industrial used oils [19]. Few municipal governments (i.e., Cuenca, Ambato, Quito, Ibarra, and El Coca) and the Provincial Government of Tungurahua have protocols to collect and recycle used oils. In Quito, the BIOfactor company [13] collects all types of oils generated by the city and neighboring cities in the north of Ecuador. The company has a re-refining capacity of 9 million L/year (in the city of Duran), which corresponds to approximately 11% of the oil used in the country. Nevertheless, only the municipal government of Cuenca has an environmental license from the Ministry of Environment (MAE) to collect and treat used oils [19]. This city (a medium-size Latin-American city, with less than half a million inhabitants), through the ETAPA company (Municipal Company of Telecommunications, Potable Water, and Sewerage), has been carrying out a program called "Collection of Used Batteries and Used Oils" since 1998, aiming to generate environmental awareness among the population and promoting an adequate management of these wastes [20].

Currently, ETAPA collects 151,200 L/month (~1534 t/year) of used oil from "used oil generators", a group of approximately 1300 car-lubricating shops, mechanic shops, car wash shops, and tire repair shops, and some miscellaneous industrial companies in Cuenca. This amount of used oil collected is approximately 57% of the total consumed in the city [20]. The remaining is sold to an informal market that employs this oil as fuel in furnaces for bricks, tile manufactures, and, less frequently, for wood treatment, concrete formwork, spraying of cars' bottoms after car-washing, as a pesticide, to reduce dust on unpaved roads, or to protect cattle from subcutaneous parasites [19]. The practice of employing used oils as fuel is common in Latin-American countries (see, for example, [21]). Evidently, some of these uses are not environmentally friendly. It is expected that, in Ecuador, Law 042 could promote better management practices of used oils and reduce pollution.

The oil collected by ETAPA is subsequently treated in its own waste treatment facility through decantation, filtration, and clarification processes to separate the water and solid impurities. Following the Ecuadorian Technical Standard NTE-INEN 2266: 2017 for the transport, storage, and handling of hazardous materials, the treated oil is sent to a cement company in Guayaquil where it is burnt as fuel [20]. However, the management system of used oils in the city requires expenses that currently are paid by both the municipal government and the used oil generators. The sale of used oils to an informal market could be explained in part by the propensity of used oil generators to reduce such additional costs. Thus, adding value to used oils could help to reduce its utilization in informal markets. Despite these problems, the collection and treatment system have been effective at avoiding spillages and harmful discharges of used oil.

The implementation of Law 042 [18] would boost the required steps towards a circular economy that prioritizes regeneration and reuse of used lubricant oil. For this purpose, a deep understanding of the properties of used oil and the potential options for adding value to this material are critical [22]. In the case of Ecuador, information on the properties of used oils is scarce. Almeida et al. [4] characterized some properties of used oils in Quito. In Cuenca, Jaramillo et al. [23] characterized two types of oils commonly used in taxis (SAE 20W50 and SAE 10W30) to determine the most common causes of engine wear. Likewise, ETAPA has determined some properties of the used oils they collect (mixtures of automotive and industrial used oils) [24]. However, there are no current works showing a complete characterization of used oils or suggestions about possible routes for re-refining these materials as part of a circular economy strategy in the conditions of Cuenca and other cities in Ecuador. The objective of this work was twofold: (a) to characterize the used lubricant oils collected in Cuenca and compare the resulting properties with those of used oils elsewhere; and (b) to recommend regeneration options for used oils, considering the characteristics of the oils and the experiences reported in other places. The work hypothesizes that, if the characteristics of used oils in Cuenca are comparable to those in other places where used oil management is adequately performed, it is possible to adapt such proven processes to the conditions of our city. This paper summarizes the main findings of the research and expects to contribute with ideas for a better management of used lubricating oils in medium- (e.g., Cuenca) and small-size cities.

Figure 1 summarizes the process to be developed in this research in order to meet the proposed objective.

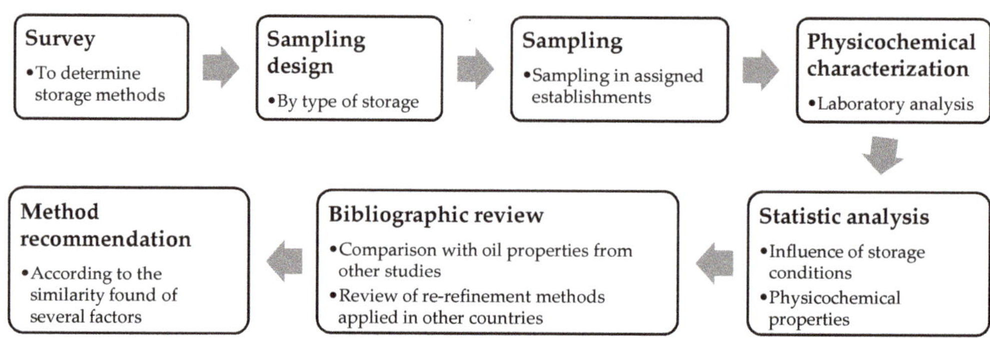

Figure 1. Process for the development of this research.

2. Materials and Methods

2.1. Materials

Samples of used lubricant oils were obtained from mechanic car shops and other used oil generators in the city of Cuenca. A preliminary step consisted of surveying a group of 265 randomly selected (out of the 1300) used oil generators registered by ETAPA to identify the collection methods and storage conditions of the used oils [25]. Based on the preliminary results, it was determined that the number of storage containers in the city is 16, for which a sample was taken from each one and the four most frequently repeated were duplicated, taken from different establishments (Samples 4 and 7, 5 and 6, 8 and 9, 14 and 15). The collection of three extra samples directly from the engines of three gasoline vehicles (i.e., used oil without being mixed or exposed to contamination from external agents) also was conducted for a comparison of the properties. Therefore, twenty-three oil samples were collected in total. The twenty samples were mixtures of various types of used automotive oils (mostly engine and gearbox lubricant oils) that have been stored under different conditions (e.g., using plastic or metallic containers), as presented in Table 1. As seen in the table, in Cuenca, the used oils are stored in the used oil generators' facilities for up to three months prior to collection by ETAPA. Samples of ~500 mL were extracted directly from the storage tanks. The collected samples were then sent to the Chemical Control Laboratory of the Guangopolo Power Plant (Termopichincha Unit—CELEC EP) for characterization.

2.2. Characterization of the Used Oil Mixtures

The twenty three samples were characterized with the purpose of determining the water content, density at 15 °C, viscosity index, base number, flash point, kinematic viscosity at 100 °C, and metals content (Al, Ba, B, Cd, Ca, Cu, Cr, Sn, P, Fe, Mg, Mn, Mo, Ni, Ag, Pb, Si, Na, Ti, V, and Zn). Water content was determined following ASTM D95 [26]. Density was determined as per ASTM D4052-11 [27]. The viscosity index was calculated from the kinematic viscosity at 40 and 100 °C and following the ASTM D 2270-93 procedure. The base number (TBN) was obtained by potentiometric titration, following ASTM D2896-11 Procedure B [28]. The flashpoint was obtained in accordance with ASTM D92-12b [29], i.e., using the Cleveland open cup method. The thermal bath method (ASTM D445-15 Procedure A) [30] was used for the kinematic viscosity index at 100 °C. The metal content was determined in accordance with ASTM 6595-00 (2011) [31]. All tests were conducted in duplicate. Table 2 show the analytical instruments' specifications for the characterization of the used oil mixtures.

Table 1. Storage conditions of the used oils in mechanics and car lubricating shops in Cuenca.

Sample Number	Container Material	Top of Tank	Pre-Filtered [a]	Covered [b]	Collection Frequency [c]
1	Directly from vehicle's engine		no	-	-
2	Directly from vehicle's engine		no	-	-
3	Directly from vehicle's engine		no	-	-
4	Metal	Closed	no	yes	monthly
5	Metal	Closed	no	yes	monthly
6	Metal	Closed	no	yes	monthly
7	Metal	Closed	no	yes	monthly
8	Metal	Closed	no	yes	bimonthly
9	Metal	Closed	no	yes	bimonthly
10	Metal	Closed	no	yes	quarterly
11	Metal	Closed	no	no	monthly
12	Metal	Closed	yes	no	monthly
13	Metal	closed	yes	yes	bimonthly
14	Metal	closed	yes	yes	monthly
15	Metal	closed	yes	yes	monthly
16	Metal	open	yes	yes	monthly
17	Metal	open	no	yes	monthly
18	HDPE	closed	no	no	monthly
19	HDPE	closed	no	yes	bimonthly
20	HDPE	closed	no	yes	quarterly
21	HDPE	closed	yes	yes	quarterly
22	HDPE	closed	yes	yes	monthly
23	HDPE	open	yes	yes	monthly

[a] Refers to a screening process to remove large particles (e.g., wipe cloth) prior to storing in the tanks. [b] Indicates if the storage container has been under roof conditions or not. [c] Frequency of collection of the used oils from the used oil generators' facilities.

Table 2. The analytical instruments' specifications.

Parameter	A. Instrument	Brand	Model	Serial Number
Water content	Distiller	Stanhope Seta	24415-2	1031631
Density at 15 °C	Digital densimeter	Mettler Toledo	Densito 30 PX	LWE13A54
Viscosity index	Thermostatic bath	Petrotest	Visco Bath	0263125001
Total base number (TBN)	Potentiometer	Metrohm	916 Ti Touch	1916001003195
Flashpoint	Equipment to flashpoint	Fisher Tag	Flash Tester	926
Metal content	Atomic emission spectrometer	Spectroil	M/F-W	626

2.3. Statistical Analysis of the Used Oil Properties

The results obtained were statistically analyzed in three steps: The first step focused on identifying the similarities among the 23 oil samples, from a physicochemical perspective. For this purpose, a cluster analysis using the SPSS program was conducted and the samples with similar characteristics were grouped [32]. Then the agglomeration method (i.e., a hierarchical classification in which algorithms are used to group objects, using a measure of Euclidean squared distance between data and the Ward linkage that measures remoteness between groups) was employed. According to [32], this method is used to analyze quantitative variables when the sample size is relatively small (<50). With this analysis, a classification tree or dendrogram was prepared (see Figure 2). Since small distances indicate homogeneous conglomerates and large distances show diversified conglomerates, it is convenient to stop the grouping process when the horizontal lines become long [32]; this procedure determines the number of clusters. In the second step of the analysis, the inference statistical technique "confidence interval" (with a 95% level of confidence) was employed [33]. In the third step, a comparison of the physicochemical characteristics of the used oils in Cuenca to the properties of used oils in other countries was conducted. Box and whisker plots were employed to summarize the data and to identify outliers (using the SPSS software).

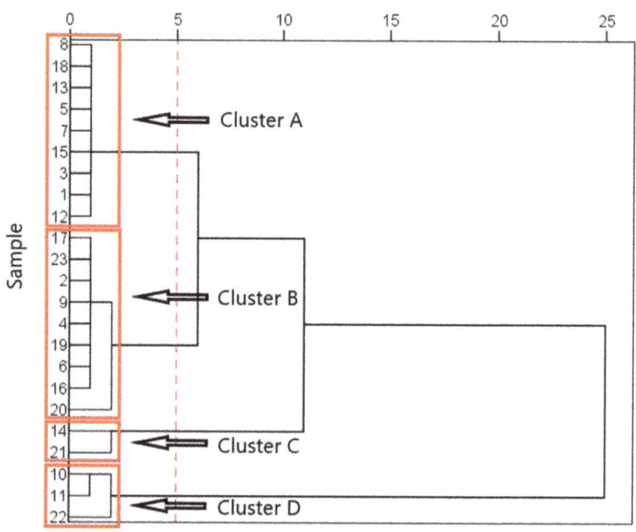

Figure 2. Cluster analysis of the physicochemical properties of the mixtures of used oils (a dendrogram using the Ward link).

2.4. Analysis of Options for Re-Refining Used Oils in Cuenca

The results of the characterization and experiences reported in the literature served to propose options of technologies that could be employed to re-refine used oils in Cuenca. This stage sought to establish the requirements that could enable the processing and utilization of used oils as part of the circular economy strategy in the city. The options that were analyzed considered five factors: (1) the amount of available used oil collected in the city alongside the possibility of expanding used oil collection in neighboring cities; (2) the quality of the used oil; (3) the yield of byproducts that can be obtained from used lubricant oils through re-refining, based on the literature and the limitations of some re-refining technologies; (4) the availability of materials for the treatment process close to the city; and (5) the costs of the regeneration technologies and the possible uses of the waste.

3. Results and Discussion

3.1. Physical Characteristics of the Used Oils, and Their Metal Content, in Cuenca

The initial test of normality of the properties of the used oils in Cuenca showed that the physical properties, such as water content, density, viscosity index, kinematic viscosity, and the presence of elements (Al, B, Cd, Cr, Fe, Mg, Mn, Mo, Ni, Ag, Na, Ti, and V), present a normal behavior. Thus, the central tendency measure is the mean. However, the TBN and the concentrations of Ba, Ca, Cu, Sn, P, Pb, Si, and Zn in the used oil did not show a normal distribution. Therefore, for these properties, we performed a data transformation through the Cox–Box method, using Minitab 19 software. After these considerations, the ranges of the values for each physical property and the metal content are shown in Table 3.

Table 3. Physical properties and metal content of the used oils in Cuenca city.

Parameter	Unit	Mean	Range [a]
Water content	%v/v	0.2	0.1–0.3
Density at 15 °C	g/mL	0.88	0.88–0.89
Viscosity index	-	132.6	128.1–137
Total Base number (TBN)	mg KOH/g	5.8	4.8–6.8
Flashpoint	°C	179.5	168–191
Kinematic viscosity at 100 °C	mm^2/s	12.2	11–13.4
Al	mg/kg	8	6.1–9.8
Ba	mg/kg	1.5	0.3–2
B	mg/kg	43	24.8–62
Cd	mg/kg	0.4	0.02–0.2
Ca	mg/kg	1988.7	1744.4–2233.1
Cu	mg/kg	13.4	2.8–23.9
Cr	mg/kg	2.9	1.5–4.1
Sn	mg/kg	9.2	0.5–2.9
P	mg/kg	910.5	782.2–1038.7
Fe	mg/kg	40.6	21.3–59.8
Mg	mg/kg	37.8	14.3–61.2
Mn	mg/kg	3.2	2.2–4.2
Mo	mg/kg	31	15.5–46.4
Ni	mg/kg	1.8	1.1–2.6
Ag	mg/kg	0.7	0.1–0.5
Pb	mg/kg	39.6	1.3–25.4
Si	mg/kg	20.2	13.9–26.4
Na	mg/kg	62.7	24–101.4
Ti	mg/kg	12	3.1–15.7
V	mg/kg	3.3	1.7–5
Zn	mg/kg	781	671–890.1

[a] 95% level of confidence.

3.2. Influence of Storage Conditions on Used Oil Properties

The result of the cluster analysis of the physicochemical properties of the twenty-three samples is presented in Figure 2. For the analysis, Euclidean squared distance and Ward's linkage were chosen to create homogeneous groups [32]. In the dendrogram (Figure 2), a distance of 5 was chosen, obtaining four clusters (according to the similarity of their physicochemical properties) called Clusters A, B, C, and D. Cluster A contains nine samples (i.e., Samples 8, 18, 13, 5, 7, 15, 3, 1, and 12), Cluster B also consists of nine samples (i.e., Samples 17, 23, 2, 9, 4, 19, 6, 16, and 20), Cluster C consists of two samples (i.e., Samples 14 and 21), and Cluster D consists of three samples (i.e., samples 10, 11, and 22). Sample 2, as seen in Figure 2, corresponds to the average values of the properties of the 23 oil samples. The corresponding properties moved away as the Euclidean distances of the other samples moved too. An ANOVA test was performed to compare the means of the groups (at the 95% confidence level). Figure 2 suggests that the storage conditions are not grouped but are different within the same cluster. Samples 1, 2, and 3, which were taken directly from vehicles' engines (Table 1), are located within Clusters A and B, with characteristics similar to the other stored samples, suggesting that the storage time (i.e., up to three months) does not affect the properties of the used oils. Clusters C and D, despite having fewer samples, also show different storage conditions. Therefore, this finding indicates that there is no correlation between the storage conditions (i.e., type of material of the containers and time) and the oil's characteristics. This can be confirmed because the properties of the oils extracted directly from the vehicle engines, which correspond to the zero storage date, were analyzed under the same environmental conditions as the other stored samples. Furthermore, due to the climatic conditions of Cuenca (which is more or less constant throughout the year), there are no temperature variations that could compromise the properties of the oil during storage.

3.3. Physical Properties of the Used Lubricating Oils in Cuenca and Comparison with the Properties of the Used Oils in Various Countries

Table 4 shows the values of the physical properties and metal content of the used oils in Cuenca (Column 2) and the new lubricant oils that are employed in the city (Column 3). The table also presents the typical values and expected limits of both the physical properties and metal contents reported in the literature, as well as the expected tendencies of each property or value compared to the new oils. It can be seen that, for used oils in Cuenca, most values are below the accepted limits, according to [34]. Moreover, the water content is below the limits reported by [34] and the rest of the results (except the concentration of some metals) are in the range of expected values [8,35]. This result suggests that the re-refining options for used oils in Cuenca should take advantage of the experiences in other places that process used oils with similar properties to guarantee the properties of the regenerated oil as required for further use.

Table 4. The physical properties and metal concentration of the used lubricant oils in Cuenca, in comparison with other oils.

Property	Oils in Cuenca		Typical Range Other Studies		Expected Limits [34]	Tendency with Respect to New Oils
	This Study	**Average New Oils [23]**	**Ref. [8]**	**Ref. [35]**		
Water content (%v/v)	0.2	0.1	-	1.1–13.7	-	increases
Flashpoint (°C)	178.1	-	75–240	120–214	>38	decreases
Density at 15 °C (g/mL)	0.9	-	-	-	-	increases
Viscosity index	134.2	138.1	-	92.5–144	-	increases
TBN (mg KOH/g)	5.7	5.3	-	4.1–4.7	-	decreases
Kinematic Viscosity 100 °C (mm^2/s)	11.8	16.4	-	9.3–14.8	-	increases
Metals content (mg/kg)						
Al	8	<1	-	4–20.3	<20	increases
Ba	1.5	64.7	5–110	-	-	decreases
B	43	10	-	30–4100	-	decreases
Cd	0.4	-	0–3.8	2–9.4	<2	increases
Ca	1988.7	1396.6	936–1670	606–1890	-	decreases
Cu	13.4	<1	-	7–30	<30	increases
Cr	2.9	<1	-	3.2–233	<10	increases
Sn	9.2	<1	-	9.7	<20	increases
P	910.5	802.2	494–973	-	-	decreases
Fe	40.6	1.3	23–99	15–348	<100	increases
Mg	37.8	12	66–120	43.5–332	-	decreases
Mn	3.2	-	0.9–18	-	-	increases
Mo	31	97.9	24–42	-	-	decreases
Ni	1.8	<1	0–5.4	2	<10	increases
Ag	0.7	<1	-	-	<3	increases
Pb	39.6	<1	7–57	2–2813	<30	increases
Si	20.2	4.3	29–76	11–21.9	<20	increases
Na	62.7	48.1	174–606	13–65	<50	increases
Ti	12	<1	0–4.9	-	-	increases
V	3.3	<1	0.2–10	-	-	increases
Zn	781	803.3	444-1280	54.9–1100	-	decreases

Table 5 shows a comparison of the physical properties, presence of metals, and concentration of additives of the used oils in Cuenca with those of used oils in other countries. Some properties are difficult to compare due to a lack of data. However, it is seen that, as with other materials, the water content is below the values reported in other countries. Properties such as viscosity are of interest since they show that the oils have not degraded too much during their use, as well as indicating a lower water content.

Metals such as Al, Cd, Cr, Cu, Fe, Pb, Mn, Mg, Ni, Ag, Sn, Ti, and V (Table 6), known as wear metals [8], arise in the used oil due to the engines' high operating temperatures and pressures. Thus, the concentrations of metals found in the used oil vary depending on the brand and the type of used oil, the conditions of the engine that originated them, and the distance traveled by the vehicle until oil change is performed [36]. Ti is outside the range of typical values for used oils with a much higher value. Oil liquid Ti is widely used to reduce the friction and wear of engine components [37], which could explain such high values for Ti content. In the case of engine wear metals, all values are below the mean of the countries reported in the literature (except for S and V, from which there are not enough results to compare in Table 5); however, they are below the typical ranges as reported by Parry [8] and Widodo et al. [35], as seen in Table 4. The results indicate that the oils have a low contamination by wear metals, especially in the case of Si, Al, Cu, Cr, and Fe. Si is considered a highly abrasive material [38]. Its presence in the used oil is mainly due to contamination by sediments or some brake fluids, as well as by dust (dirt) entering the engine. The concentration of additives such as Ba, B, Ca, Mg, Mo, Na, P, and Zn (Table 7) is reduced with respect their initial concentrations. P and Zn are elements that are generally found in anti-wear additives and antioxidants used to protect metal segments under engines' extreme operating conditions. Ca, Mg, and Na sulfonate are mainly associated with the detergents/dispersants used for engine cleaning. Some engine oils contain MoS2 to reduce wear at high temperatures and pressures. B and Ba are used in some synthetic oil formulations and act as anti-wear additives [22,36,39]. Regarding B, there are not enough data to compare with other countries (Table 5), but it is within a typical range according to Widodo. et al. as seen in Table 4. Some metals that are components of the engines also serve as additives in lubricating oils. This is the case of Al, Cu, Mg, Ti, and Mo. If the concentration of an element in the used oil increases with respect to the concentration of the virgin oil, it is considered a wear metal, otherwise it is considered an additive. Ba and Mg content is below the mean of the typical values reported in the literature (Table 7), which suggests that part of these elements was lost during the operation of the engines. The Mo, Na, P, and Zn concentrations are also below the mean values (right column), suggesting loss of these metals, but they are still within the typical ranges (see Table 4). Ca has a higher value than the expected values [8,9]. Its content depends on the types or brands of oils present in the mixture.

In general, it is observed that the characteristics of the oils used in Cuenca are within the ranges of values obtained from other countries. The results of this research are close to those reported for used oils in other countries in South America (e.g., Colombia, Chile, and Venezuela), except for the higher water content of the used oils in Chile. This could possibly result from similar climatic conditions, age of the car fleet, and types of oils consumed. Likewise, in Peru, a higher presence of wear metals in used oils is seen, although those used oils retain more additives. The literature also shows that, in Australia, used oils are less contaminated by wear metals, but there is greater loss of additives. Works conducted in the US on the concentration of metals in used oils show values close to used oils in Cuenca. In the case of Spain and Portugal, there is less metal contamination in the used oil, except for Fe. Iraq, Nigeria, and Poland present approximately similar results, although those oils appear less contaminated by metals. The same happens in the case of used oils in the UAE, except for the high content of Pb. The physical characteristics of the used oils in Egypt, Pakistan, and Saudi Arabia show some similarities to the used oils analyzed in this work. However, the metals content cannot be compared due to a lack of information. A high water content and low viscosity index is evident in South Africa's used oil. Kazakhstan, Ukraine, and Ghana show the highest metal-contaminated automotive used oil.

Table 5. The physical properties, concentration of wear metals, and additives of the used lubricating oils in Cuenca, in comparison with the properties of the oils used in various countries.

Element (ppm)	This Work	Quito-Ecuador [4]	Chile [40]	Colombia [5]	Venezuela [41]	Perú [42]	Spain [43]	Portugal [22]	UAE [44]	USA [45]	Kazakhstan [46]	Ghana [2]	Nigeria [47]	South Africa [48]	Ukraine [49]	Australia [50]	Poland [51]	Iraq [52]	Iraq [53]	Pakistan [54]	Saudi Arabia [55]	Egypt [56]	Mean
Al	8	-	-	17	-	375	-	-	4	-	57.9	-	-	20.1	-	64	-	-	-	-	-	-	89.7
Ba	1.5	-	-	-	113	9	297	-	<1	2.9	-	-	-	-	-	-	-	-	-	-	-	-	84.6
B	43	-	-	-	-	-	-	-	3	-	-	-	-	-	-	-	-	-	-	-	-	-	3
Cd	0.4	-	0.1	-	-	-	-	-	<1	1.7	-	-	-	-	-	0.04	-	1	-	-	-	-	0.8
Ca	1988.7	-	-	285.3	190	1250	4.5	1805	1667	-	171.9	606.5	-	-	1445.7	-	-	-	-	-	-	-	825.1
Cu	13.4	-	9.7	14.3	16	15	29	-	7	36	37.9	21.3	-	-	74.1	7	18	4.6	-	-	-	-	22.3
Cr	2.9	-	5	-	<1	-	-	-	4	3.3	18.3	-	1.3	-	52.6	0.2	1.4	1.5	-	-	-	-	8.9
Sn	9.2	-	5	-	-	-	-	-	<1	-	-	-	-	-	-	-	-	1.6	-	-	-	-	2.5
P	910.5	-	-	18.4	-	1550	931	739	632	-	-	-	-	-	2797.1	485	-	-	-	-	-	-	1021.8
Fe	40.6	-	-	-	46	240	205	212	15	-	-	89.5	-	21.3	696.4	41	53.5	72	-	-	-	-	153.8
Mg	37.8	-	-	-	1.5	-	309	-	38	-	436.1	-	-	-	549.8	24	-	81	-	-	-	-	205.6
Mn	3.2	-	-	-	-	-	-	-	1	-	-	-	5.3	-	8.8	1	1.8	1.5	-	-	-	-	3.2
Mo	31	-	-	-	-	-	-	-	7	-	-	-	-	-	48.15	-	462.9	-	-	-	-	-	172.7
Ni	1.8	-	1.4	-	-	-	-	5	<1	1.5	-	2	0.8	-	5.8	0.2	0.4	-	-	-	-	-	2
Ag	0.7	-	-	-	-	-	-	-	<1	-	-	-	-	-	-	-	-	-	-	-	-	-	<1
Pb	39.6	-	15.5																				45.4
Si	20.2	-	-	2.3	-	-	-	-	11	-	-	-	-	-	140	-	-	-	-	-	-	-	51.1
Na	62.7	-	-	216.8	43.5	-	118	-	20	-	-	9.5	-	-	-	-	-	-	-	-	-	-	73.3
Ti	12	-	-	-	-	-	-	-	2	-	-	-	-	-	-	-	-	-	-	-	-	-	2
V	3.3	-	2	-	-	-	-	-	<1	-	-	-	-	-	-	-	-	-	-	-	-	-	1.5
Zn	781	-	340	825.5	651	1025	1	1000	780	1152	403	-	-	-	4361	254	1106.7	1280	-	-	-	-	1013.8
Water content	0.2	0.4	10	0.5	0.6	5.1	-	-	0.8	-	0.03	-	-	11.5	0.2	-	-	0.9	1.5	-	-	0.3	2.7
Density at 15 °C	0.9	-	-	0.9	-	-	-	-	0.9	-	0.9	-	0.9	0.9	0.9	-	-	0.9	0.9	0.9	0.9	0.9	0.9
Viscosity index	134.2	122	-	-	-	115	-	103.1	-	-	-	-	-	25.5	110	-	-	89.1	-	110.6	80	115.1	96.7
Base number (TBN)	5.7	18.6	-	-	-	-	1.7	-	4.1	-	-	-	-	-	3.5	-	-	0.1	5.8	4.6	13.5	-	6.5
Flashpoint	178.1	90.8	-	-	182.2	-	348	-	152	-	190	-	135	125	511	-	-	158	178	183	101	128	190.9
Kinematic viscosity at 100 °C	11.8	29.8	55	81.1	-	17	-	7.8	-	-	9.5	-	-	6.5	13.9	-	-	13.5	-12	15.5	14.1	12.93	22.2

3.4. Analysis of Options for Re-Refining Used Oils in Cuenca

The treatment of used lubricant oils involves three processes: recovery, reprocessing, and regeneration (or re-refining) [8]. Reprocessing is carried out by the elimination of contaminants from used oils and may include distillation and chemical treatment, as well as a combination of methods with those used for recovery. With the re-refining process, the highest degree of contaminant removal is reached, obtaining the base oil to manufacture new lubricating oils. Table 6 shows the methods commonly employed for re-refining of used oils at the laboratory scale and Figure 3 summarizes the sequence of these steps. Regardless of the re-refining process, it begins with a pretreatment step that depends on the characteristics of the oil. According to Fong et al. [5], good quality oils can be obtained from used oils and their production costs are relatively lower than the costs of producing oils through petroleum refining. A gallon of used oil (3.78 L) provides 2.5 L of lubricating oil, which, otherwise, should require 42 gallons (>158 L) of crude oil [57]. Besides, the by-products of re-refining can be converted into valued end products for other manufacturing processes, such as asphalt. The yields of the products and byproducts can cover the cost of buying the chemicals needed to keep the operation of the re-refining plant and, therefore, to make the process profitable [5]. An oil of higher quality than the corresponding original oil cannot be obtained using these processes. For this reason, for cost-effective recycling of used oil, it is important to separate the different types of oils in the places used oils are generated because, from the mixture of high-quality and low-quality used oils, only low-quality oils are obtained [8]. Speight and Exall [58] (cited by [35]) indicated that, in the re-refinement process, solid particles and water are separated from the used oil by physical treatments to obtain an oil comparable to the original, but the contaminant metals could not be completely removed.

Table 6. Steps for re-refining used oils and the technologies employed at the laboratory scale.

	Technology	Country and Reference
Pre-Treatment	Filtration, centrifugation, decantation, and sedimentation	South Africa [48], Egypt [59]
	Distillation	
	Filtration	Egypt [56]
	Sedimentation, magnetization, heating, and agitation	Nigeria [60]
	Technology	**Country and Reference**
Regeneration	Sedimentation	UAE [44]
	Acid treatment	Romania [61], Iran [62], Nigeria [47], Colombia [5]
	Caustic treatment	Ghana [2]
	Activated carbon/clay treatment	South Africa [48]
	Distillation/clay	
	Acid clay treatment	
	Irradiation and ultrasonic adsorption	Kazakhstan [46]
	Solvent extraction	Egypt [56], Iraq [53], Portugal [22], Egypt [59], Nigeria [60], Spain [43], UAE [44]

Table 6. *Cont.*

Fractionation of the Bases	Vacuum distillation	Ukraine [49], Egypt [56], Iraq [53], Ghana [2], UAE [44]
	Atmospheric and vacuum distillations.	Spain [43]
Finishing	Adsorption, neutralization, sedimentation, and filtration.	Romania [61], Nigeria [47]
	Filtration and heating	South Africa [48]
	Filtration	Colombia [5]
	Adsorption	Kazakhstan [46], Egypt [56], Iraq [53], Nigeria [60], Iran [62], UAE [44]

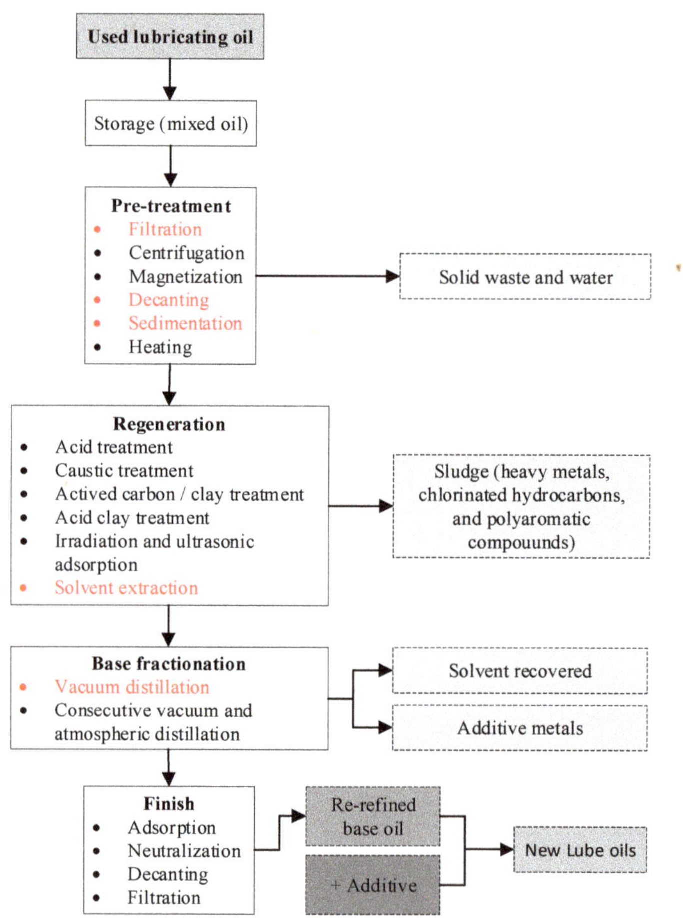

Figure 3. Main stages of conventional technologies for the re-refining of used oils. Adapted from [35].

Re-refining consists of four stages: pre-treatment (i.e., removal of water and solid particles), regeneration (i.e., elimination of degradation products), fractionation of the bases (i.e., separation of light hydrocarbons), and finishing (i.e., improvement of color and smell of the treated oil) [35,59,63]. Some of these processes still have some disadvantages, such as the poor performance of oil regeneration or the generation of other environmentally hazardous contaminants [35]. Following the logic of the circular economy, the ideal regeneration method for used lubricating oils should guarantee the function the product had at the beginning, as many times as possible after a treatment is conducted.

For the finishing step, excellent results have been obtained with the use of bentonite [44,50,60,61]. This is the process that appears most suitable for used oil treatment in Cuenca, due to the large amounts of bentonite available in the province of Napo [64], located ~400 km from Cuenca. Bentonite can be used as bleaching land to clarify and reduce (by adsorption purification) the intense color of the oils [65]. It is expected that used oil generators (e.g., mechanic and car lubricating shops) in the city will play a vital role in the collection system as they are responsible for the handling, separating, and storing operations for different types of used oils before delivering them to authorized collectors to ensure proper treatment [22].

In Ecuador, in the city of Quito, the re-refining of used lubricating oils is carried out, in this process methods such as filtration, extraction by solvents, and vacuum distillation are applied, while in the city of Cuenca only filtration, decantation, and sedimentation is performed; in Figure 3 these processes are marked in red.

At an industrial scale, the main technologies employed for re-refining used oils are (a) the acid/clay process; (b) activated clay process; (c) thin-film evaporation (TFE), based in vacuum distillation; (d) solvent extraction process; and (e) hydro-treatment process (hydrogen extraction) [58,65,66]. The steps for each re-refining technology and the yields that can be recovered from the process are presented in Figure 4.

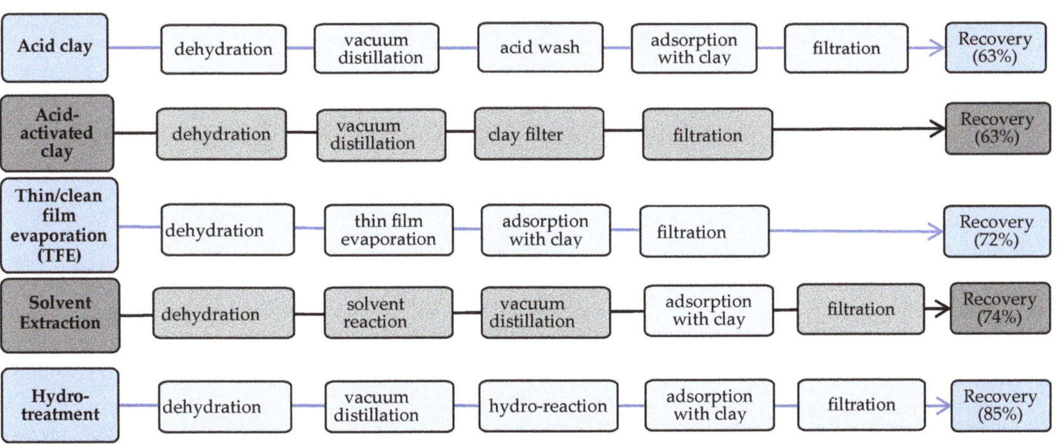

Figure 4. Summary of the technologies employed at the industrial level for re-refining used lubricant oil. Adapted from [66,67].

Table 7. The main technologies employed at an industrial scale for re-refining used lubricating oils.

Re-Refining Method	Process	Advantages	Disadvantages
Acid–clay [68,69]	Dehydration: removal of impurities, water, antifreeze, and solvents by preliminary distillation at low temperature. Vacuum distillation. Sulfuric acid: reverse reaction with oxygen, sulfur, and nitrogen to form sludge. Additional refining to eliminate paraffinic and naphthenic hydrocarbons. Adsorption with clay to reduce odor and dark color.	Proven technology. Low production cost. Simple and straightforward process. Does not require qualified operators. Adequate for small capacities. Low capital investment, profitable for small-scale plants. Low energy consumption. Produces good quality base oils.	Generates polluting waste. Causes corrosion of equipment. Low yield due to loss of oil in mud and clay. Not compatible with pollution control regulations. Technology is prohibited in many countries
Activated clay [68,69]	Dehydration. Vacuum distillation with conventional vacuum column. Adsorption with activated clay at 120 °C for 2 h. Filtered.	No acid is required. Simple process. Suitable for small capacity plants. Produces good quality base oils.	High consumption of clay. Low performance. Inconsistent quality. Environmental problems to eliminate large amounts of spent clay. Depends on a particular type of clay that may not be readily available
Thin film evaporation (TFE) [59,68]	Dehydration. Evaporation of thin-film (via vacuum distillation). Process involves the removal of volatile materials and the separation of the high boiling distillate from residues containing heavy metals. Two options: (a) hydraulic finishing to remove chlorine, nitrogen, oxygen, and sulfur compounds or (b) adsorption with clay to remove impurities (e.g., heavy metals and decomposition products)	Suitable for high capacity plants The thin-film evaporator is capable of operating under high vacuum and is usually used for high value products It does not cause contamination. Produces very good quality base oils	Operates at high temperature and high vacuum. The plant must have high capacity to be economically viable. High energy consumption
Solvent extraction [67,68,70]	Dehydration. Removal of impurities by mixing solvents that do not include sulfuric acid, the solvent produces insoluble and suspended substances (e.g., asphalt, metal compounds, and resin). The sludge removed as non-hazardous waste. Vacuum distillation with conventional vacuum column. Adsorption with clay to reduce odor and dark color.	The solvent is recyclable. Does not cause contamination. Functional recovery of base oils. Produces good quality base oils. Propane is used as a solvent to remove additives, metals, and tar. It operates at room temperature	Economical only for high plant capacity. Operates at higher pressure. Requires operating systems and qualified personnel. May involve operating losses of solvent. Propane can cause fires/explosion. High energy consumption
Hydro-treatment [66–68]	Dehydration. Vacuum distillation with conventional vacuum column. The oil treated with hydrogen in the presence of a catalyst removing sulfur, nitrogen, and oxygen. The combination of this process with solvent extraction strengthens the oil quality. Adsorption with clay.	Corrects the color and smell of the oil. The distillation process is commonly used to remove sulfur, nitrogen, metals, or unsaturated hydrocarbons	Expensive. Unsafe. Not suitable at small scale. High energy consumption. Operates at high temperature and pressure.

Details on each technology and the corresponding advantages and disadvantages are presented in Table 7. Other new technologies are currently available, especially in Europe and China, including the combined finishing of TFE and clay, TFE and solvent finishing, TFE and hydro-treatment, thermal deasphalting (TDA), clay finishing, and TDA and hydro-treatment. Kupareva et al. [66] presented an exhaustive review of the technologies employed at the industrial scale for re-refining used oils in Europe and CleanOil in China, as well as a list of the companies/countries using those technologies (see Table 8). Most of those commercial plants use the technologies described above, with small modifications in some cases.

The selection of a specific technology for re-refining used oil at an industrial level involves several criteria [67,68], including (a) technical and sustainability aspects; (b) health, safety, and environmental impacts; and (c) economic considerations. The thin-film evaporation (TFE) technology appears to offer the best results (in terms of product quality), followed by the solvent extraction, the hydrotreatment, and the activated acid–clay and acid–clay technologies. However, the TFE technology is economically viable only at large scales. The minimum capacity needed for a plant with this technology, according to Kupareva, is around 40,000 t/year (see Table 8), which is several times higher than the ~1534 t/year of used oil collected in Cuenca. Therefore, TFE technology is not currently viable in Cuenca.

The hydrotreatment and the activated acid–clay and acid–clay technologies, however, are incapable of producing products with adequate quality while possibly creating environmental problems [67] (see Table 7). Thus, these technologies are not appropriate compared to the TFE or the solvent extraction technologies. For these reasons, in this study, we believe the solvent extraction technology is the best option for re-refining used oils in Cuenca. The solvent extraction process is more adequate for processing relatively low volumes of used oils and the quality of the product complies with engine requirements. The oil recovery rate is relatively high (around 74%) and the process accepts both synthetic and mineral oils used in vehicles [66]. The solvents can be recovered and reused in the process. Moreover, this technology is widely used in other countries [66]. Since the physical properties of the used oils in Cuenca resemble those of countries such as the UAE [44], Iraq [53], and Egypt [56], which recover oils through solvent extraction, the use of this technology is justified and currently feasible.

Table 8. The main re-refining processes currently used in European countries and China (adapted from [66,71]).

Technology	Raw Material	Investment Costs	Product Quality	Yield%	Waste	Plant Capacity (t/year)
Solvent Extraction						
MRD process (Germany, Denmark and the Netherland)	It can contain up to 5% vegetable oils, complete preservation of synthetic oils.	Relatively low operating and capital costs.	Base oils of good quality.	91	It can use as fuel.	120,000
Interline (Spain and United Kingdom)	Motor mineral oil and industrial oils.	Relatively low operating and capital costs.	Low quality of base oil (API I).	79	Asphalt Modifier Production	Spain: 36,000, United Kingdom: 50,000
Vacuum distillation/TFE + hydro-treatment, solvent extraction finishing or chemical treatment						
Vaxon (Denmark and Spain)	Mineral motor and industrial oils and some stable synthetics in the presence of a strong base.	Financially attractive.	The average quality of the oil produced.	65–70	Asphalt extender, thermal energy production, or as a bitumen mix component.	Denmark: 40,000, Spain: 42,000
EcoHuile (Sotulub) (France)	Mineral oils and some stable synthetics in the presence of alkaline additives.	It does not require a final step; this decreases the investment cost of the process.	A base oil of poor quality.	82–92		125,000
Hydrotreatment						
Hylube (Germany)	Accept used oil from different sources.	The capital investment and operational costs of the process are minimizing by eliminating superfluous equipment.	Pretty high oil quality (API II).	85	Stable and acceptable for asphalt mixing.	
TFE + hydro-treatment						
CEP process (Finland)	It accepts used oil from different sources: industrial and motor, mineral and synthetic oils.	High operating and capital costs.	High quality base oils (API II).	70		60,000
TDA+ hydro-treatment						
Revivoil (Italy, Poland and Spain)	Use motor oil types.	Operation and capital costs higher.	Products of comparable quality to virgin base oils meet the requirements of API II.	72	Asphalt and bituminous membrane extender. Hydro-treatment catalyst regenerated and reused in the process.	Italy: 130,000, Poland: 80,000, Spain: 59,000

Table 8. *Cont.*

Technology	Raw Material	Investment Costs	Product Quality	Yield%	Waste	Plant Capacity (t/year)
Solvent extraction + hydro-treatment						
Snamprogetti (Italy)	Use motor oil types.	Relatively expensive due to PDA (propane deasphalting) and hydro-treatment.	High quality base oils.	74–80	Asphalt production.	84,000
Cyclon process (Greece)	Use motor oil types.	Operation and capital costs higher.	High-quality base oils.	72	Light hydrocarbon fuel used in boiler oil heaters.	40,000
CleanOil ORT (China)	Use motor oil types.	Relatively expensive due to PDA (propane deasphalting), vacuum distillation and hydro-treatment.	High quality Group II/II + base oils, exceed performance standards (API)	90	Asphalt flux	20,000

The pretreatment of used oil that ETAPA currently carries out (water and sludge removal through sedimentation, centrifugation, and filtration) is similar to that used in the UAE [44] and Egypt [56]. Thus, the preliminary steps required for the solvent extraction process are already being performed. In the re-refining process through solvent extraction, a 1:3 oil-to-solvent ratio is proposed (where the solvents are blends of butanol + toluene + methanol or butanol + KOH), following the processes used in the two aforementioned countries [44,56]. The solvent extraction process is followed by a vacuum distillation step for the recovery of solvents that will be reused in the process to remove metals (especially Pb and Fe) [44]. For the finishing step, different adsorbents can be used. However, better results are obtained with alumina, bentonite, and activated bentonite. The use of almond and palm kernel powders can also be effective [44]. At the moment, ETAPA's Used Oil Collection Program sends the filtered and ceded oil sludge to the ECOTECNO Foundation attached to HOLCIM, where a thermal destruction of the waste is carried out with the endorsement of the local and regional Environmental Control Entity. The waste generated by the re-refining plants in Europe and China are used as a component in asphalt preparation and as fuel in boilers [66]. These could be an option for re-using the wastes of the re-refining processes since the city has a small asphalt production plant that could process all this by-product. The development of a re-refining plant with a capacity to process higher amounts of used oils than that currently collected in the city is suggested. Our estimates indicate that there could be up to 7330 t/year of used oil to be processed in the proposed plant (which is around 10% of the total lubricant oil consumed in Ecuador; see Section 1). This amount would result from (a) a complete recovery of used oil in Cuenca; and (b) the contribution of neighboring cities in the South of Ecuador (e.g., Loja, Azogues, Machala, and other smaller cities) that are located at distances less than 220 km from Cuenca (Figure 5). According to [72], the added total number of cars in these cities is approximately double that in Cuenca. However, further work is necessary to better define the collection, transport, and storage logistics of used oil in those cities to comply with the requirements of a re-refining plant and Law 042. The benefit for Ecuador is enormous since the import of up to 7330 t/year of lubricant could be avoided in these conditions. In addition to environmental and social benefits, the country could save around US$ 30 million/year, according to data from the Central Bank of Ecuador [73], due to avoiding the import of lubricant oils.

Figure 5. Map of Ecuador showing Cuenca and some neighboring cities that could supply used lubricant oils for the operation of a re-refining plant.

4. Conclusions

Disposing of used lubricating oils by direct incineration causes damage to the environment while losing a non-renewable resource. A circular economy requires the reuse of used lubricating oils to reduce waste and environmental damage while promoting social benefits. While the current used oil collection system in Cuenca is primarily focused on avoiding spills and the collection of used oils, strategies to re-refine and therefore add value to these materials are important steps to consider. The work has found that the storage conditions of the used oil in the generators' premises do not significantly affect its properties. This result shows that the system currently used by used oil generators in Cuenca can be used for further processing and no changes to the system or logistics are required. The characterization of used oils in the city has shown that most of the properties are comparable to those of used oils in other countries (for example, the United Arab Emirates, Iraq, and Egypt). Some critical properties, such as the water content, are below those reported in the literature, which helps the re-refining process. Consequently, technologies to recover and add value to used oils in Cuenca must consider the experiences of other nations in the processing of similar used oils. European industries show great progress on the subject, as well as good recovery rates and quality of the regenerated product. The union of several methods improves the final product even more, but the economic investment increases. However, small- and medium-sized cities such as Cuenca do not generate large amounts of used oil in order to have plants with the same capacities as those in Europe. Therefore, the selection of treatment technology must also consider aspects such as the available market, operating costs, transportation, energy, and the quality of oil to be obtained. Due to the conditions in Cuenca, a good option is the solvent extraction process since it can be adapted by small-scale plants, providing adequate quality and performance of the products, as well as reducing negative environmental impacts. For the finish (improvement of color and odor), bentonite, a material that is available near the city, can be used. Furthermore, in Cuenca the initial step is already being carried out; that is, the filtration/dehydration and sedimentation process. Re-refining used oils in Cuenca and other medium-sized cities, whether in Latin America or elsewhere, is necessary to comply with the principles of a circular economy, taking advantage of non-renewable resources and contributing to the economy, society, and environment.

Author Contributions: Conceptualization, C.S.-A., J.C.-B. and D.A.-A.; methodology, C.S.-A., J.C.-B., M.R.P.-S. and D.A.-A.; sampling, J.C.-B. and D.A.-A.; software, C.S.-A. and F.G.-Á.; validation, C.S.-A., J.C.-B. and M.R.P.-S.; formal analysis C.S.-A., J.C.-B. and M.R.P.-S.; investigation, C.S.-A., J.C.-B., F.G.-Á. and D.A.-A.; data curation, C.S.-A. and F.G.-Á.; writing—original draft preparation, C.S.-A., J.C.-B. and F.G.-Á.; writing—review and editing, C.S.-A. and M.R.P.-S.; funding acquisition, J.C.-B. and D.A.-A. All authors have read and agreed to the published version of the manuscript.

Funding: This research was funded by the Research Department of the University of Cuenca (DIUC) through the project "Análisis y definición de estrategias y escenarios para el desarrollo de sistemas de mantenimiento industrial orientado a la eficiencia energética y amigable con el ambiente en la ciudad de Cuenca".

Institutional Review Board Statement: Not applicable.

Informed Consent Statement: Not applicable.

Data Availability Statement: The data presented in this study are available in this article.

Acknowledgments: The authors acknowledge Raul Pelaez-Garcia for English editing.

Conflicts of Interest: The authors declare no conflict of interest. The funders had no role in the design of the study; in the collection, analyses, or interpretation of data; in the writing of the manuscript, or in the decision to publish the results.

References

1. References Agencia de Regulación y Control Hidrocarburifero [ARCH]. Reglamento Elaboración Aceites, Lubricantes Para Uso de Automotores. Available online: https://www.controlhidrocarburos.gob.ec/wp-content/uploads/MARCO-LEGAL-2016/Registro-Oficial-Suplemento-621-Res.-ARCH-3.pdf (accessed on 12 June 2019).
2. Mensah-Brown, H. Re-refining and recycling of used lubricating oil: An option for foreign exchange and natural resource conservation in Ghana. *ARPN J. Eng. Appl. Sci.* **2015**, *10*, 797–801.
3. El-Fadel, M.; Khoury, R. Strategies for vehicle waste-oil management: A case study. *Resour. Conserv. Recycl.* **2002**, *33*, 75–91. [CrossRef]
4. Almeida Streitwieser, D.; Jativa Guzmán, F.; Aguirre Ortega, B. Conversión de aceite lubricante usado de automóviles a Diésel #2. *Rev. Digit. VI Congr. Cienc. Tecnol.* 2011. Available online: http://repositorio.espe.edu.ec/handle/21000/3483 (accessed on 6 February 2019).
5. Fong Silva, W.; Quiñones Bolaños, E.; Tejada Tovar, C. Physical-chemical caracterization of spent engine oils for its recycling Caracterización físico-química de aceites usados de motores para su reciclaje. *Prospectiva* **2017**, *15*, 135–144. [CrossRef]
6. Ministerio del Medio Ambiente [MMA]. Evaluación Económica, Ambiental y Social de la Implementación de la REP en Chile Available online: https://mma.gob.cl/wp-content/uploads/2015/07/Publicacion-impactos-2012.pdf (accessed on 28 December 2019).
7. Botas, J.A.; Moreno, J.; Espada, J.J.; Serrano, D.P.; Dufour, J. Recycling of used lubricating oil: Evaluation of environmental and energy performance by LCA. *Resour. Conserv. Recycl.* **2017**, *125*, 315–323. [CrossRef]
8. Parry, B.J. Oil Recycling. In *Kirk-Othmer Encyclopedia of Chemical Technology*; Terrapure Environmental: North Vancouver, BC, Canada, 2016; pp. 1–16. ISBN 0471238961.
9. Martelli-Tristão, J.A.; Valentini-Tristão, V.T.; Frederico, E. O processo de reciclagem do óleo lubrificante. *Rev. Ibero-Am. Ciênc. Ambient.* **2017**, *8*, 224–238. [CrossRef]
10. CONAMA. A Importância da Destinação Correta do Óleo Lubrificante Usado—A Logística Reversa Como Instrumento na Aplicação da Resolução CONAMA No 362/2005. Available online: https://www.linkedin.com/pulse/importância-da-destinaçao-correta-do-óleo-usado-logística-ramos/ (accessed on 29 November 2019).
11. SINDIRREFINO—Processo Industrial/Rerrefino. Available online: https://www.sindirrefino.org.br/institucional/a-instituicao (accessed on 17 December 2020).
12. Compañía Industrial Lima S.A. [CILSA]. Protegemos el Medio Ambiente. Available online: https://www.cilsaperu.com/leer.html (accessed on 30 May 2020).
13. BIOFACTOR S.A. Available online: http://biofactorsa.com/ (accessed on 30 May 2020).
14. Nueva Energía S.A. Available online: https://nuevaenergiasa.com.ar/ (accessed on 30 May 2020).
15. Muñoz, E.; Montoya, D.; Muñoz, A. La Re-Refinación: Una Alternativa a la Mano Para la Disposición Adecuada de Aceites usados (1). Available online: http://nuevagaceta.co/inicio/re-refinacion-alternativa-disposicion-adecuada-de-aceites-usados (accessed on 30 May 2020).
16. BID. BID Lab Apoya Creación de Primera Planta de Recuperación de Aceite en Centroamérica | IADB. Available online: https://www.iadb.org/es/noticias/bid-lab-apoya-creacion-de-primera-planta-de-recuperacion-de-aceite-en-centroamerica (accessed on 29 November 2019).
17. Banco Central del Ecuador [BCE]. Evolución de la Balanza Comercial Enero-Julio/2018. Available online: https://contenido.bce.fin.ec/documentos/Estadisticas/SectorExterno/BalanzaPagos/balanzaComercial/ebc201812.pdf (accessed on 25 January 2019).
18. Ministerio del Ambiente y Agua [MAE]. Instructivo para la Gestión Integral de Aceites Lubricantes Usados y Envases Vacíos—Apel. Available online: http://apel.ec/biblioteca/mae-instructivo-para-la-gestion-integral-de-aceites-lubricantes-usados-y-envases-vacios/ (accessed on 30 May 2020).
19. Diario el Comercio Convenios en cinco ciudades del Ecuador para Reciclar Aceites Usados. Available online: https://www.elcomercio.com/actualidad/convenios-ciudades-ecuador-reciclar-aceites.html (accessed on 29 November 2019).
20. ETAPA. EP—Servicios de Telefonía, Televisión, Internet, Agua Potable, Alcantarillado de Cuenca—Ecuador > Información > Gestión Ambiental > Gestión Ambiental Urbana > Programa de Recolección de Aceites. Available online: https://www.etapa.net.ec/informacion/gestion-ambiental/gestion-ambiental-urbana/programa-de-recoleccion-de-aceites (accessed on 31 March 2020).
21. Navarro-Nuñez, W. Estado Situacional del Manejo del Aceite Lubricante Usado en la Ciudad de Ayacucho y Propuesta de Disposición Final. Master's Thesis, Universidad de Piura, Piura, Perú, 2014.
22. Pinheiro, C.T.; Ascensão, V.R.; Cardoso, C.M.; Quina, M.J.; Gando-Ferreira, L.M. An overview of waste lubricant oil management system: Physicochemical characterization contribution for its improvement. *J. Clean. Prod.* **2017**, *150*, 301–308. [CrossRef]
23. Jaramillo, D.; Redrován, L.; Urgilés, D. Analisis Técnico de la Vida Útil de un Lubricante de Aceite Mineral, para Motores de Combustion Interna a Gasolina de los Vehículos de Servicios de Taxis en la Ciudad de Cuenca. Bachelor's Thesis, Universidad Politécnica Salesiana, Cuenca, Ecuador, 2011.
24. Fundación PRO-AMBIENTE Manejo Ambientalmente Racional de Aceites Lubricantes Usados. Available online: https://es.slideshare.net/JGNathyvidad/aceites-usados-39623939 (accessed on 30 November 2019).
25. Peñafiel Chiriboga, S.L. Caracterización del Manejo de Aceites de Desecho de Automóviles e Hidráulicos de Origen Industrial en la Ciudad de Cuenca. Bachelor's Thesis, Universidad de Cuenca, Cuenca, Ecuador, 2017.

26. ASTM International Standard. *Practice for Calculating Viscosity Index from Kinematic Viscosity*; D 2270-04; ASTM International: West Conshohocken, PA, USA, 2009; pp. 1–6.
27. ASTM International Standard. *Test Method for Density, Relative Density, and API Gravity of Liquids by Digital Density Meter*; D4052-11; ASTM International: West Conshohocken, PA, USA, 2011; pp. 1–8.
28. ASTM International Standard. *Test Method for Acid Number of Petroleum Products by Potentiometric Perchloric Acid Titration*; D 2896-03; ASTM International: West Conshohocken, PA, USA, 2003; pp. 1–7.
29. ASTM International Standard. *Test Method for Flash and Fire Points by Cleveland Open Cup Tester*; D 92-05; ASTM International: West Conshohocken, PA, USA, 2005; pp. 1–10.
30. ASTM International Standard. *Practice for Calculating Viscosity Index from Kinematic Viscosity at 40 and 100 °C*; D 2270-93 (Reapproved 1998); ASTM International: West Conshohocken, PA, USA, 1993; pp. 1–6.
31. ASTM International Standard. *Test Method for Determination of Wear Metals and Contaminants in Used Lubricating Oils or Used Hydraulic Fluids by Rotating Disc Electrode Atomic Emission Spectrometry*; D6595-16; ASTM International: West Conshohocken, PA, USA, 2011; pp. 1–6.
32. De la Fuente Fernandez, S. Análisis de Conglomerados. Available online: https://www.academia.edu/32046069/Análisis_Conglomerados_Santiago_de_la_Fuente_Fernández (accessed on 18 September 2018).
33. Hines, W.; Montgomery, D.; Borror, C. *Probabilidad y Estadística para Ingeniería*, 3rd ed.; Compañía Editorial Continental: Mexico City, Mexico, 2005; ISBN 970240553X.
34. Widman, R. Interpretando el Reporte de Análisis de Aceite. Available online: http://www.mantenimientomundial.com/notas/w46.pdf (accessed on 6 February 2019).
35. Widodo, S.; Ariono, D.; Khoiruddin, K.; Hakim, A.N.; Wenten, I.G. Recent advances in waste lube oils processing technologies. *Environ. Prog. Sustain. Energy* **2018**, *37*, 1867–1881. [CrossRef]
36. Gómez Estrada, Y. Contribución al Desarrollo y Mejora para la Cuantificación de la Degradación en Aceites Lubricantes Usados de MCIA a Través de la Técnica de Espectrometría Infrarroja por Transformada de Fourier (FT-IR). Ph.D. Thesis, Universitat Politècnica de València, Valencia, Spain, 2013.
37. Terpel Terpel Oiltec 20W 50 Titanio. Available online: https://www.terpel.com/Plantillas/Terpel/Descargables/Lubricantes/TerpelOiltec20W-50Titanio.pdf (accessed on 30 June 2020).
38. Noria Corporation. Q & A: ¿Qué es el TBN y qué Indica en el Análisis de Aceite? Noria Latín América. Available online: http://noria.mx/lublearn/q-a-que-es-el-tbn-y-que-indica-en-el-analisis-de-aceite/ (accessed on 30 November 2019).
39. Mang, T.; Dresel, W. *Lubricants and Lubrication*, 2nd ed.; WILEY-VCH: Weinheim, Germany, 2007; ISBN 9783527610341.
40. Arellano, L. *Proyecto Planta de Reciclaje de Aceites Usados por Extracción por Solvente*; Asesorías y Proyectos Ribotta: Valparaíso, Chile, 2008.
41. Amesty, R.; De Turris, A.; Rojas, D.; Hurtado, A.; Medrano, J.; López, Y. Characterization of Oil Lubricant Automotive for Reuse. *Redieluz* **2015**, *5*, 43–48.
42. Villanueva Torres, C. Diseño de una Planta Piloto para Desarrollar Tecnología de Extracción con Solvente para Tratamiento de Aceites Usados. Bachelor's Thesis, Universidad Nacional de Ingeniería, Lima, Perú, 2005.
43. Moya Díaz, L. Desde el Aceite Lubricante Usado Hasta su Puesta en el Mercado Tras Su Regeneración. Master's Thesis, Fundación Escuela de Organización Industrial, Madrid, Spain, 2010.
44. Abdel-Jabbar, N.M.; Al Zubaidy, E.A.H.; Mehrvar, M. Waste lubricating oil treatment by adsorption process using different adsorbents. *World Acad. Sci. Eng. Technol.* **2010**, *62*, 9–12.
45. Boughton, B.; Horvath, A. Environmental Assessment of Used Oil Management Methods. *Environ. Sci. Technol.* **2004**, *38*, 353–358. [CrossRef]
46. Syrmanova, K.K.; Kovaleva, A.Y.; Kaldybekova, Z.B.; Botabayev, N.Y.; Botashev, Y.T.; Beloborodov, B.Y. Chemistry and recycling technology of used motor oil. *Orient. J. Chem.* **2017**, *33*, 3195–3199. [CrossRef]
47. Owolabi, R.; Amosa, M.K. Some Physico-Chemical and Adsorptive Reclamation Strategies of Spent Automobile Engine Lubricating Oil. *J. Eng. Res.* **2017**, *22*, 98–106.
48. Abdulkareem, A.S.; Afolabi, A.S.; Ahanonu, S.O.; Mokrani, T. Effect of treatment methods on used lubricating oil for recycling purposes. Energy Sources, Part A Recover. *Util. Environ. Eff.* **2014**, *36*, 966–973. [CrossRef]
49. Korchak, B.; Hrynyshyn, O.; Chervinskyy, T.; Polyuzhin, I. Application of vacuum distillation for the used mineral oils recycling. *Chem. Chem. Technol.* **2018**, *12*, 365–371. [CrossRef]
50. Ramadass, K.; Megharaj, M.; Venkateswarlu, K.; Naidu, R. Ecological implications of motor oil pollution: Earthworm survival and soil health. *Soil Biol. Biochem.* **2015**, *85*, 72–81. [CrossRef]
51. Zając, G.; Szyszlak-Bargłowicz, J.; Słowik, T.; Kuranc, A.; Kamińska, A. Designation of chosen heavy metals in used engine oils using the XRF method. *Polish J. Environ. Stud.* **2015**, *24*, 2277–2283. [CrossRef]
52. Hamawand, I.; Yusaf, T.; Rafat, S. Recycling of waste engine oils using a new washing agent. *Energies* **2013**, *6*, 1023–1049. [CrossRef]
53. Mohammed, R.R.; Ibrahim, I.A.R.; Taha, A.H.; McKay, G. Waste lubricating oil treatment by extraction and adsorption. *Chem. Eng. J.* **2013**, *220*, 343–351. [CrossRef]
54. Muhammad, Q.; Tariq, M.A.; Mazhar, H. Physico-chemical characteristics of Pakistani used engine oils. *J. Pet. Technol. Altern. Fuels* **2016**, *7*, 13–17. [CrossRef]

55. Elkhaleefa, A.M. Waste Engine Oil Characterization and Atmospheric Distillation to Produce Gas Oil. *Int. J. Eng. Adv. Technol.* **2016**, *5*, 6–8.
56. Osman, D.I.; Attia, S.K.; Taman, A.R. Recycling of used engine oil by different solvent. *Egypt. J. Pet.* **2018**, *27*, 221–225. [CrossRef]
57. U.S. Environmenteal Protection Agency (EPA). Used Oil Management Program. Available online: https://archive.epa.gov/wastes/conserve/materials/usedoil/web/html/index.html (accessed on 29 November 2019).
58. Speight, J.G.; Exall, D.I. *Refining Used Lubricating Oils*; CRC Press: Boca Raton, FL, USA; Taylor and Francis: Boca Raton, FL, USA, 2014; ISBN 9781466551497.
59. Emam, E.A.; Shoaib, A.M. Re-refining of Used Lube Oil, II- by Solvent/Clay and Acid/Clay-Percolation Processes. *ARPN J. Sci. Technol.* **2012**, *2*, 1034–1041.
60. Isah, A.G.; Abdulkadir, M.; Onifade, K.R.; Musa, U.; Garba, M.U.; Bawa, A.A.; Sani, Y. Regeneration of used engine oil. *Lect. Notes Eng. Comput. Sci.* **2013**, *1 LNECS*, 5–8.
61. Stan, C.; Andreescu, C.; Toma, M. Some aspects of the regeneration of used motor oil. *Procedia Manuf.* **2018**, *22*, 709–713. [CrossRef]
62. Salem, S.; Salem, A.; Agha Babaei, A. Preparation and characterization of nano porous bentonite for regeneration of semi-treated waste engine oil: Applied aspects for enhanced recovery. *Chem. Eng. J.* **2015**, *260*, 368–376. [CrossRef]
63. García González, M.T. Regeneración de Aceites Lubricantes Usados Mediante Extracción. Available online: https://dialnet.unirioja.es/servlet/tesis?codigo=105028 (accessed on 30 November 2019).
64. Uribe, R. Investigaciones de Materias Primas Minerales No Metálicas en el Ecuador. *Rev. Politec.* **2015**, *36*, 34. [CrossRef]
65. Haro, C.; De La Torre, E.; Aragón, C.; Guevara, A. Regeneración de arcillas de blanqueo empleadas en la decoloración de aceites vegetales comestibles. *Rev. EPN* **2014**, *34*, 1–8.
66. Kupareva, A.; Mäki-Arvela, P.; Murzin, D.Y. Technology for rerefining used lube oils applied in Europe: A review. *J. Chem. Technol. Biotechnol.* **2013**, *88*, 1780–1793. [CrossRef]
67. Maghsoodi, A.I.; Afezalkotob, A.; Ari, I.A.; Maghsoodi, S.I.; Hafezalkotob, A. Selection of waste lubricant oil regenerative technology using entropy-weighted risk-based fuzzy axiomatic design approach. *Informatica* **2018**, *29*, 41–74. [CrossRef]
68. Jafari, A.J.; Hassanpour, M. Analysis and comparison of used lubricants, regenerative technologies in the world. *Resour. Conserv. Recycl.* **2014**, *103*, 179–191. [CrossRef]
69. Nwachukwu, A. Review and assessment of mechanic village potentials for small scale used engine oil recycling business. *Afr. J. Environ. Sci. Technol.* **2012**, *6*, 464–475. [CrossRef]
70. Oladimeji, T.E.; Sonibare, J.A.; Omoleye, J.A.; Emetere, M.E.; Elehinafe, F.B. A review on treatment methods of used lubricating oil. *Int. J. Civ. Eng. Technol.* **2018**, *9*, 506–514.
71. CleanOil Limited—China's Most Advanced and Cleanest Oil Re-Refiner. Available online: http://www.cleanoil.com.hk/ (accessed on 26 December 2020).
72. AEADE SECTOR AUTOMOTOR en Cifras. Available online: http://www.aeade.net/wp-content/uploads/2017/03/Sector-en-cifras_6-Marzo-2017.pdf (accessed on 30 December 2019).
73. Ecuador Evolución de la Balanza Comercial Enero—Diciembre/2019. Subgerencia Program. y Regul. Available online: https://contenido.bce.fin.ec/documentos/Estadisticas/SectorExterno/BalanzaPagos/balanzaComercial/ebc202002.pdf (accessed on 16 February 2019).

recycling

Article

Evaluation of the Use of Recycled Vegetable Oil as a Collector Reagent in the Flotation of Copper Sulfide Minerals Using Seawater

Felipe Arcos and Lina Uribe *

School of Mining Engineering, University of Talca, Talca 3460000, Chile; farcos14@alumnos.utalca.cl
* Correspondence: luribe@utalca.cl; Tel.: +56-220-1798

Abstract: Considering sustainable mining, the use of seawater in mineral processing to replace conventional water is an attractive alternative, especially in cases where this resource is limited. However, the use of this aqueous medium generates a series of challenges; specifically, in the seawater flotation process, it is necessary to adapt traditional reagents to the aqueous medium or to propose new reagents that achieve better performance and are environmentally friendly. In this research, the technical feasibility of using recycled vegetable oil (RVO) as a collector of copper sulfide minerals in the flotation process using seawater was studied. The study considered the analysis of the metallurgical indexes when different concentrations of collector and foaming reagent were used, considering as collectors the RVO, potassium amyl xanthate (PAX) and mixtures of these, in addition to the methyl isobutyl carbinol (MIBC) as foaming agent. In addition, it was evidenced that the best metallurgical indexes were achieved using 40 g/t of RVO and 15 g/t of MIBC, which corresponded to an enrichment ratio of 6.29, a concentration ratio of 7.01, a copper recovery of 90.06% and a selectivity index with respect to pyrite of 4.03 and with respect to silica of 12.89. Finally, in relation to the study of the RVO and PAX collector mixtures, it was found that a mixture of 60 g/t of RVO and 40 g/t of PAX in the absence of foaming agent presented the best results in terms of copper recovery (98.66%) and the selectivity index with respect to pyrite (2.88) and silica (14.65), improving PAX selectivity and recovery compared to the use of RVO as the only collector. According to these results, it is possible to conclude that the addition of RVO improved the selectivity in the rougher flotation for copper sulfides in seawater. This could be an interesting opportunity for the industry to minimize the costs of the flotation process and generate a lower environmental impact.

Keywords: flotation; seawater; collectors; vegetable oil; recycled vegetable oil

check for
updates

Citation: Arcos, F.; Uribe, L. Evaluation of the Use of Recycled Vegetable Oil as a Collector Reagent in the Flotation of Copper Sulfide Minerals Using Seawater. *Recycling* 2021, 6, 5. https://doi.org/10.3390/recycling6010005

Received: 14 December 2020
Accepted: 4 January 2021
Published: 11 January 2021

Publisher's Note: MDPI stays neutral with regard to jurisdictional claims in published maps and institutional affiliations.

Copyright: © 2021 by the authors. Licensee MDPI, Basel, Switzerland. This article is an open access article distributed under the terms and conditions of the Creative Commons Attribution (CC BY) license (https://creativecommons.org/licenses/by/4.0/).

1. Introduction

Froth flotation is one of the most widely used physicochemical methods of concentrating minerals in the mining industry today because of the high capacity and efficiency it offers in the process of obtaining metals [1]. Separation of minerals by froth flotation techniques depends primarily on the differences in the wettability of particles. The most important reagents used are the collectors, which selectively adsorb onto the surface of the mineral particles, making them hydrophobic and facilitating their adhesion to gas bubbles injected into the process that rise through the slurry to form a concentrate rich in the mineral of interest [2].

Among the different collectors currently available in the copper mining industry, the xanthates are widely used in the flotation of minerals such as common sulfides (Cu, Mo, Pb, Zn, Co and Ni) and native metals (Cu, Ag, Au, etc.). The xanthates are ionic collectors and contain a hydrophilic polar group (affinity with water) attached to a chain of hydrocarbons that turns out to be the hydrophobic part. Due to their strong ionic powers, xanthates have negative effects to the biota and health hazards to human and animals [3–5], and they become a major problem at the time of handling and disposal of the waste generated by the flotation [6]. It was estimated that a total of 52,000 tons of xanthates were consumed annually by the mining industry worldwide [7]. Most of the mining industry across the globe use xanthate. The efforts to reduce the harmful effects of these are focused on the search to develop a "green" collector, or environmentally friendly, that reduces or inhibits the generation of toxic substances.

On the other hand, because of the water scarcity in some mining areas around the world, some projects consider the use of seawater to process coper ores by flotation in countries such as Chile, Australia and Indonesia [8]. In Chile, Las Luces plant has been using sweater for their operations since 1994 [9], and more recently Esperanza mine from Antofagasta Minerals [10]. However, even though the use of seawater in mineral process can be seen as interesting alternative to solve the scarcity problems, its use in mineral processing leads to significant challenges such as corrosion in different equipment such as pipes, mills and flotation cells; and the reduction of the performance in the concentration process [11]. Within this framework, as possible solutions to these problems, it has been proposed to use some coatings of materials with high chemical and mechanical resistance [12], to use alloys of materials that better resist the corrosion phenomenon [13] and to study new operating conditions in the different processes associate, in order to achieve metallurgical recoveries and grades of copper without affecting the efficiency of by-products (Mo and Au) [14].

Specifically, when seawater is used to carry out this concentration process, a series of problems is generated due to the high concentration of ions that it possesses (Na^+, K^+ Mg^{2+}, Ca^{2+}, $SO_4{}^{2-}$ and Cl^-). Mainly, there is the pH adjustment as seawater acts as a buffer solution, which generates a significant increase in lime consumption to achieve the pH values required in the process to depress pyrite [15]. Additionally, seawater generates a depressant effect on minerals such as molybdenite at pH values above 9.5. This is mainly due to the presence of calcium and magnesium cations in solution, which generates colloidal precipitates at pH levels above this value, preventing the recovery of this mineral [16]. It is because of these reasons that, new reagents are needed, in order to recover copper minerals (Chalcopyrite) in seawater, without affecting the recovery of secondary minerals, such as molybdenite. In addition, these new reagents must be able to depress pyrite to the pH of seawater.

In this sense, the use of vegetable oils could be an interesting alternative to replace and/or reduce the use of xanthates in copper sulfide ore flotation. There exists evidence that shows that the collecting properties of these oils can be competitive compared to the commercial collectors used in the sulfides flotation and native minerals [17–21]. Owusu et al. studied the use of commercial edible oils (canola, derived from rapeseed, and palm oil) as potential flotation reagents to add to the existing array of xanthates normally deployed for the selective flotation of copper sulfide minerals [18]. In addition, Bauer et al. [20] and Greene et al. [21] patented the use of various oils as collectors for the recovery of chalcopyrite and molybdenite from an ore containing 0.579% Cu and 0.010% Mo. Furthermore, Benn et al. [22] proposed the use of rapeseed oil as a selective collector for galena (PbS) recovery against pyrite. A laboratory flotation test using a complex sulfide ore containing Pb, Zn, Cu and Fe sulfides using 240 g/t of rapeseed oil as collector gave a higher recovery of lead, copper and zinc minerals than when typical, conventional collectors (xanthate, dithiophosphate, mercaptan, etc.) were used. Pyrite recovery in this study was similar to that obtained using the conventional collectors.

Taking into account these considerations, this research proposes to study a recycled vegetable oil as a collector of copper sulfide minerals in the flotation process in seawater in order to contribute to reduce the toxic collector used in minerals processing and to find a different type of application of this waste This study considered the analysis of the metallurgical indexes when different concentrations of collector and foaming reagent were used, considering as collectors the RVO, potassium amyl xanthate (PAX) and mixtures of these, in addition to the methyl isobutyl carbinol (MIBC) as foaming agent.

2. Material and Methods

2.1. Mineral Sample

The mineral sample used in this investigation was supplied by Minera Paicaví S.A. from the Chépica mine, located in the town of Pencahue, Chile. For the purposes of this work, each sample of 600 g of ore were crushed, ground in a ball mill and sieved to a P80 size of 212 μm, which corresponds to an appropriate size for rougher flotation.

Tables 1 and 2 present the results of the XRF and XRD analysis performed on the mineral sample. In the first one, it can be observed that, according to XRF analysis, the copper content was 2.20% and iron was 12.97%, while, regarding its mineralogy, in the second table, it is observed that the content of these elements are mainly associated to chalcopyrite and pyrite and that gangue minerals are mainly quartz and phyllosilicates such as muscovite and chlorite.

Table 1. Results of the XRF analysis of the mineral sample.

Element	Symbol	%
Silicon	Si	41.00
Sulfur	S	23.60
Potassium	K	1.140
Titanium	Ti	0.031
Vanadium	V	0.011
Chrome	Cr	0.017
Magnesium	Mg	0.033
Iron	Fe	12.97
Copper	Cu	2.196
Zinc	Zn	0.307
Arsenic	As	0.008
Silver	Ag	0.003
Platinum	Pt	0.003
Lead	Pb	0.112
Molybdenum	Mo	0.002

Table 2. Results of the XRD analysis of the mineral sample.

Mineral Phase	Molecular Formula	%
Quartz	SiO_2	73.9
Muscovite	$KAl_2(Si_3AlO_{10})(OH)_2$	5.5
Chlorite	$(Mg,Fe++)5Al(Si_3Al)O_{10}(OH)_8$	1.9
Pyrite	FeS_2	13.0
Chalcopyrite	$CuFeS_2$	5.7

2.2. Chemical Reagents and Aqueous Media

The collector reagents used in this research were potassium amyl xanthate (PAX) supplied by SOLVAY Chile and purified according the methodology described by Montalti et al. [23]. Sunflower recycled RVO was collected by Universidad de Talca and was previously filtered and was introduced in the flotation cell as supplied. The foaming agent used was methyl isobutyl carbonyl (MIBC Aerofroth 70 supplied by SOLVAY, Antofagasta, Chile). The only modifier considered in this project was lime (CaO) to adjust the pH to 8. MIBC was made down to 1 mL/L, by weighing 1 g into a clean 100 mL volumetric flask and making up to the mark with distilled water.

The supply of seawater used in this research was from the coast of the Maule region in Chile, specifically from the town of Iloca, highlighting that the water was extracted at a point far from the shore so that its chemical composition was not affected or physically contaminated. The collected seawater had a pH of 7.6 and salinity of 35 g/L. During the flotation stage, all experiments were done at room temperature of 20 °C.

2.3. Experimental Procedure

2.3.1. Reduction of Mineral Sample Size

The first comminution process was performed in the jaw crusher in order to obtain a P80 size of 2000 µm. Then, a wet milling process was performed, for which 67% of solids by weight (using 600 g of ore sample and 300 g of water) ore slurry was prepared. A first addition of reagents was made during this stage, specifically of lime (0.7 g) to regulate the pH to the desired value (pH 8). In addition, to determine the optimum grinding time, kinetic tests were carried out beforehand, allowing the grinding time to be set at 50 min.

2.3.2. Flotation Test

The flotation process was carried out in a Edemet cell of 1.5 L by adding a pulp containing 600 g of mineral and 1200 g of water, forming 33% of solids by weight, maintaining a pH 8, adding wash water after each palletization and dosing the collecting and foaming reagents according to the test to be carried out for the investigation. The reagents were added at the same time during conditioning stage, which lasted 5 min and was done at an agitation speed of 1100 rpm. At this stage, the pH value was also adjusted to the desired value by the addition of lime. Once this was achieved, the flotation stage took place, which lasted 16 min and was carried out at a stirring speed of 1000 rpm, injecting an airflow of 5 L/min maintaining a manual palletization every 10 s. All tests were done in duplicate with an average standard error of 5%.

Finally, once each flotation test was completed, both the concentrate and the process tail were dried. First, an Edemet pressure filter was used to remove the excess water and then the samples obtained were dried inside the drying oven for approximately 24 h at a temperature of 100 °C. Subsequently, the corresponding analyses were carried out using XRF to determine the content of Cu, Fe, S and Si in each sample and finally to calculate the metallurgical indexes for each test.

Rougher flotation tests were performed by evaluating the dosages of RVO or PAX collector to 20, 40, 60, 80 and 100 g/t. After that, mixtures of RVO/PAX at 80/20, 60/40, 40/60 and 20/80 g/t, respectively, were evaluated. Finally, to evaluate the effect of MIBC frother in the presence of RVO, the best dosages of collectors (PAX or RVO) or mixtures of them were kept constant, according to the metallurgical results, to evaluate the dosages of MIBC at concentration levels between 0 and 20 g/t.

2.3.3. Calculations and Methods of Analysis Used

- Metallurgical Indexes

The most important metallurgical indexes studied were the grade and the recovery of Cu and Fe in the concentrate, since these data provide an indication of the efficiency of the flotation. Similarly, enrichment ratio, selectivity indexes, concentration ratio and concentration yield were studied. Table 3 shows the formulas used to calculate these indexes.

Table 3. Formulas to calculate metallurgical performance [24].

Metallurgical Performance	Formula		Nomenclature
Recovery, %	$R = \frac{Cc}{Aa}$	(1)	C and A are the masses of concentrate and feed, respectively. c and a are the grades of the species of interest in the concentrate and in the feed, respectively.
Enrichment Ratio	$Re = \frac{c}{a}$	(2)	c and a are the grades of the species of interest in the concentrate and in the feed, respectively.
Selectivity	$IS = \sqrt{\frac{M*n}{m*N}}$	(3)	M and m are the grades of the species of interest in the concentrate and in the tailings, respectively. N and n are the grades of the gangue in the concentrate and in the tailings, respectively.
Concentration Ratio	$R_c = \frac{A}{C}$	(4)	C and A are the masses of concentrate and feed, respectively.
Concentration Efficiency	$V = \frac{C}{A} = \frac{1}{R_c}$	(5)	C and A are the masses of concentrate and feed, respectively.

- Applied Assumptions

To analyze the amount of chalcopyrite and pyrite present in the concentrates, the following mathematical relations were made based on the stoichiometric composition of chalcopyrite and pyrite, after having obtained the percentage of copper and iron in the concentrate by means of XRF.

This application is based on the assumption that all the copper present is associated with the chalcopyrite ore using a certain amount of iron. On the other hand, the remaining iron content is related to pyrite [18].

Equation (6) is used to estimate the percentage of chalcopyrite present in the concentrate.

$$\%Cpy = \%Cu\left(\frac{183.3}{63.5}\right) \tag{6}$$

Equation (7) refers to the percentage of iron present as pyrite as estimated above.

$$\%Fe^* = \%Fe - \%Cu\left(\frac{55.8}{63.5}\right) \tag{7}$$

Finally, Equation (8) represents the percentage of pyrite present in the concentrate.

$$\%Py = \%Fe^*\left(\frac{119.8}{55.8}\right) \tag{8}$$

3. Results and Discussion

3.1. Results of Collective Flotation Tests as a Function of PAX and RVO Concentration

Figure 1a,b shows the results obtained from the recovery and grade of copper and iron, as a function of PAX (Figure 1a) and RVO (Figure 1b) concentration. Figure 1a clearly shows that, in the absence of PAX, the lowest recovery and highest copper grade is obtained, which was to be expected due to the fact that under these conditions only the most hydrophobic particles of the ore, such as copper sulfides, should float. Subsequently, the addition of 20 g/t of PAX collector produces an increase in recovery because this reagent gives hydrophobic characteristics to the ore particles, achieving at 40 g/t the highest recovery and copper grade. Finally, at higher concentrations of this reagent, it can be seen that recovery and copper grade decrease considerably, suggesting an overdose of collecting reagent Accordingly, it should be noted that a 40 g/t concentration of PAX collector allows for the best copper recoveries and grades in seawater, under the operating conditions set forth above. On the other hand, in relation to the recovery and iron grades obtained at different PAX concentrations, a similar trend can be observed in the curves obtained with copper, suggesting that the iron present in the concentrate is mainly associated to chalcopyrite and pyrite minerals.

Figure 1b shows the results obtained for recovery and grade of copper and iron, as a function of the different studied RVO collector concentrations. In the figure, it can be seen that, in the case of copper, the recovery curve shows a significant increase when RVO is added to the process, regardless of its concentration, going from having a recovery of 58.8% in the absence of collector reagent to achieving an average recovery of close to 90% at the different RVO concentrations studied. Similarly, when analyzing the copper grade, it is observed that, at concentrations between 20 and 80 g/t of RVO collector, the highest copper grades are achieved, which is close to 9%, while, in the absence of collector and at a maximum concentration of 100 g/t, this grade decreases to values of 6.84% and 5.77%, respectively. On the other hand, with respect to the recovery and grade of iron, it is possible to note that, at 80 g/t, the highest recovery and grade was achieved with 65.7% and 29.42%, respectively.

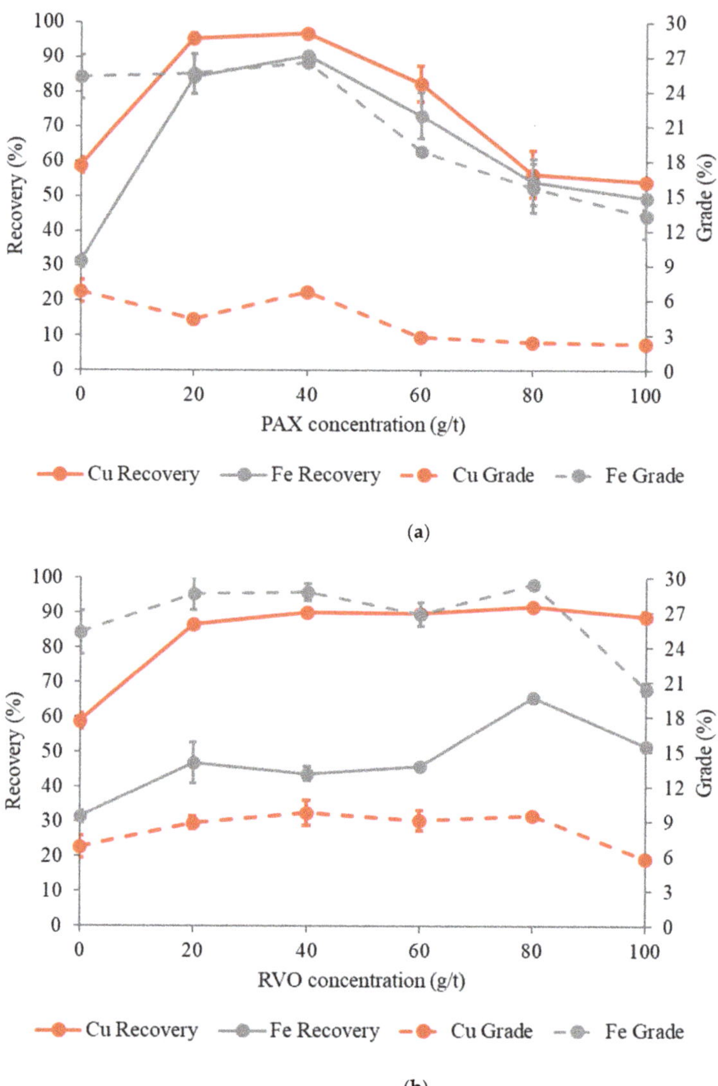

(a)

(b)

Figure 1. Recovery and grade of copper and iron obtained based on the PAX (**a**) and RVO (**b**) concentration, using 15 g/t of MIBC and seawater.

It is important to point out that, within the interval between 20 and 80 g/t of RVO, we have both the highest grades and the highest copper recoveries obtained with this collector, considerably improving the results obtained in its absence, which indicates that this reagent has a positive effect on the process and a great potential to be considered as a collector reagent in the flotation of copper sulfides with seawater.

Figure 2a,b shows the recovery and grade of both minerals (Cpy and Py) obtained as a function of the PAX and RVO concentration, respectively. In the figures, it can be seen that, for both minerals and collectors studied, there is a similar trend to the grades and recoveries obtained from each element, with the highest recoveries and grades of chalcopyrite at a concentration of 40 g/t. However, when comparing the results obtained in the presence of 40 g/t of PAX and RVO separately (Figure 2a,b), it is noteworthy that the chalcopyrite grades in the presence of RVO (28.2%) are significantly higher than those obtained with PAX (19.4%). On the other hand, it is also shown that pyrite grades remain between 40% and 45% and the increase in the concentration of the RVO collector does not significantly affect the grade of this mineral. On the basis of the above, it can be stated that the same amount of pyrite is concentrated in the presence of RVO as with PAX, but there is a considerable increase in the percentage of chalcopyrite. The above indicates that RVO adsorbs preferentially on copper particles and that in seawater this reagent delivers favorable results in terms of both recovery and copper grades.

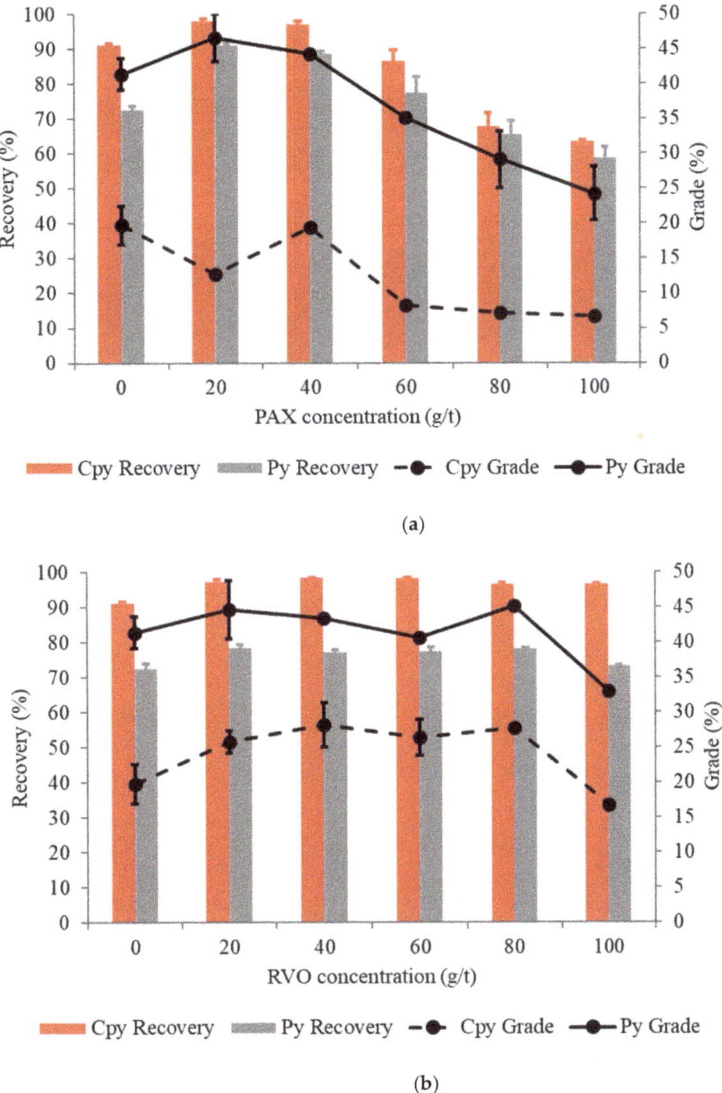

Figure 2. Recovery and grade of chalcopyrite and pyrite obtained based on the PAX (**a**) and RVO (**b**) concentration, using 15 g/t of MIBC and seawater.

Tables 4 and 5 show the metallurgical indexes calculated for the PAX and RVO collectors in flotation tests. Within these indexes, we find the recovery; grade; enrichment ratio; selectivity index with respect to copper–silica and chalcopyrite–pyrite, because these species are the main gangue that are required to be removed during the flotation process and the concentration ratio and yield.

Table 4. Metallurgical indexes obtained based on the PAX concentration, using 15 g/t of MIBC and seawater.

PAX Dosage (g/t)	R (%)	Cu Grade (%)	Re	IS (Cu-Si)	IS (Cpy-Py)	Rc	V
0	58.78	6.84	4.69	6.27	1.95	8.02	0.12
20	95.41	4.41	2.99	14.92	2.06	3.15	0.32
40	96.85	6.70	1.93	7.62	1.98	1.99	0.50
60	82.32	2.85	1.93	3.42	1.36	2.36	0.42
80	56.40	2.45	1.47	1.65	1.05	2.63	0.38
100	54.05	2.29	1.34	1.47	1.10	2.48	0.40

Table 5. Metallurgical indexes obtained based on the RVO concentration, using 15 g/t of MIBC and seawater.

RVO Dosage (g/t)	R (%)	Cu Grade (%)	Re	IS (Cu-Si)	IS (Cpy-Py)	Rc	V
0	58.78	6.84	4.69	6.27	1.95	8.02	0.12
20	86.50	8.91	5.62	11.14	3.09	6.53	0.16
40	90.06	9.74	6.29	12.89	4.03	7.01	0.14
60	89.75	9.10	5.87	11.05	3.78	6.56	0.15
80	91.74	9.55	3.17	8.19	2.78	3.46	0.29
100	88.80	5.77	3.77	6.59	3.08	4.26	0.24

From the results presented in Table 4, it can be seen that the best metallurgical indexes with respect to copper were achieved at 20 and 40 g/t of PAX collector, being 20 g/t the case in which the highest enrichment ratio, concentration ratio and selectivity of copper with respect to the main gangue, silica and pyrite and 40 g/t the case in which the highest recovery and grade of copper is achieved. With regard to the capacity indexes obtained at these conditions, it is noted that the highest concentration ratio occurs at 20 g/t and, consequently, the highest concentration yield at 40 g/t. On the other hand, in relation to the RVO collector from the results presented in Table 5, it can be seen that the best metallurgical indexes with respect to copper were achieved at 40 and 80 g/t of RVO collector, being 40 g/t the case in which the highest grade, enrichment ratio, concentration ratio and selectivity of copper with respect to the main gangue, silica and pyrite and 80 g/t the case in which the highest copper recovery is achieved.

Comparing the results obtained in the presence of PAX and RVO, it is possible to establish that, at 40 g/t of RVO collector, the flotation process achieves better metallurgical indexes than those obtained with PAX, in which it can be clearly indicated that the copper grade, the enrichment ratio, the selectivity indexes and the concentration ratio are better when using RVO than PAX, being the copper recovery the only index favored in the presence of PAX.

3.2. Results of Collective Flotation Tests Using Mixtures of RVO/PAX Collectors in Different Proportions

Tests were performed by mixing both collector reagents with the aim of studying possible configurations of the collectors that would have favorable effects on metallurgical indexes.

Figure 3 shows, in the same way as the previous cases, the recovery curves and copper and iron grades obtained as a function of the different RVO/PAX mixtures studied. In the figure, it can be seen that, in the case of copper, the recovery obtained for the 80/20, 60/40, 40/60 and 20/80 reagent mixtures is approximately constant, reaching a recovery close to 98%. Meanwhile, in cases where separate reagents were used, it can be seen that the recovery of copper is lower with 100 g/t of PAX than when using the same concentration of RVO, with 88.80% and 54.05%, respectively. By analyzing the copper grade, it can be seen that the highest grades are obtained in the 80/20 and 60/40 RVO/PAX blends, which corresponded to 5.51% and 7.40%.

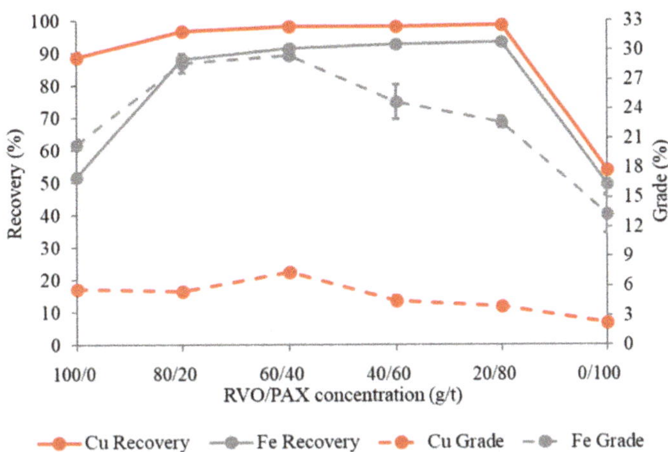

Figure 3. Recovery and grade of copper and iron obtained based on the RVO/PAX concentration, using 15 g/t of MIBC and seawater.

On the other hand, with respect to the recovery and iron grades obtained as a function of the RVO/PAX mixtures, it can be seen that the curves show a behavior similar to that of copper, reaching a recovery for the 80/20, 60/40, 40/60 and 20/80 cases of approximately 90%, and, with respect to the iron grade, it can be seen that the highest grades of this element are obtained in the 80/20 and 60/40 RVO/PAX mixtures, which were close to 30%.

When comparing the results of the recoveries and grades obtained in the mixtures with those obtained only with RVO (Figure 1b), it can be seen that, in the mixture, the presence of PAX favors the recovery of copper obtained with RVO but affects the copper grade obtained in the concentrate.

With respect to the mineralogy associated with the process, Figure 4 shows the recovery and grade of chalcopyrite and pyrite obtained for each mixture according to the assumptions indicated above. In the figure, it can be seen that the highest recovery and chalcopyrite grade was obtained with an RVO/PAX mix of 60/40, obtaining 98.6% and 21.4%, respectively.

Comparing the results obtained with the mixtures against those obtained only with RVO (Figures 2 and 4), it can be observed that the presence of PAX generates a slight increase in the recovery of chalcopyrite but a decrease in the grade of this mineral and, in the same way, significantly increases the presence of pyrite in the concentrates, being the mixture RVO/PAX 60/40 the one that presents the best results.

Table 6 presents the metallurgical indexes calculated for the flotation test set performed. It shows that, in general, the best metallurgical indexes with respect to copper were achieved with the RVO/PAX 60/40 mixture, especially in terms of process selectivity.

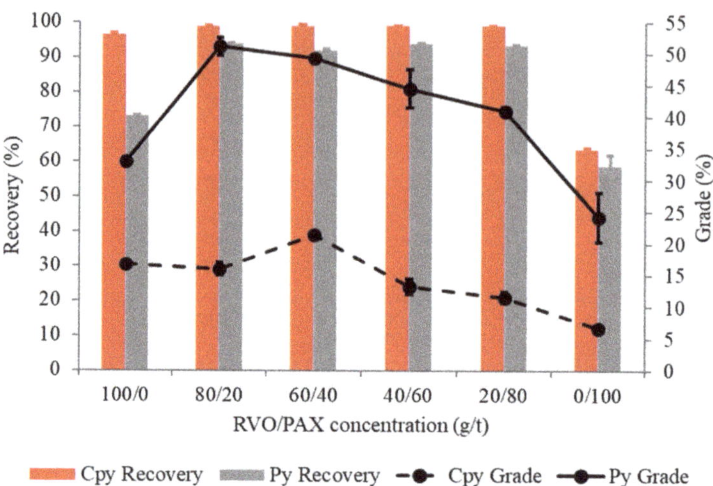

Figure 4. Recovery and grade of chalcopyrite and pyrite obtained based on the RVO/PAX concentration, using 15 g/t of MIBC and seawater.

Table 6. Metallurgical indexes obtained based on the RVO/PAX concentration, using 15 g/t of MIBC and seawater.

RVO/PAX Dosage (g/t)	R (%)	Cu Grade (%)	Re	IS (Cu-Si)	IS (Cpy-Py)	Rc	V
100/0	88.80	5.77	3.77	6.59	3.08	4.26	0.24
80/20	96.94	5.51	3.22	13.80	2.24	3.33	0.30
60/40	98.24	7.40	2.25	13.78	2.53	2.29	0.44
40/60	98.41	4.56	2.36	13.38	2.37	2.41	0.42
20/80	98.75	4.00	2.11	12.91	2.53	2.14	0.47
0/100	54.05	2.29	1.34	1.47	1.10	2.48	0.40

When comparing the results under this condition (Table 6) with the best results obtained with the separate action of each collector (Tables 4 and 5), it can be established that the use of reagent mixtures improves the copper recovery compared to that obtained with both collectors separately, while the grade and the enrichment ratio were improved when compared with PAX alone. On the other hand, compared to the selectivity indexes with respect to the main gangue, it can be seen that the selectivity of copper, with respect to silica, increased when using mixtures, while, with respect to pyrite, it was only increased when comparing it with PAX. Finally, in. relation to the concentration ratio, the use of mixtures increased this index with respect to the value obtained with PAX, but, when compared with those obtained with RVO alone, these values are significantly higher, generating the concentrates with the lowest amount of mass in the presence of RVO.

In general, the reagent mixtures generated a positive effect, achieving 98.24% and 7.40% maximum copper recovery and grade, respectively, when using an RVO/PAX 60/40 mixture. Although it is true that the grade is higher than that obtained with PAX and lower compared with the cases of RVO, the results obtained with this mixture with respect to the other metallurgical indexes are attractive considering that the studied process corresponds to a stage of collective flotation.

3.3. Seawater Flotation Results as a Function of MIBC Concentration

3.3.1. Results of Collective Flotation Tests as a Function of MIBC Concentration in Presence of 40 g/t PAX and RVO, Separately

Figure 5 presents the results of copper and iron recovery and grade obtained as a function of MIBC concentration. In the figure, it is possible to observe that the recovery curves for both elements showed a similar trend and a constant behavior when varying the MIBC concentration, while, in the case of copper and iron grades, they increase as the concentration of the foaming agent increases. A minimum recovery of copper in the absence of foaming agent is obtained, corresponding to 96.6% and a maximum of 97.8% at 20 g/t MIBC. On the other hand, in relation to the recovery of iron, a behavior similar to that of copper is obtained, with a minimum recovery of 89.7% in the absence of foaming agent and a maximum recovery of 90.8% using 10 g/t of MIBC. With respect to the grades, in the case of copper, the lowest value was 5.43% and occurs at 5 g/t MIBC, from which it begins to increase progressively and slightly depending on the concentration of foaming agent, reaching the maximum value of 7.23% at 20 g/t. On the other hand, when analyzing the iron grade obtained in the concentrate, it is emphasized that there is a similar behavior to the previous one, where the lowest grade obtained for iron was 23% at 5 g/t MIBC, while the highest was 28.7% at 20 g/t foaming agent. In this way, and according to the previous evidence, it can be established that for this test with 40 g/t of PAX collector, the best results in terms of grade and copper recovery are obtained with 20 g/t of MIBC foaming agent.

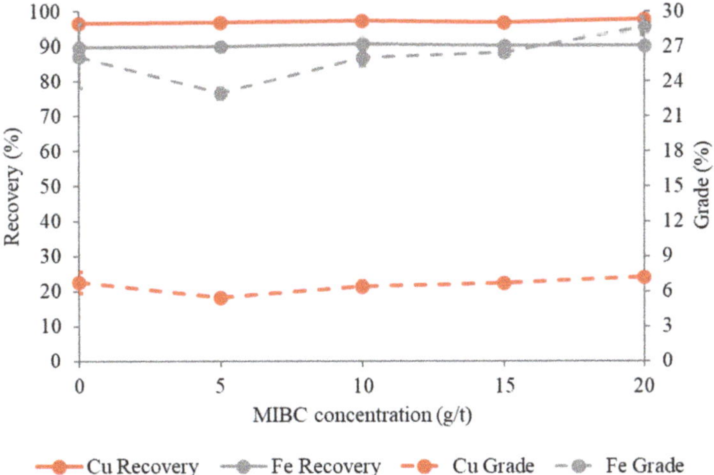

Figure 5. Recovery and grade of copper and iron obtained based on the MIBC concentration, using 40 g/t PAX and seawater.

Figure 6 shows the recovery and grade curves of copper and iron obtained at the different concentrations of foaming agent for a 40 g/t RVO concentration. In the figure, it is possible to emphasize that the recovery of copper has a stable behavior that oscillates between 87% and 90% in the interval that goes from 0 to 15 g/t, presenting the minor recovery at 20 g/t, which was of 84.7%. On the other hand, in relation to the recovery of iron, it can be seen that the curve obtained showed a behavior similar to that of copper, i.e., recovery remains stable at around 56%, while at 15 g/t, unlike the case of copper which slightly increased its recovery, for iron it decreases markedly to around 44% approximately. In relation to the grades obtained as a function of the MIBC concentration, in the case of copper, it can be observed that in the absence of foaming agent a grade of 9.74% is achieved, which then, by adding low concentrations of foaming agents, corresponding to 5 and 10 g/t, decreases to values of 8.33% and 7.92%, respectively. However, at higher concentrations, corresponding to 15 and 20 g/t, a considerable increase is achieved, reaching values of 9.74% and 9.51%, respectively. With respect to the iron grade, a similar behavior can be seen to that obtained in the case of copper which at 0 g/t has a value of 28.37%; at concentrations of 5 and 10 g/t decreases to 27.79% and 27.26%, respectively; and at higher concentrations the grade is reversed, obtaining the highest value at 15 g/t of MIBC, corresponding to 28.79%.

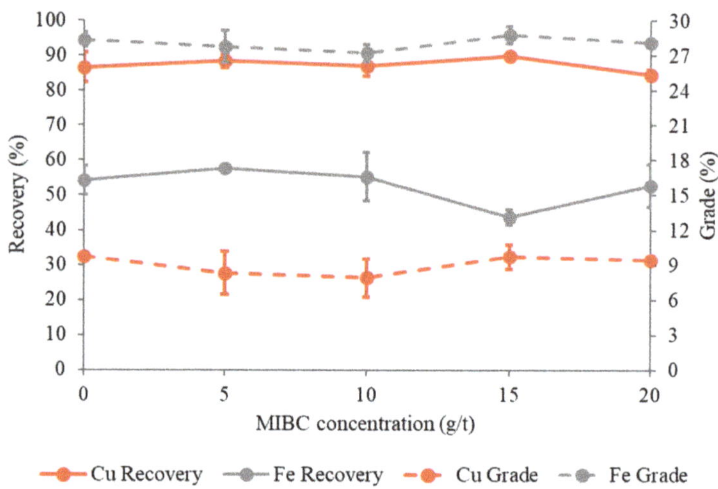

Figure 6. Recovery and grade of copper and iron obtained based on the MIBC concentration, using 40 g/t RVO and seawater.

On the other hand, Figure 7a,b shows the percentages of chalcopyrite and pyrite obtained as a function of the concentration of foaming agent. It shows a similar behavior to that obtained with copper and iron grades, where, as the concentration of foaming agent increases, the grades of both minerals increase, the highest grades being obtained at a dose of 20 g/t MIBC, 20.9% for chalcopyrite. With regard to the recovery of these minerals, this index shows little variation depending on the concentration of the foaming agent, reaching a maximum value of 98.3% at 20 g/t of MIBC, thus improving the results obtained before 15 g/t.

In addition, Figure 7b shows the recovery and grade of chalcopyrite and pyrite obtained in the presence of RVO. It can be seen that chalcopyrite is directly related to copper grade, so at 0 and 15 g/t it has its highest grade with a value of 28.2%. On the other hand, it is observed that the pyrite grade in the concentrate remains practically constant between 42% and 43%. With regard to the recoveries of both minerals, it is observed that they remain practically constant throughout the range of foaming agent concentrations studied, obtaining the highest recovery of chalcopyrite at 15 g/t MIBC with a value of 98.2%, while the lowest recovery of pyrite is reached at 20 g/t.

Table 7 presents the metallurgical indexes obtained as a function of the concentration of foaming agent. It shows that, at 20 g/t MIBC, the best metallurgical indexes are indeed obtained. It should be pointed out that the Cu-Si selectivity index increases considerably in the presence of this dose of MIBC, reaching 11.99, as well as for the Cpy-Py case, where a value of 2.43 is obtained. Finally, with respect to the capacity indexes, the concentration ratio is higher in comparison to the other MIBC concentrations, indicating that the concentrate obtained in the process has a lower mass quantity, making the process more efficient in both collective and selective terms. Consequently, the lowest concentration yield is obtained of all the conditions studied, confirming the fact of having a lower mass quantity in the concentrate with respect to the process feed. On the other hand, in relation to the RVO collector from the results presented in Table 8, it is observed that the best indexes are obtained at a concentration of 15 g/t, highlighting both the enrichment ratio of 6.29, the selectivity index Cu-Si of 12.89 and Cpy-Py of 4.03, as well as the concentration ratio of

7.01, indicating a high selective capacity in the process in this condition of operation and dosage of reagents in the presence of seawater.

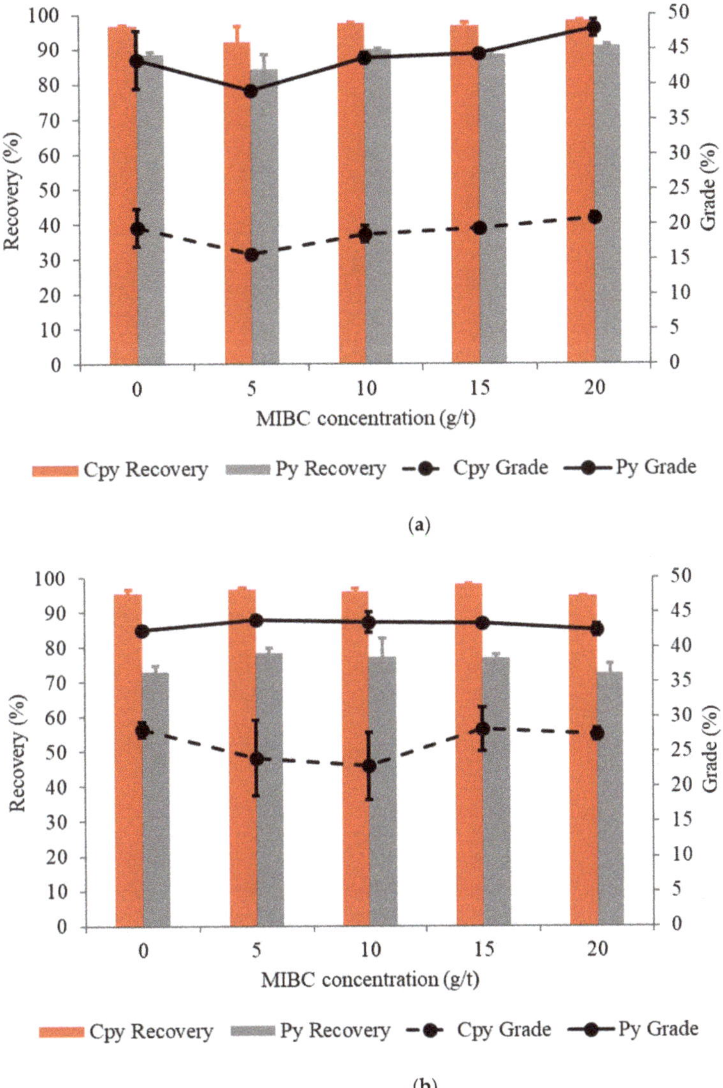

Figure 7. Recovery and grade of chalcopyrite and pyrite obtained based on the MIBC concentration, using 40 g/t PAX (**a**) and RVO (**b**), separately, and seawater.

Table 7. Metallurgical indexes obtained based on the MIBC concentration, using 40 g/t PAX and seawater.

MIBC Dosage (g/t)	R (%)	Cu Grade (%)	Re	IS (Cu-Si)	IS (Cpy-Py)	Rc	V
0	96.62	6.74	1.99	7.77	1.99	2.06	0.49
5	90.98	5.42	1.92	5.50	1.48	2.13	0.47
10	97.44	6.42	2.04	9.24	2.16	2.10	0.48
15	96.85	6.70	1.93	7.62	1.98	1.99	0.50
20	97.84	7.23	2.27	11.99	2.43	2.33	0.43

Table 8. Metallurgical indexes obtained based on the MIBC concentration, using 40 g/t RVO and seawater.

MIBC Dosage (g/t)	R (%)	Cu Grade (%)	Re	IS (Cu-Si)	IS (Cpy-Py)	Rc	V
0	86.76	9.74	3.56	8.06	2.76	8.02	0.12
5	88.61	8.33	3.95	8.59	2.78	4.48	0.22
10	87.10	7.92	4.00	8.10	2.63	4.61	0.22
15	90.06	9.74	6.29	12.89	4.03	7.01	0.14
20	84.70	9.51	3.56	7.11	2.58	4.23	0.24

In the same context, the selectivity index of chalcopyrite on pyrite obtained at 15 g/t MIBC stands out due to its high value, which was 4. According to the literature, this represents a good metallurgical result for the flotation process, which, integrated with the rest of the indexes studied, makes this case a good operating condition in terms of feasibility for its use in industrial processes. Finally, the concentration yield is highlighted which indicates that, of the total mass fed to the flotation, only 14% is transferred to the concentrate stream, suggesting high selectivity for the process.

3.3.2. Results of Collective Flotation Tests as a Function of MIBC Concentration in Presence of RVO/PAX 60/40 Mixture

Figure 8 presents the results of copper and iron recovery and grade obtained as a function of MIBC concentration using a 60/40 g/t RVO/PAX mixture as collector. In general, it is possible to observe that all the curves obtained have a practically stable behavior as a function of the MIBC concentration. With regard to copper recovery, it can be seen that its behavior is homogeneous throughout the whole range of frother agent concentration, i.e., it remains constant with a value close to 98%. On the other hand, the recovery of iron also behaves in a stable way, maintaining a recovery that ranges between 90.78% and 92.4%.

With respect to the grades obtained, it can be observed that, with respect to the copper grade, it remains stable with a value close to 7%, reaching its maximum value in the absence of MIBC corresponding to 7.58% of copper in the concentrate and a minimum value at 5 g/t corresponding to 7.05%. On the other hand, analysis of the iron grade may indicate that it exhibits constant behavior as a function of the MIBC concentration, the maximum and minimum iron grade being obtained at 15 and 5 g/t, respectively, with values of 29.49% and 28.80% for each of the aforementioned doses.

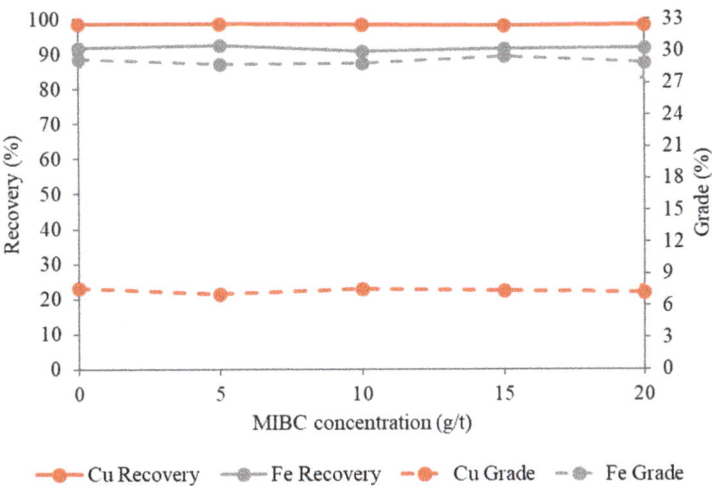

Figure 8. Recovery and grade of copper and iron obtained based on the MIBC concentration, using RVO/PAX 60/40 and seawater.

Figure 9 shows the grade and recovery of chalcopyrite and pyrite obtained at the different concentrations of foaming agent studied. It is possible to observe in the figure that the highest chalcopyrite content, as well as the highest copper grade, is achieved in the absence of foaming agent and corresponds to 21.9% together with 48.5% pyrite. Meanwhile, the lowest grade of the latter ore was 47.7% and is reached at a concentration of 10 g/t MIBC, which coincides with one of the lowest values for iron grade. With regard to the recovery of these minerals, in general, it is maintained without great alterations depending on the concentration of the foaming agent; however, the maximum recoveries of chalcopyrite are reached at 0 and 5 g/t with values of approximately 98.9%, while the maximum recovery of pyrite occurs at 5 g/t with a value of 92.5%.

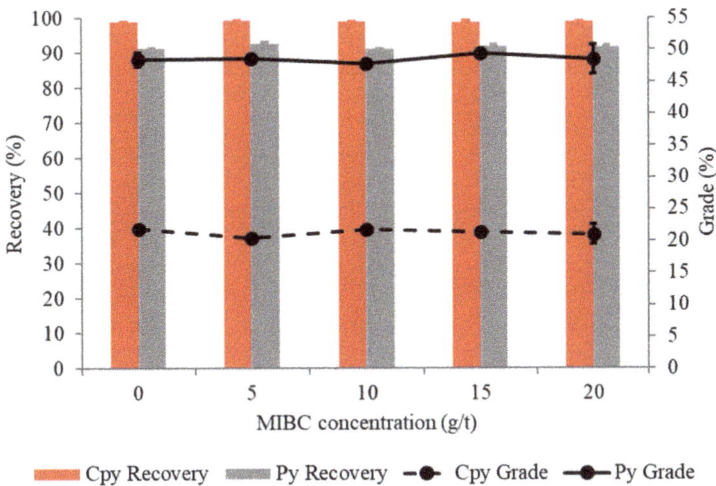

Figure 9. Recovery and grade of chalcopyrite and pyrite obtained based on the MIBC concentration, using RVO/PAX 60/40 and seawater.

Table 9 presents the metallurgical indexes obtained according to the different concentrations of foaming agent. It shows that, with respect to copper recovery and grade, there is a very slight variation between each test depending on the concentration of the foaming agent and the highest result is obtained in the absence of MIBC. Compared to the enrichment ratio, a similar behavior is observed, with the highest enrichment at 15 g/t MIBC. On the other hand, compared to the selectivity indexes, the Cu-Si selectivity index presented its highest values at 5 g/t of frother, while for the Cpy-Py case the highest value was obtained in the absence of MIBC. Finally, in relation to the capacity indexes, it is observed that the highest concentration ratio was achieved in the absence of foaming agent along with the best performance.

Table 9. Metallurgical indexes obtained based on the MIBC concentration, using RVO/PAX 60/40 and seawater.

MIBC Dosage (g/t)	R (%)	Cu Grade (%)	Re	IS (Cu-Si)	IS (Cpy-Py)	Rc	V
0	98.66	7.58	2.14	14.65	2.88	4.26	0.24
5	98.67	7.05	2.22	15.06	2.74	2.26	0.44
10	98.40	7.50	2.23	13.75	2.79	2.27	0.44
15	98.24	7.40	2.25	13.78	2.53	2.29	0.44
20	98.43	7.25	2.16	12.85	2.64	2.20	0.45

From the results obtained for the different concentrations of MIBC for the best separate collectors (Tables 7 and 8) and the mixture of these (Table 9), it can be seen that, except for recovery, the best metallurgical indexes obtained are obtained when using 40 g/t of RVO and 15 g/t of MIBC. On the other hand, the use of RVO/PAX mixtures allows the flotation process to be carried out in the absence of foaming agent, achieving even better metallurgical indexes than those obtained with conventional PAX collectors. This phenomenon could be explained by the presence of RVO in the process, which could play an additional role as frother, given its physical and chemical properties. In this aspect, it is necessary to emphasize that the use of mixtures achieves a great performance in collective terms, achieving the highest recovery of all the conditions studied and the highest selectivity with respect to the Cu-Si case, as well as a mass reduction of approximately four times of the feed to the concentrate stream.

4. Conclusions

This research proposed to study an RVO as a collector of copper sulfide minerals in the flotation process in seawater in order to contribute to reducing the toxic collector used in minerals processing and to find a different type of application of this waste. Based on the metallurgical analysis, it is possible to note that the RVO collector was more selective for chalcopyrite than PAX but gives a lower recovery. In addition, by analyzing the mixture of both collectors, in the absence of frother, the best results in collective terms and selectivity were obtained. According to these preliminary results, it is possible to conclude that the addition of RVO improved the selectivity of the conventional collector PAX, which is highly collective, in the rougher flotation of copper sulfide minerals in seawater. This could be an interesting opportunity for the industry in order to minimize the costs of the flotation process and generate a lower environmental impact and significant health advantages since it could help to reduce the use of toxic xanthates. However, future research is needed in order to find the best method of addition of the RVO to the process, to study new mixtures with other collectors and to analyze the frother characteristics. In addition, it could be important to understand the effect of the presence of RVO in subproducts as molybdenite.

Author Contributions: Conceptualization, methodology, data curation, visualization, validation, formal analysis, investigation, writing-original draft preparation writing-review and editing, F.A and L.U. All authors have read and agreed to the published version of the manuscript.

Funding: This research received no external funding.

Data Availability Statement: The data presented in this study are available on request from the corresponding author.

Acknowledgments: The authors wish to thank the support from the ANID/FONDAP/15130015 project.

Conflicts of Interest: The authors declare not conflict of interest.

References

1. COCHILCO. *Sulfuros Primarios: Desafíos y Oportunidades*; COCHILCO: Santiago, Chile, 2017.
2. Rezaei, R. Removal of the residual xanthate from flotation plant tailings using modified bentonite. *Miner. Eng.* **2018**, *119*, 1–10. [CrossRef]
3. Li, X.F. Teratogenic toxicity of butyl xanthate to frog embryos. *Acta Scient. Circum.* **1990**, *10*, 213–216.
4. Lam, K.S. Biodegradation of Xanthate by Microbes Isolated from a Tailings Lagoon and a Potential Role for Biofilm and Plant/Microbe Associations. Ph.D. Thesis, Western Sydney University, Penrith, Australia, 1999.
5. Zhao, Y.H.; Xie, M.H.; Luo, X.P. Investigation on removal of xanthate in flotation effluent. *Metal Min.* **2006**, *6*, 75–77.
6. Boening, D.W. Aquatic toxicity and environmental fate of xanthates. *Miner. Eng.* **1998**, *50*, 65–68.
7. Harris, G.H. Xanthates. In *Kirk-Othmer Encyclopedia of Chemical Technology*; John Wiley & Sons: Hoboken, NJ, USA, 2000. [CrossRef]
8. Wang, B.; Peng, Y. The effect of saline water on mineral flotation—A critical review. *Miner. Eng.* **2014**, *66–68*, 13–24. [CrossRef]
9. Moreno, P.A.; Aral, H.; Cuevas, J.; Monardes, A.; Adaro, M.; Norgate, T.; Bruckard, W. The use of seawater as process water at Las Luces copper-molybdenum beneficiation plant in Taltal (Chile). *Miner. Eng.* **2011**, *24*, 852–858. [CrossRef]
10. Wang, B.; Peng, Y. The behaviour of minerals matter in fine coal flotation using saline water. *Fuel* **2013**, *109*, 309–315. [CrossRef]
11. Morales, F. Estudio del Efecto de las Interacciones del Sistema 'agua de mar–cal' en Procesamiento de Minerales. Master's Thesis, Universidad de Chile, Santiago, Chile, 2017.
12. Minería Chilena. Available online: https://www.mch.cl/reportajes/el-combate-contra-la-corrosion/# (accessed on 15 October 2019).
13. Marulanda, J.; Pérez, D.; Remolina-Millán, A. Resistencia a la corrosión en ambiente salino de un acero al carbono recubierto con aluminio por rociado térmico y pintura poli aspártica. *Rev. Ion Investig. Optim. Nuevos Procesos Ing.* **2017**, *30*, 21–31.
14. Uribe, L. Efecto del Agua de mar en la Recuperación de Minerales de Cobre-Molibdeno por Procesos de Flotación. Ph.D. Thesis, Universidad de Concepción, Concepción, Chile, 2017.
15. Castro, S. Challenges in flotation of Cu-Mo Sulfide ores in seawater. In *Water in Mineral Processing*; Society for Mining, Metallurgy, and Exploration: Englewood, CO, USA, 2012; pp. 29–40.
16. Laskowski, J.S.; Castro, S.; Ramos, O. Effect of seawater main components on frothability in the flotation of Cu-Mo sulfide ore. *Physicochem. Probl. Miner. Process.* **2013**, *50*, 17–29.
17. Williams, C.; Peng, Y.; Dunne, R. Eucalyptus oils as green collectors in gold flotation. *Miner. Eng.* **2013**, *42*, 62–67. [CrossRef]
18. Owusu, C. The use of canola oil as an environmentally friendly flotation collector in sulphide mineral processing. *Miner. Eng.* **2016**, *98*, 127–136. [CrossRef]
19. Brandao, P.R.G.; Caires, L.G.; Queiroz, D.S.B. Vegetable lipid oil-based collectors in the flotation of apatite ores. *Miner. Eng.* **1994**, *7*, 917–925. [CrossRef]
20. Bauer, K.; Greene, M.G.; Di Reber, N.R.; Young, S.K.; Young, T.L. Flotation of Sulfide Minerals with Oils. Australia. Patent WO 2000 009 268 A1, 24 February 2000.
21. Greene, M.G.; Walton, K.; Dimas, P.A.; Laney, D.G.; Young, S.K.; Young, T.L.; Reber, N.R.; Reber, N.R., Jr. Collectors for Flotation of Molybdenum-Containing Ores. U.S. Patent Application No. 0,145,605A1, 14 June 2012.
22. Benn, F.W.; Dattilo, M.; Cornell, W.L. Flotation of Lead Sulfides Using Rapeseed Oil. U.S. Patent 5,544,760, 13 August 1996.
23. Montalti, M.; Fornasiero, D.; Ralston, J. Ultraviolet-visible spectroscopic study of the kinetics of adsorption of ethyl xanthate on pyrite. *J. Colloid Interface Sci.* **1991**, *143*, 440–450. [CrossRef]
24. Gaudin, A.M. *Flotation*; McGraw-Hill: New York, NY, USA, 1957.

recycling

Review

Prospects of Red King Crab Hepatopancreas Processing: Fundamental and Applied Biochemistry

Tatyana Ponomareva, Maria Timchenko , Michael Filippov, Sergey Lapaev and Evgeny Sogorin *

Federal Research Center "Pushchino Scientific Center for Biological Research of the RAS", 142290 Pushchino, Russia; tatyanap91875@gmail.com (T.P.); matimchenko@gmail.com (M.T.); filippmv@gmail.com (M.F.); a3t1v4i1n5@gmail.com (S.L.)
* Correspondence: evgenysogorin@gmail.com; Tel.: +7-915-132-5419

Abstract: Since the early 1980s, a large number of studies on enzymes from the red king crab hepatopancreas were conducted. They have been relevant both from a fundamental point of view in terms of studying the enzymes of marine organisms and in terms of rational natural resource management aimed to obtain new valuable products from the processing of crab fishing waste. Most of these works were performed by Russian scientists due to the area and amount of waste of red king crab processing in Russia (or the Soviet Union). However, the close phylogenetic kinship and the similar ecological niches of commercial crab species and the production scale of the catch provide the bases for the successful transfer of experience in the processing of the red king crab hepatopancreas to other commercial crab species caught worldwide. This review describes the value of recycled commercial crab species, discusses processing problems, and suggests possible solutions for these issues. The main emphasis is made on hepatopancreatic enzymes as the most salubrious products of red king crab waste processing.

Keywords: commercial crab species; red king crab; waste processing; hepatopancreas; proteases; hyaluronidase

Citation: Ponomareva, T.; Timchenko, M.; Filippov, M.; Lapaev, S.; Sogorin, E. Prospects of Red King Crab Hepatopancreas Processing: Fundamental and Applied Biochemistry. *Recycling* **2021**, *6*, 3. https://dx.doi.org/10.3390/recycling6010003

Received: 19 October 2020
Accepted: 29 December 2020
Published: 2 January 2021

Publisher's Note: MDPI stays neutral with regard to jurisdictional claims in published maps and institutional affiliations.

Copyright: © 2021 by the authors. Licensee MDPI, Basel, Switzerland. This article is an open access article distributed under the terms and conditions of the Creative Commons Attribution (CC BY) license (https://creativecommons.org/licenses/by/4.0/).

1. Introduction

The global growth in consumer demand for food products based on commercial species of marine organisms (fish, crab, squid, etc.) has stimulated fishers to increase production. However, the inability to increase the catch volume endlessly and related waste recycling problems have put forward the consideration of advanced processing of secondary raw materials to obtain new commercially valuable products. Crabs are a favorite catch of fishers worldwide due to the high price of their meat. The main northern commercial species of crab include the following: Red king crab (*Paralithodes camtschaticus*), Blue king crab (*Paralithodes platypus*), Spiny brown king crab (*Paralithodes brevipes*), Golden king crab (*Lithodes aequispinu*), Opilio snow crab (*Chionoecetes opilio*), Tanner snow crab (*Chionoecetes bairdi*), Triangle tanner crab (*Chionoecetes angulatus*), Red snow crab (*Chionoecetes japonicus*), Chinese mitten crab (*Eriocheir sinensis*), Hair crab (*Erimacrus isenbeckii*) [1]. The first four species belong to the infraorder of half-tailed or false crabs (Anomura) of the suborder Pleocyemata, order Decapoda. The rest of them belongs to the infraorder of true crabs (Brachyura) (Figure 1) [2].

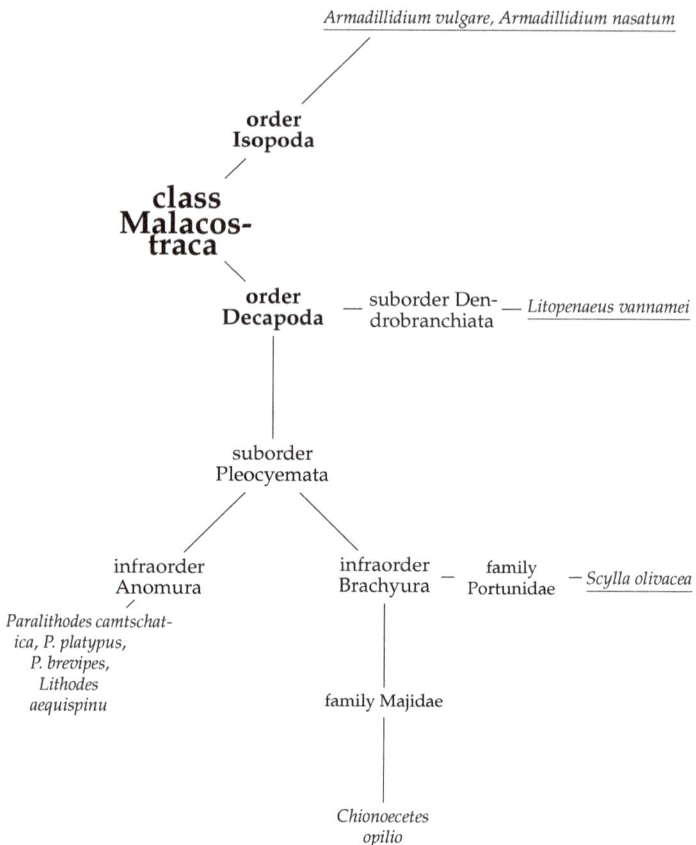

Figure 1. Phylogeny of some members of the class Malacostraca. Species with known primary structures of hyaluronidases are underlined (see Section 3.4.3).

All crabs have a massive cephalothorax covered with carapace from above, a flat abdomen, and are bent under the cephalothorax. Brachyura representatives move with the help of four pairs of pectoral legs, and the fifth pair of limbs are claws. A specific feature of the Anomura is the asymmetric structure of the body (the right claw is larger than the left) and the presence of only three pairs of walking legs (one of the five pairs is hidden under the carapace and is used for the regular cleaning of gills) [1,3–5].

Red king crab, or Kamchatka crab, is famous around the world. In recent decades, the catch of this crab by Russian fishers has reached 15,000–20,000 tonnes per year. The original habitat of red king crab ranges from Karaginsky Island (Russia) off the east coast of Kamchatka and Shelikhov Bay (Russia) in the Sea of Okhotsk to Hokkaido Island (Japan) and the Korea Strait (between Korea and Japan). The crab is also found on the west coast of North America from Cape Barrow to the Queen Charlotte Archipelago in the south [1,6].

Red king crab was successfully introduced to the Barents Sea in the 1960s and 1970s. Optimal temperature conditions, the absence of natural predators, and sufficient amount of food led to the spread of acclimatized red king crab from the coast of the Kola Peninsula to Norway and north to Svalbard (Figure 2). Individuals of this king crab population are larger, grow faster, and mature earlier than individuals of the far-eastern population [7]. The rapid growth of the northern king crab population is an environmental problem [7,8]. The growth rate of the resource potential of this population made it possible to open commercial fishing in Norway in 2002 and in Russia in 2004.

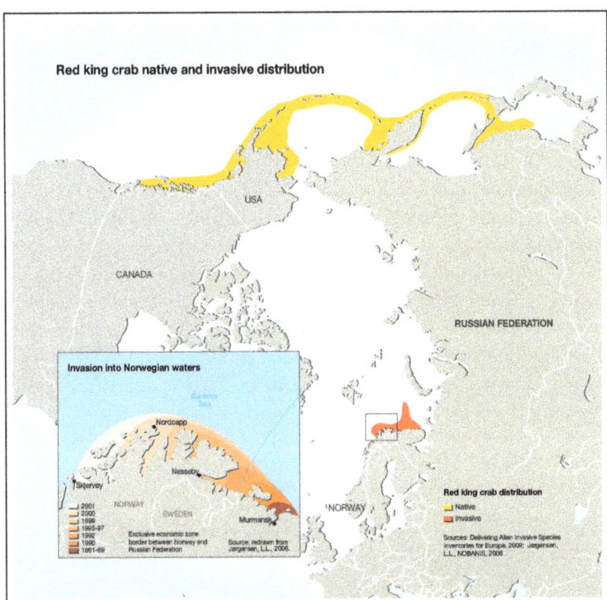

Figure 2. Red king crab native and invasive distribution.

In Russia, the crab catch is often processed immediately on board. One of the methods for processing the crab on a ship is as follows: the just-caught crab is placed on a hook; the limbs are additionally cleaned, boiled, and frozen; and the broken carapace with entrails is dumped immediately into the sea as a waste product. On average, this waste can account for 50% of the catch mass [9,10]. The mass fraction of carapace from these wastes is approximately 60%; the rest comprises the entrails (including the digestive organ, the hepatopancreas) [6]. In terms of protein content, these wastes are identical to crab meat, and even superior in terms of the content of minerals, lipids, and carbohydrates. Crab flour used as animal feed can be obtained from crab processing waste. In addition, the crab shell is an excellent source of raw materials for the production of chitin and chitosan, which can be used to meet the demands of the food industry and medicine [10]. Chitin, chitosan, and their derivatives represent promising matrices for the development of novel biomaterials with antioxidant, bactericidal, hypotensive, antiallergic, antiinflammatory and antitumor activities [11]. Recently, crab shell has also been used for the production of α-glucosidase inhibitors and anticancer agent prodigiosin via microbial fermentation [12,13]. Due to their peculiar properties such as nontoxicity, biocompatibility, and biodegradability, chitin and chitosan are used as potential excipients and as biological active agents in cosmetology [14].

The main focus of the review is on the research and use of the crab hepatopancreas enzyme complex due to the strong industry interest in this enzyme source. Russia is the leader in the catch of this fishing object. This is reflected in the number of studies on the processing of this waste by Russian scientists. Most of these works were published only in Russian, which significantly limited the availability of information on that research results and development for the world community. The goal of this review is to solve this problem.

2. Materials and Methods

The review has been systematically approached based on electronic databases including Scopus, Web of Science, Google Scholar. Most of the resources are the articles of highly specialized journals without translation and abstracts of regional symposia and conferences specialized in the water resources of Russia (Soviet Union), which are not available in electronic form.

Figure "Red king crab native and invasive distribution" was generously provided by GRID-Arendal (nonprofit environmental communications centre based in Norway). This graphic item may be reproduced in any form for educational or nonprofit purposes without special permission from GRID-Arendal, provided acknowledgement of the source is made [15,16].

The chapter "Technologies for processing hepatopancreas" presents patented technologies for processing waste from red king crab and other closely related commercial crabs. This information corresponds to the study of the technical level of patent research on this topic.

Experimental data from research articles were used to compile tables. Among other things, the molecular masses of the proteins under study and the methods used in these works are indicated. The table shows the scatter of these data from different works in order to demonstrate the measurement by different researchers. The correct molecular weight should be estimated by the results of mass spectrometry and interpretation of the results or calculations for the amino acid sequence of proteins.

3. Results

3.1. Crab Hepatopancreas

The hepatopancreas is an organ of the digestive system that functions as both the liver and pancreas [17]. In red king crab, the hepatopancreas makes up about 90% of the intestines of the carapace and 5–10% of the total weight of the animal [18]. The Decapoda hepatopancreas secretes a wide variety of highly active enzymes.

Under the action of the digestive enzymes of the hepatopancreas, food is broken down into easily digestible substances. The hepatopancreas of crabs of the family Lithodidae (infraorder Anomura, for example, red king crab) consists of brown mass of fragile liver tubules filling the main part of the body cavity. The integrity of these tubules is easily destroyed, even by a slight mechanical action, and the enzymes enter the body cavity, where they start the process of autolysis. The hepatopancreas of true crabs (Brachyura, for example, opilio crab) is shapeless orange-brown mass [1].

3.2. Use of Red King Crab Hepatopancreatic Enzymes

The hepatopancreas of the digestive system of commercial crabs is a valuable source of a complex of enzymes with various activities: collagenase, protease, hyaluronidase, lipase, nuclease, etc. The complex of proteolytic enzymes of the red king crab hepatopancreas is of interest in various industries. For example, the prospect of using hepatopancreas enzyme preparation in the hydrolysis of soy protein was recently highlighted [19]. A red king crab hepatopancreas enzyme preparation was successfully used to separate roe from the connective tissue of ovaries of commercial fish [20], to hydrolyze minced fish meat to obtain a dietary food ingredient [21], to hydrolyze crustacean processing waste products to obtain components for microbiological nutrient media [22], and to isolate chondroitin sulfate from marine wastes, namely from tissues of marine organisms [23]. Based on collagenolytic proteinases, wound healing and wound cleansing preparations [24–26], including wound dressings [27,28], have been designed.

3.3. Hepatopancreas Recycling Technologies

Many technologies are used to process the hepatopancreas of commercial crab species to obtain enzymatic preparations. Most technologies were developed for the hepatopancreas of red king crab, but some were transferred to the processing of the hepatopancreas from other commercial species (for example, snow crab and blue crab) [29,30]. Most often, the initial raw material is the hepatopancreas (Figure 3), which is homogenized in salt buffers or by osmotic shock under hypotonic conditions (excess distilled water). The integrity of the hepatopancreas tissue is easily destroyed during the freezing/thawing process; therefore, no considerable effort is required for mechanical homogenization. However, in some cases, a colloid mill can be used [31]. Sometimes, triton X-100 or sodium dodecyl sulfate (SDS) is added to the homogenization buffer [29,32]. The homogenate is incubated at room temperature for several hours. Under such conditions, cell autolysis occurs, which increases the level of protein extraction; however, this can lead to the inactivation of target enzymes. A significant problem in further processing is caused by lipids. Ballast substances and lipids are removed from the homogenate by centrifugation or flocculation followed by chitosan filtration [30–34] or immediately by filtration through the hollow fiber [35,36]. The filtrate is concentrated by ultrafiltration through hollow fibers with a pore size of 15–30 kDa and then dried by freeze-drying or spray-drying. The choice of hollow fibers with a specific pore size (considering the variation in real pore sizes) should be based on the known molecular weights of the target enzymes. To obtain more purified preparations before drying, proteins are precipitated with ammonium sulfate and/or tert-butanol, and then ion exchange, hydrophobic, or affinity chromatography is performed [37–40]. In the simplest case, after homogenization, proteins are precipitated with an excess of cooled acetone ("acetone powder") [41]. This method is not suitable for upscaling, since the use of a large volume of acetone is unsafe for both humans and the environment. Most of the developed technologies have been successfully tested on tens and hundreds of kilograms of hepatopancreas as raw materials [29–31,33–36]. The yield of dry final substance in such protocols varies from 0.6 to 1.3% based on the feedstock [33–36], and the collagenolytic activity is 500–3000 units Mandl/mg (the substrate is collagen type I, at 37 °C) [35,36], even reaching 11,000 Mandl/mg (the substrate is collagen type III, at 42 °C) [29], whereas the amount of protein in dry matter is 80–98% [29,33,35,36].

The basic conditions for enzymes obtained with the mentioned technologies include the maintenance of neutral pH of the enzyme solution; at pH values below 5.5 and above 9.5, the activity of the proteolytic complex and its individual components decreases dramatically. At pH values below 3 and above 10.5, parts of the enzymes are irreversibly inactivated [33,42].

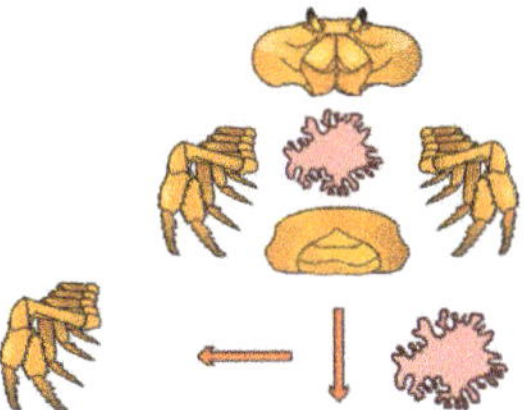

Homogenization: freezing/thawing process, osmotic shock, detergent, cell autolysis, or mechanical homogenization

Ballast substances and lipids removing: centrifugation, flocculation, or filtration through the hollow fiber

crude preparation is ready 

Additional purification: precipitation of proteins (ammonium sulfate, acetone, etc.), chromatography

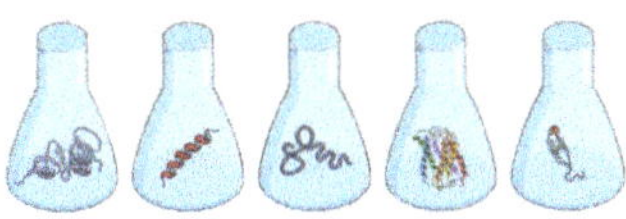

Figure 3. Concept of hepatopancreas processing.

3.4. Enzymes of the Red King Crab Hepatopancreas

3.4.1. Proteolytic Enzymes

Currently, 10 proteolytic enzymes of hepatopancreas—collagenolytic serine proteinase PC (PC—*Paralithodes camtschaticus*), collagenolytic serine proteinase PC 2, trypsin-like proteinase A, chymotrypsin-like proteinase C, aminopeptidase PC, carboxypeptidase PC trypsin PC, elastase, cathepsin L, and metalloproteinase—were described in the literature in detail (Table 1). Most of these are small proteins up to 30 kDa (based on the denaturing gel electrophoresis data) with an isoelectric point (pI) of 2.5–4.4. The optimal operating conditions for these enzymes include neutral pH and temperature range from 25 to 55 °C. The collagenolytic serine proteinases are of particular interest due to their ability to degrade native collagen of types I–III. Most likely, these two enzymes are the basis of the active substance in existing wound healing and wound cleaning preparations. Collagenolytic activity has also been assigned to other hepatopancreas proteinases. However, the temperature of the reaction mixture in these studies was often 42 °C, at which native collagen partially denatures [43]; therefore, the measured activity may refer not to true collagenase but to gelatinase activity. Despite the large number of scientific works devoted to hepatopancreas enzymes, the complete nucleotide sequence of mRNA is known only for collagenolytic serine proteinase PC, trypsin PC, cathepsin L, and metalloproteinase (Table 1).

Table 1. Hepatopancreas proteolytic enzymes. [1] expected molecular weight based on the mRNA nucleotide sequence, [2] mass spectrometry data. PC—*Paralithodes camtschaticus*.

Proteinase	kDa (Based on Electrophoresis)	Opt. pH	Opt. °C	pI	Substrate Specificity and Other
Collagenolytic serine proteinase PC [44]	29 23.5 [1] [45]	7.5	47–55	3	Preferably hydrolyzes peptide bonds, the carbonyl formed by Arg, Lys, and hydrophobic amino acids. Hydrolyzes native collagen type I even at 4 °C [46]. The mRNA nucleotide sequence was determined [45], European Molecular Biology Laboratory (EMBL) Nucleotide Sequence Database: AF461035.
Collagenolytic serine proteinase PC 2 [47]	25	8.5	38–40	?	Preferably hydrolyzes peptide bonds, carbonyl formed by positively charged amino acids. Hydrolyzes native collagen types I–III.
Trypsin-like proteinase A [48]	27 30 [49]	7.9	55 [49]	2.5	Preferably hydrolyzes peptide bonds, the carbonyl formed by Arg and Lys. Proteolytic activity is not inhibited by ethylenediaminetetraacetic acid (EDTA), and partially inhibited by soybean trypsin inhibitor.
Chymotrypsin-like proteinase C [50]	24	9	55 [49]	2.9 [49]	Preferably hydrolyzes peptide bonds, carbonyl formed by hydrophobic amino acids (Phe, Val, and Leu). Not inhibited by Tos-Phe-CH$_2$Cl (chymotrypsin inhibitor).
Aminopeptidase PC [51]	110	6	36–40	4.1	Effectively cleaves N-terminal amino acids: Arg, Lys, Leu, Phe, and Met. Most likely it is a homodimeric, Zn-containing enzyme.

Table 1. *Cont.*

Proteinase	kDa (Based on Electrophoresis)	Opt. pH	Opt. °C	pI	Substrate Specificity and Other
Carboxypeptidase PC [52]	34	6.5	55	3.1	Effectively cleaves C-terminal amino acids: Arg, Lys, Phe, Tyr, Leu, and Ile. The enzyme is inhibited by 0.5 mM Ag^+, Zn^{2+}, Cd^{2+} and 1 mM EDTA, whereas it is activated by Co^{2+} and Ca^{2+}.
Trypsin PC [53]	29 23 [54] 24.8 [1] [45] 24.8 [2] [54]	7.5–8	55	<2.5	Preferably hydrolyzes peptide bonds, carbonyl formed by Arg and Lys. The mRNA nucleotide sequence was determined [45], EMBL: AF461036.
Elastase [55,56]	28.5	8–8.5	50 [57]	4.5	Hydrolyzes native elastin (inhibited by elastinal). NaCl, $MnCl_2$, $CdCl_2$ at a concentration of 1–100 mM stimulate elastase activity, whereas it is inhibited by $HgCl_2$ (100 mM).
Cathepsin L [58]	29 24 [1]	8	25	?	Enzyme has cathepsin activity, hydrolyzes Z-Phe-Arg-pNA substrate. Hydrolyzes collagen types X and VI. $HgCl_2$, E64, and leupeptin inhibit cathepsin activity; soybean trypsin inhibitor practically does not suppress activity. The mRNA nucleotide sequence was determined, EMBL: HQ437281
Metalloproteinase [59]	22.2 [1]	8–8.5	45	4.43	Destroys peptide bonds formed by both acidic and hydrophobic amino acids. Hydrolyzes azocollagen. Proteolytic activity is maintained at 1–3 M NaCl and is inhibited by isopropanol, o-phenanthroline, and EDTA. Zn-containing enzyme. The mRNA nucleotide sequence was determined, EMBL: AF492483

3.4.2. Nucleases and Other Enzymes of Hepatopancreas

The red king crab hepatopancreas is a source of enzymes such as nucleases (Table 2). Ca^{2+}- and Mg^{2+}-dependent DNase, which is characterized by high thermal stability, has been studied in detail: the enzyme remains absolutely active after 3 h of incubation at 60 °C, whereas incubation for 30 min at 100 °C resulted in only 7% loss of activity [60]. Based on the circular dichroism (CD) spectrum, the protein appears to be a compact globule and consists mainly of β-layers (75%) [61]. The pI value is 4 and the optimal reaction temperature range is 50–60 °C [62].

This DNase has a pronounced specificity for the secondary structure of DNA and predominantly cleaves double-stranded substrates (DNA and RNA–DNA duplexes, while the RNA remains intact). The minimum duplex size for cleavage by DNase is 9 bp; the enzyme does not hydrolyze RNA [62]. The DNase cleaves phosphodiester bonds with the formation of 5′-phosphate and 3′-OH terminal groups. Ca^{2+} and Mg^{2+} together have a positive synergistic effect on the rate of the hydrolysis reaction. The unique properties of DNase enable it to be used effectively for the rapid analysis of single nucleotide polymorphisms and the normalization of nucleic acid libraries [63–66]. In the hepatopancreas of red king crab, other nucleases (RNases) and phosphoesterases were found (Table 2).

Table 2. Nucleases and other enzymes of hepatopancreas. [1] denaturing gel chromatography data, [2] denaturing electrophoresis, [3] expected molecular weight based on the mRNA nucleotide sequence.

Enzymes	kDa (Based on Gel-Chromatography)	Opt. pH	Known Properties
Ca^{2+}- and Mg^{2+}-dependent DNase [60]	53 47 [1] 42 [2] 41.5 [3] [62]	7–8 6.6 [62]	The primary structure was determined. There are two sequences in UniProtKB, Q8I9M9 (2003) and B6ZLK3 (2009), differing by two amino acids.
Alkaline RNase (AlkR) [67]	19	7.2–7.5	Broad specificity. Poorly hydrolyzes poly(AUC). $MgCl_2$ at a concentration of 10–50 mM stimulates the activity of enzyme. 0.1 M NaCl inhibits the enzyme activity by 50%.
Acid RNase (AcR and AcR') [67]	33 and 70	5.5	Does not hydrolyze poly(C) and poly(AUC). $MgCl_2$ inhibits its activity, 0.25 M NaCl inhibits its activity by 50%. Most probably, these are monomeric and dimeric forms of the same protein.
Two acidic phosphomonoesterases [68]	80 and 82	5.5	Do not hydrolyze $(3',5')$cAMP (cyclic adenosine monophosphate); 1.5 M NaCl inhibits the enzyme activity by 20% (protein 80 kDa); 1.1M NaCl inhibits the enzyme activity by 50% (protein 82 kDa)
Alkaline phosphomonoesterase [68]	80	7.2–7.5	Does not hydrolyze $(3',5')$cAMP. 0.4 M NaCl inhibits the enzyme activity by 50%.
Acidic phosphodiesterase [68]	57	4.8–5	50% inhibition at 1.4 M NaCl.
Alkaline phosphodiesterase [68]	51	7.2–7.5	No inhibition up to 1.4 M NaCl is observed.

The enzyme preparation from the hepatopancreas contains several other enzymatic activities, which allows this enzyme complex to be used for the depolymerization of β glycosidic bonds in chitosan [69–71]. Lipase activity has also been observed [33]; however other activities of hepatopancreas enzymes have not been sufficiently studied in comparison with proteolytic and nuclease activities.

3.4.3. Hyaluronidase Activity of Hepatopancreas Homogenate

Currently, the structure of hyaluronidases of the Malacostraca class has not been studied in sufficient detail. In the open database of UniProt protein sequences, there are only four representatives of this class for which amino acid sequences of hyaluronidases are available: two representative of Decapods, including Whiteleg shrimp (*Litopenaeus vannamei*, UniProt A0A423SH46) and orange mud crab (*Scylla olivacea*, two UniProt proteins A0A0P4VVV1 and A0A0N7ZAX3), and two representative of Isopoda, which are pill-bug *Armadillidium vulgare* (only one of two proteins is characterized as UniProt A0A444ST78) and *Armadillidium nasatum* (UniProt A0A5N5TJL6) (Figure 1). Earlier, hyaluronidase activity was found in the complex of enzymes of the red king crab hepatopancreas homogenate [72].

In our previous work, the kinetics of the hydrolysis of hyaluronic acid in cosmetic fillers with the red king crab hepatopancreas homogenate was studied for the first time [73]. Using the methods of turbidimetric analysis, atomic force microscopy, and nuclear magnetic resonance spectroscopy, the kinetics of hydrolysis and the structural transformation of hyaluronic acid under the action of homogenate enzymes were investigated. We found that the obtained homogenate has activity comparable to commercially available hyaluronidase preparations. In this work, we demonstrated the possibility of using an enzymatic preparation based on hepatopancreas homogenate for the treatment of complications after the injection of fillers based on hyaluronic acid. Further studies of the effectiveness and safety of hyaluronidase from hepatopancreas on model animals will enable us to develop new drugs for the treatment of complications of filler injections in the near future.

3.5. Other Valuable Non-Protein Components of the Red King Crab Hepatopancreas

The crab hepatopancreas is a source of enzymes and other valuable products. For example, a preparation with the properties of an inhibitor of serine proteinases was obtained and characterized from the hepatopancreas, and its effect on the process of human blood plasma coagulation was studied [74]. The tested inhibitor showed an anticoagulant effect that was more pronounced when combined with heparin. Procedures hwere developed to obtain an inhibitor both from the raw material [75] and in the recombinant form [76].

The hepatopancreas contains a large amount of fat, which varies from 10 to 26% [18,77,78]. In the study of the fractional composition of crab fat, it was found to contain triglycerides at a rate of up to 55%, as well as polyunsaturated fatty acids, including ω-3 (14–24% of the total of all fatty acids). Hepatopancreas fat does not contain toxic substances and can be used as a food supplement or as an ingredient for cosmetic products.

4. Discussion
4.1. Prospects of Processing Waste from Other Commercial Crab Species

The commercial species of crabs include representatives of the false (Anomura) and true crabs (Brachyura) of Pleocyemata. Brachyura includes opilio crab (*Chionoecetes opilio*), which is also a commercially important catch for marine fisheries. Crabs of this suborder have similar enzymatic activity in their digestive system. For example, the proteolytic activities of enzyme preparations from the hepatopancreas of red king crab and opilio crab are practically the same [21]. The zymogram showed that the hepatopancreas of opilio crab contains at least 10 proteolytic enzymes.

The activity of the proteolytic complex was comparable to the commercially available collagenase of the gas bacillus *Clostridium histolyticum* [79,80]. In these works, proteolytic enzymes of the hepatopancreas of opilio crab, but not red king crab, were isolated and biochemically characterized, and the N-terminal amino acid sequences of proteolytic enzymes were obtained. The authors noted their mistake in their next work [42]. Unfortunately, these incorrect data were published in the UniProt database. For example, the amino acid sequence UniProtKB-P20734 (COGC_PARCM) actually belongs not to *Paralithodes camtschaticus*, but to *Chionoecetes opilio*. The sequence of UniProtKB-P34153 (COG1_CHIOP) and UniProtKB-P34156 (COG4_CHIOP) is not derived from the hepatopancreas proteins of *Chionoecetes opilio*, but from *Paralithodes camtschaticus*. The rest of the N-terminal sequences of proteolytic enzymes of these two crabs in [79,80] completely coincide in the UniProt database, which once again confirms the biochemical similarity of the digestive systems of the representative of Pleocyemata.

The close phylogenetic kinship and similar ecological niches of the commercial crab species, as well as the industrial scale of the catch, provide grounds for the successful transfer of the experience of the processing of the king crab hepatopancreas to other crab species to obtain new valuable products. For example, the enzymatic complex of hepatopancreas of opilio crab was successfully used in the production of protein hydrolysates from cod waste processing, as well as to improve the consistency and juiciness of cod fillets at the stage of the salting of the semi-finished product [21,30,81].

Russia is a leader in terms of the catch level of red king crab, and this catch is increasing every year (Figure 4a). Most of this catch comes from fishers in the Russian Far East, however, the total catch of Russia and Norway in the Barents and Norwegian Seas is expected in the near future to equal and even exceed the catch level in the Russian Far East [82]. In the United States (Alaska), red king crab is also caught, but to a lesser extent. Fishers in Alaska have caught 200–400 tonnes per year from 2013 to 2020 [83]. However, at the same time the United States catches about 5000–10,000 tonnes of crabs that are collectively known as king crabs [82]. Other commercial species of crabs are also promising species for advanced processing. For example, Food and Agriculture Organization (FAO) data indicate that the global catch of opilio crab is more than 100,000 tonnes [84] (Figure 4b). We confidently state that high catch levels of large commercially significant crabs, and therefore vast amount of waste, provide an opportunity to develop waste processing technologies and include them in the industrial process. The most promising direction of processing should be considered the processing of the hepatopancreas as a source of new high-margin products.

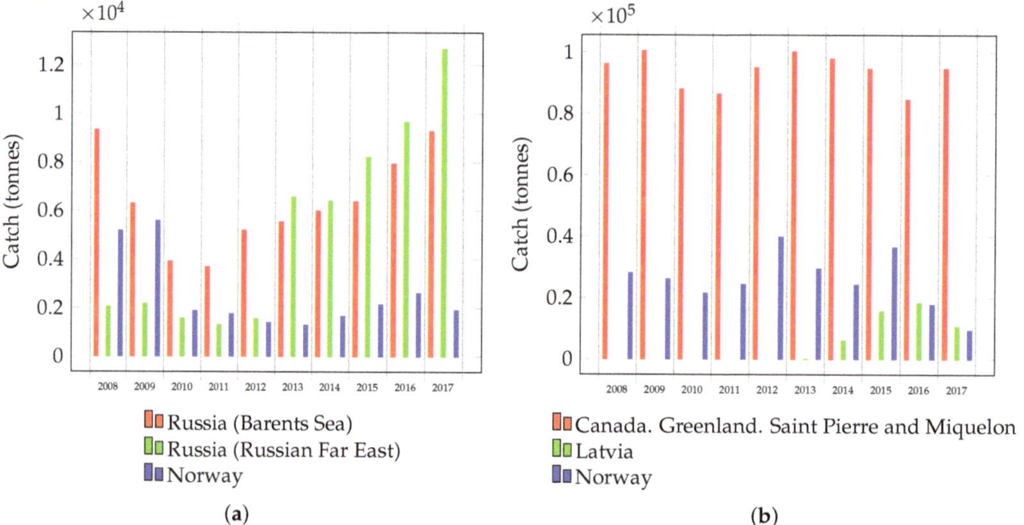

Figure 4. Crab catch: (**a**) red king crab in the Barents Sea (Russia and Norway) and the Russian Far East (Russia) and (**b**) the global catch of opilio crab.

4.2. Development Strategy for Waste Processing

Crab is processed using different approaches. If the caught crab is not boiled, complex waste processing can be used to maximize the yield of new valuable products. After limb separation, the rest of the body can be completely processed: the carapace can be used for chitosan production, the hepatopancreas for the production of multicomponent enzymatic complexes and specific purified enzymes, the fat from the hepatopancreas for dietary nutrition or biofuel production, and the gills and other internal organs as a feed supplement for birds, fish, and other animals. This approach adopts the principle of non-waste recycling.

The methods of processing the crab hepatopancreas to obtain enzyme preparations can be divided into two strategies: obtaining complexes of various enzymes or further purification of specific enzymes (or class of enzymes) from this complex. The enzyme complex is appropriate for the treatment of multicomponent substrates where the simultaneous degradation of proteins, nucleic acids, and other polymers into monomers is required, for example, in the production of easily digestible food products from animal and plant tissues. The activity of the enzyme complex in the preparations will differ for the hepatopancreas from sample to sample. In this regard, the technology for using such enzymatic preparations should allow for fluctuations in the activities in different batches of the preparation. The second strategy, based on the isolation of one enzyme from the complex, has advantages, since the final product can be used to produce a high-margin product (for example, a pharmaceutical product or a reagent for scientific research). For this purpose, it is important to have a simple and scalable method for the purification of a specific enzyme. Notably, all the currently existing technologies for processing the crab hepatopancreas do not meet the specified requirements. The main vector of development in this field could be an approach including several stages of tangential flow filtration and affinity chromatography using a cheap carrier.

All these works discussed in this review consider the processing of a specific type of waste (carapace/hepatopancreas/fat). However, the waste is most often a mixture, which is difficult to sort. To implement the principle of non-waste recycling, it is necessary to develop new approaches to recycling this mixture, which will convert the waste into new valuable products. The relevance of deep processing is provided by the sustainable continual increase in catches of fisheries. On the one hand, a growth in catch leads to an increase in waste, which might cause several environmental problems. On the other hand, insufficient control of the catch will lead to a decrease in the crab population, which will entail a sharp decline in the economic performance of the industry in the coming years. Thus, all types of processing could reduce the environmental burden, as well as help satisfy the financial appetites of the industry by selling new valuable products.

The main problem with the complex processing of wastes from commercial marine species is the complexity of their collection and storage under sailing conditions on board a trawler and delivery to coast or further processing. The solution of this problem could be, for example, the creation of a maximally automated integrated waste processing unit immediately on a catching vessel, which could eliminate the problem of the time spent by fishers on the manual removal of an organ, as well as its storage and transportation. Another method of solving this problem is to establish an adequate cost of the hepatopancreas for fishers, allowing a stable supply of raw materials onshore on an industrial scale. The task could be simplified if the non-waste processing of live crabs is organized onshore. Further studies of the hepatopancreas of commercial crab species in fundamental scientific terms considering its potential applications will increase the value of crab processing waste, possibly leading to a time when this will equal or even exceed the cost of crab meat.

5. Outlook

Despite a long period of scientific research, deep processing of crab has not been launched yet. The main reason for this is laboratory protocols being not adapted to catch conditions. Also these technologies should provide deep recycling based on the principle of waste-free recycling taking into account financial benefits. It also requires financial support for R&D projects to develop technologies for creating a new high-value products for an industrial scale. The solution to the problem lies in close collaboration between scientists, developers, and fishers. In our review, we discuss all these parties and hope that we touch upon the interests of all the listed stakeholders in order to unite their efforts in the deep processing of wastes from commercial crab species.

Author Contributions: T.P. and E.S. conceived of the presented idea, supervised the findings of this work. E.S. designed the figures, and wrote the manuscript with support from T.P., M.T., M.F., and S.L. S.L. developed the theoretical formalism, performed the analytic calculations and performed the statistical data. M.T., M.F., and S.L. processed the literature data, performed the analysis. All authors discussed the results and contributed to the final manuscript. All authors have read and agreed to the published version of the manuscript.

Funding: This research received no external funding.

Institutional Review Board Statement: At the time of writing this review, no both human or another living thing were harmed (i.e., did not participate in study).

Informed Consent Statement: At the time of writing this review, no both human or another living thing were harmed (i.e., did not participate in study).

Data Availability Statement: Most of the data is available on internet platforms, some are available only on paper.

Conflicts of Interest: The authors declare no conflict of interest. The funders had no role in the design of the study; in the collection, analyses, or interpretation of data; in the writing of the manuscript, or in the decision to publish the results.

References and Notes

1. Slizkin, A.; Safronov, S. Commercial crabs of Kamchatka waters. *Petropavlovsk Kamchatskii Sev. Patsifika* **2000**, *180*, 12.
2. Taxonomy Browser (Root)-NCBI-NIH. Available online: https://www.ncbi.nlm.nih.gov/Taxonomy/Browser/wwwtax.cgi?mode=Undef&id=6681&lvl=3&lin=f&keep=1&srchmode=1&unlock (accessed on 16 December 2020).
3. Tavares, M. True crabs. In *FAO Species Identification Sheets for Fishery Purposes: Western Central Atlantic (Fishing Area 31)*; FAO: Rome, Italy, 2003; pp. 327–352.
4. Garth, J.S.; Abbott, D.P. Brachyura: The True Crabs. In *Intertidal Invertebrates of California*; Stanford University Press: Stanford, CA, USA, 1980; pp. 594–630.
5. Steneck, R.S. Functional Groups. In *Encyclopedia of Biodiversity*, 2nd ed.; Levin, S.A., Ed.; Academic Press: Waltham, MA, USA, 2001; pp. 609–623. [CrossRef]
6. Maksimova, S.; Poleschuk, D.; Surovtseva, E.; Vereshchagina, K.; Milovanov, A. *King crab* wastes potential as the technological valuable raw materials. *Food Ind.* **2019**, *4*. [CrossRef]
7. Dvoretsky, A.G.; Dvoretsky, V.G. Red king crab (*Paralithodes camtschaticus*) fisheries in Russian waters: Historical review and present status. *Rev. Fish Biol. Fish.* **2018**, *28*, 331–353. [CrossRef]
8. Matishov, G. *Marine Ecosystems and Communities in the Conditions of Current Climate Changes*; MMBI KSC RAS: St. Petersburg, Russia, 2014.
9. Podkorytova, A.; Strokova, N.; Semikova, N. Complex processing of *Kamchatka crab* in the production of food products and biologically active substances. *Tr. VNIRO* **2018**, *172*, 198–212. [CrossRef]
10. Nemtsev, S. *Comprehensive Technology for Obtaining Chitin and Chitosan from Crustacean Shell*; VNIRO: Moscow, Russia, 2006.
11. Šimat, V.; Elabed, N.; Kulawik, P.; Ceylan, Z.; Jamroz, E.; Yazgan, H.; Čagalj, M.; Regenstein, J.M.; Özogul, F. Recent Advances in Marine-Based Nutraceuticals and Their Health Benefits. *Mar. Drugs* **2020**, *18*, 627. [CrossRef] [PubMed]
12. Nguyen, V.B.; Wang, S.L. Reclamation of marine chitinous materials for the production of α-glucosidase inhibitors via microbial conversion. *Mar. Drugs* **2017**, *15*, 350. [CrossRef] [PubMed]
13. Nguyen, V.B.; Nguyen, D.N.; Nguyen, A.D.; Ngo, V.A.; Ton, T.Q.; Doan, C.T.; Pham, T.P.; Tran, T.P.H.; Wang, S.L. Utilization of Crab Waste for Cost-Effective Bioproduction of Prodigiosin. *Mar. Drugs* **2020**, *18*, 523. [CrossRef]
14. Casadidio, C.; Peregrina, D.V.; Gigliobianco, M.R.; Deng, S.; Censi, R.; Di Martino, P. Chitin and chitosans: Characteristics, eco-friendly processes, and applications in cosmetic science. *Mar. Drugs* **2019**, *17*, 369. [CrossRef]
15. Hulme, P.E. *DAISIE (Delivering Alien Invasive Species Inventories for Europe). Handbook of Alien Species in Europe*; Springer: Dordrecht, The Netherlands, 2009.
16. Jørgensen, L.L. NOBANIS—Invasive Alien Species Fact Sheet—*Paralithodes camtschaticus*.—From: Online Database of the European Network on Invasive Alien Species—NOBANIS www.nobanis.org. 2013. Available online: https://www.nobanis.org/globalassets/speciesinfo/p/paralithodes-camtschatica/paralithodes_camtschaticus.pdf (accessed on 19 October 2020).
17. Gibson, R. The decapod hepatopancreas. *Oceanogr. Mar. Biol. Ann. Rev.* **1979**, *17*, 285–346.
18. Kas'janov, S.; Kuklev, D.; Kucheravenko, K.; Blinov, J.; Akulin, V. Method of Extracting Fat from Crab Liver. RU 2162648 C2, Filed 7 April 1999, and Issued 10 February 2001. Patent Holder Pacific Branch of VNIRO (TINRO).
19. Muranova, T.; Zinchenko, D.; Melanyina, L.; Miroshnikov, A. Hydrolysis of soybean proteins with Red king crab hepatopancreas enzyme complex. *Appl. Biochem. Microbiol.* **2018**, *54*, 74–81. [CrossRef]
20. Sova, V. Method for Preparing Caviar from Salmon Species Fish. RU 2111681 C1, Filed 12 March 2003, and Issued 10 April 2004. Patent Holder Sova, V.
21. Lyzhov, I.; Rysakova, K.; Mukhin, V.; Novikov, V.; Shironina, A. Applying of snow crab hepatopancreas for obtaining of protein hydrolysates from cod filleting wastes. In Proceedings of the Biological Resources of the White Sea and Inland Waters of the European North, Materials of the XXVIII International Conference, 5–8 October 2009, Petrozavodsk, Russia; pp. 337–340.
22. Mukhin, V.; Novikov, V. Enzymatic hydrolysis of proteins from crustaceans of the Barents Sea. *Appl. Biochem. Microbiol.* **2001**, *37*, 538–542. [CrossRef]
23. Novikov, V.; Portsel, M. Method of Producing Chondroitin Sulphate From Sea Hydrobiont Tissue. RU 2458134 C1, Filed 27 December 2010, and Issued 10 August 2012. Patent Holder POLAR Branch of VNIRO (PINRO).
24. Glyantsev, S.; Vishnevsky, A.; Adamyan, A.; Vishnevsky, A.; Sakharov, I. Crab collagenase in wound debridement. *J. Wound Care* **1997**, *6*, 13–16. [CrossRef] [PubMed]
25. Sakharov, I.; Glyanzev, S.; Litvin, F.; Savvina, T. Potent debriding ability of collagenolytic protease isolated from the hepatopancreas of the king crab *Paralithodes camtschatica*. *Arch. Dermatol. Res.* **1993**, *285*, 32–35. [CrossRef] [PubMed]

6. Rudenskaya, G. Brachyurins—Serine collagenolytic enzymes from crabs. *Bioorg. Khim.* **2003**, *29*, 117–128.
7. Belov, A.; Filatov, V.; Belova, E.; Filatov, N. Medical Bandage Containing Proteolytic Enzyme Complex Including Collagenolytic Proteases from Crab Hepatopancreas. RU 2268751 C2, Filed 25 April 2003, and Issued 27 January 2006. Patent Holders Belov, A.; Filatov, V.; Belova, E.; Filatov, N.
8. Belov, A.; Filatov, V.; Belova, E. A Medical Dressing Containing a Complex of Enzymes from the Hepatopancreas of a Crab, and a Method for Its Production. RU 2323748 C2, Filed 21 February 2006, and Issued 10 May 2008. Patent Holders Belov, A.; Filatov, V.; Belova, E.
9. Sova, V.; Strongyn, A.; Klimovava, O.; Stadnikov, V. A Collagenolytic Activity Exhibiting Enzyme Complex and the Method for Isolation and Purification Thereof. EP 0402321 A1, Filed 8 June 1990, and Issued 12 December 1990, Patent Holder SEATEC.
0. Artjukov, A.; Sakharov, I. Method for Production of Collagenase. RU 2039819 C1, Filed 29 March 1990, and Issued 20 July 1995. Patent holders Artjukov, A.; Sakharov, I.
1. Majorov, A.; Albulov, A.; Samujlenko, A. Method of Collagenase Preparing. RU 2112036 C1, Filed 9 October 1992, and Issued 27 May 1998. Patent Holders Majorov, A.; Albulov, A.; Samujlenko, A.
2. Pivnenko, T.; Zhdanjuk, V.; Ehpshtejn, L. Method of Preparing of Proteolytic Complex. RU 2034028 C1, Filed 7 February 1992, and Issued 30 April 1995. Patent Holder "Tikhookeanskij nauchno-issledovatel'skij institut rybnogo khozjajstva i okeanografii".
3. Artjukov, A.; Menzorova, N.; Kozlovskaja, E.; Kofanova, N.; Kozlovskij, A.; Rasskazov, V. Enzyme Preparation from Hepatopancreas of Commercial Crab Species and Method for Production of the Same. RU 2280076 C1, Filed 6 December 2004, and Issued 20 July 2006. Patent Holder "Tikhookeanskij institut bioorganicheskoj khimii Dal'nevostochnogo otdelenija Rossijskoj Akademii nauk".
4. Kireev, V.; Titov, O.; Sheveleva, O. Method for Preparing Enzyme Preparation from Crab Hepatopancreas. RU 2285041 C1, Filed 14 April 2005, and Issued 10 October 2006. Patent Holder "FGUP Poljarnyj nauchno-issledovatel'skij institut morskogo rybnogo khozjajstva i okeanografii im. N.M. Knipovicha (FGUP PINRO)".
5. Stadnikov, V.; Erkhov, S. Method of Preparing Enzyme Preparation Showing Collagenolytic Activity. RU 2096456 C1, Filed 10 November 1996, and Issued 20 November 1997. Patent Holders Stadnikov, V.; Erkhov, S.
6. Erkhov, S. Biologically Active Substance, Raw Material and Method of Its Preparing. RU 2132876 C1, Filed 13 October 1998, and Issued 10 July 1999. Patent Holder Erkhov, S.
7. Sakharov, I.; Artyukov, A.; Berezin, V. Method of Purification of Collagenase. SU 1699350 A3, filed 29 March 1990, and issued 15 December 1991.
8. Isaev, V.; Rudenskaja, G.; Kupenko, O.; Stepanov, V.; Popova, I.; Didenko, J. Method of Collagenase Preparation Producing. RU 2008353 C1, Filed 5 August 1991 and Issued 28 February 1994. Patent Holder "NAUCHNO-PROIZVODSTVENNOE PREDPRIJATIE "TRINITA"".
9. Sandakhchiev, L.; Danilov, A.; Malygin, E.; Zinov'ev, V.; Ovechkina, L.; Zakabunin, A.; Mistjurin, J. Method of Collase Purification. RU 2121503 C1, Filed 5 Spetember 1996, and Issued 10 November 1998. Patent Holder "Gosudarstvennyj nauchnyj tsentr virusologii i biotekhnologii "Vektor"".
0. Isaev, V.; Rudenskaja, G.; Shmojlov, A. Method for Preparing Collagenase Preparation. RU 2225441 C1, Filed 5 March 2003, and Issued 10 March 2004. Patent Holder "Zakrytoe aktsionernoe obshchestvo Nauchno-proizvodstvennoe predprijatie "Trinita"".
1. Sakharov, I.; Dzhunkovskaya, A.; Artyukov, A.; Sova, V.; Sakandelidze, O.; Kozlovskaya, E. Method of Producing Collagenase. SU 1343591 A1, Filed 13 December 1985, and Issued 30 April 1992.
2. Klimova, O.; Chebotarev, V. Collagenolytic complex of proteases from the hepatopancreas of the *Kamchatka crab*: Enzymologic activity of the individual components. *Biulleten'eksperimental'noi Biologii I Meditsiny* **1999**, *128*, 391–396. [CrossRef]
3. Štancl, J.; Skočilas, J.; Landfeld, A.; Žitný, R.; Houška, M. Electrical and thermodynamic properties of a collagen solution. *Acta Polytech.* **2017**, *57*, 229–234. [CrossRef]
4. Rudenskaia, G.; Isaev, V.; Stepanov, V.; Dunaevskii, I.; Baratova, L.; Kalebina, T.; Nurminskaia, M. Isolation and properties of serine PC proteinase from the *Kamchatka crab, Paralithodes camtschatica*—A proteolytic enzyme with broad specificity. *Biochem. Mosc.* **1996**, *61*, 1119.
5. Rudenskaya, G.N.; Kislitsin, Y.A.; Rebrikov, D.V. Collagenolytic serine protease PC and trypsin PC from king crab *Paralithodes camtschaticus*: CDNA cloning and primary structure of the enzymes. *BMC Struct. Biol.* **2004**, *4*, 2. [CrossRef]
6. Papisova, A.; Semenova, S.; Kislitsyn, Y.; Rudenskaya, G. Peculiarities of substrate hydrolysis by endopeptidases from hepatopancreas of king crab. *Russ. J. Bioorg. Chem.* **2008**, *34*, 428–434. [CrossRef]
7. Semenova, S.; Rudenskaya, G.; Lyutova, L.; Nikitina, O. Isolation and properties of collagenolytic serine proteinase isoenzyme from king crab *Paralithodes camtschatica*. *Biochem. Mos.* **2008**, *73*, 1125–1133. [CrossRef] [PubMed]
8. Sakharov, I.; Litvin, F.; Artyukov, A.; Kofanova, N. Purification and characterization of collagenolytic protease from the *Paralithodes camtschatica* hepatopancreas. *Biochem. Mos.* **1988**, *53*, 1844–1849.
9. Sakharov, I.Y.; Litvin, F.E.; Artyukov, A.A. Purification and characterization of two serine collagenolytic proteases from crab *Paralithodes camtschatica*. *Comp. Biochem. Physiol. Part B Comp. Biochem.* **1994**, *108*, 561–568. [CrossRef]
0. Sakharov, I.Y.; Litvin, F.E.; Artyukov, A.A. Some physico-chemical properties of collagenolytic protease C of the crab (*Paralithodes camtschatica*). *Biochem. Mosc.* **1992**, *57*, 40—45.
1. Rudenskaya, G.; Shmoilov, A.; Isaev, V.; Ksenofontov, A.; Shvets, S. Aminopeptidase PC from the hepatopancreas of the *Kamchatka crab Paralithodes camtschatica*. *Biochem. Mos.* **2000**, *65*, 164–170.

52. Rudenskaia, G.; Kupenko, O.; Isaev, V.; Stepanov, V.; Dunaevskii, I. Isolation and properties of carboxypeptidase from the *Kamchatka crab Paralithodes camtshatica*. *Russ. J. Bioorg. Chem.* **1995**, *21*, 249–255.
53. Rudenskaia, G.; Isaev, V.; Kalebina, T.; Stepanov, V.; Maltsev, K.; Shvets, S.; Lukianova, N.; Kislitsin, I.; Miroshnikov, A. Isolation of trypsin PC from the *Kamchatka crab Paralithodes camtschatica* and its properties. *Russ. J. Bioorg. Chem.* **1998**, *24*, 112–118.
54. Kislitsyn, I.; Rebrikov, D.; Dunaevskii, I.; Rudenskaia, G. Isolation and primary structure of trypsin from the red king crab *Paralithodes camtschaticus*. *Russ. J. Bioorg. Chem.* **2003**, *29*, 269–276. [CrossRef]
55. Dzhunkovskaia, A.; Zakharova, N. Pancreatic elastase is a new serine protease isolated from the hepatopancreas of the king crab. In Proceedings of the Abstracts of the All-Union Meeting Biologically Active Substances of Aquatic Organisms—New Medicinal Therapeutic and Prophylactic and Technical Preparations, Vladivostok, Russia, 23–27 September 1991; p. 8.
56. Sakharov, I.; Dzhunkovkaia, A. Elastase from the hepatopancreas of the king crab. *Biochem. Mos.* **1993**, *58*, 1445–1453.
57. Sakharov, I.Y.; Dzunkovskaya, A.V.; Artyukov, A.A.; Zakharova, N.N. Purification and some properties of elastase from hepatopancreas of king crab *Paralithodes camtschatica*. *Comp. Biochem. Physiol. Part B Comp. Biochem.* **1993**, *106*, 681–684. [CrossRef]
58. Isaev, V.; Balashova MV, S.D.; Shagin, D.; Shagina, I.; Eremeev, N.; Rudenskaya, G. New psychrophilic cathepsin L from the hepatopancreas of the red king crab (*Paralithodes camtschaticus*). *Sci. Rev. Biol. Sci.* **2016**, *6*, 81–89.
59. Semenova, S.A.; Rudenskaya, G.N.; Rebrikov, D.V.; Isaev, V.A. cDNA cloning, purification and properties of *Paralithodes camtschatica* metalloprotease. *Protein Pept. Lett.* **2006**, *13*, 571–575. [CrossRef] [PubMed]
60. Menzorova, N.; Markova, A.; Rasskazov, V. Highly stable Ca^{2+}, Mg^{2+}-dependent DNAase from crab hepatopancreas. *Biochem. Mos.* **1994**, *59*, 449–456.
61. Vakorina, T.; Menzorova, N.; Rasskazov, V. Study of conformational stability of DNAse from hepatopancreas of the crab *Paralithodes camtschatica*. *Biochem. Mos.* **1997**, *62*, 1642–1647.
62. Anisimova, V.E.; Rebrikov, D.V.; Shagin, D.A.; Kozhemyako, V.B.; Menzorova, N.I.; Staroverov, D.B.; Ziganshin, R.; Vagner, L.L.; Rasskazov, V.A.; Lukyanov, S.A.; Shcheglov, A.S. Isolation, characterization and molecular cloning of duplex-specific nuclease from the hepatopancreas of the *Kamchatka crab*. *BMC Biochem.* **2008**, *9*, 14. [CrossRef]
63. Shagin, D.A.; Rebrikov, D.V.; Kozhemyako, V.B.; Altshuler, I.M.; Shcheglov, A.S.; Zhulidov, P.A.; Bogdanova, E.A.; Staroverov, D.B.; Rasskazov, V.A.; Lukyanov, S. A novel method for SNP detection using a new duplex-specific nuclease from crab hepatopancreas. *Genome Res.* **2002**, *12*, 1935–1942. [CrossRef]
64. Zhulidov, P.A.; Bogdanova, E.A.; Shcheglov, A.S.; Vagner, L.L.; Khaspekov, G.L.; Kozhemyako, V.B.; Matz, M.V.; Meleshkevitch, E.; Moroz, L.L.; Lukyanov, S.A. Simple cDNA normalization using kamchatka crab duplex-specific nuclease. *Nucleic Acids Res.* **2004**, *32*, e37. [CrossRef]
65. Lukyanov, S.A.; Rebrikov, D.V.; Shagin, D.A. Methods and Compositions for Selectively Cleaving Dna Containing Duplex Nucleic Acids in a Complex Nucleic Acid Mixture, and Nuclease Compositions for Use in Practicing the Same. US 7435794 B2 Filed 5 December 2004, and Issued 14 October 2008.
66. Shagina, I.; Bogdanova, E.; Al'tshuler, I.; Luk'ianov, S.; Shagin, D. Application of the duplex-specific nuclease for fast analysis of single nucleotide polymorphisms and detection of target DNA in complex PCR products. *Bioorg. Khim.* **2011**, *37*, 522–529.
67. Menzorova, N.I.; Sibirtsev, J.T.; Rasskazov, V.A. Ribonuclease from the Hepatopancreas of the Red King Crab *Paralithodes camtschatica*. *Appl. Biochem. Microbiol.* **2009**, *45*, 369–373. [CrossRef]
68. Menzorova, N.; Ivleva, A.; Sibirtsev, Y.T.; Rasskazov, V. Phosphatases and phosphodiesterases isolated from the red king crab (*Paralithodes camtschatica*) hepatopancreas. *Appl. Biochem. Microbiol.* **2008**, *44*, 93–97. [CrossRef]
69. Novikov, V.; Mukhin, V. Chitosan Depolymerization by Enzymes from the Hepatopancreas of the Crab *Paralithodes camtschaticus*. *Appl. Biochem. Microbiol.* **2003**, *39*, 464–468. [CrossRef]
70. Novikov, V.; Mukhin, V.; Rysakova, K. Properties of chitinolytic enzymes from the hepatopancreas of the red king crab (*Paralithodes camtschaticus*). *Appl. Biochem. Microbiol.* **2007**, *43*, 159–163. [CrossRef]
71. Rysakova, K.; Novikov, V.; Mukhin, V.; Serafimchik, E. Glycolytic activity of enzyme preparation from the red king crab (*Paralithodes camtschaticus*) hepatopancreas. *Appl. Biochem. Microbiol.* **2008**, *44*, 281–286. [CrossRef]
72. Turkovski, I.; Paramonov, B.; Antonov, S.; Kozlov, D.; Klimova, O.; Pomorski, K. Comparative evaluation of the depth of collagen and hyaluronic acid hydrolysis in vitro by collagenase and hyaluronidase preparations. *Bull. Exp. Biol. Med.* **2008**, *146*, 89–90. [CrossRef]
73. Ponomareva, T.; Sliadovskii, D.; Timchenko, M.; Molchanov, M.; Timchenko, A.; Sogorin, E. The effect of hepatopancreas homogenate of the Red king crab on HA-based filler. *PeerJ* **2020**, *8*, e8579. [CrossRef]
74. Balashova, M.; Liutova, L.; Rudenskaia, I.; Isaev, V.; Andina, S.; Kozlov, L.; Rudenskaia, G. Anticoagulative and anticomplementary activity of endogenous inhibitor preparation from hepatopancreas of red king crab (*Paralithosed camtschaticus*) towards human blood. *Biomed. Khim.* **2012**, *58*, 176–188. [CrossRef]
75. Isaev, V.; Rudenskaya, G.; Rudenskaya, Y.; Lyutova, L.; Balashova, M. Method for Producing a Collagenase Inhibitor with Anticoagulative Action from the King Crab Hepatopancreas. RU 2403284 C1, Filed 25 January 2009, and Issued 10 November 2010. Patent Holder "Zakrytoe Aktsionernoe Obshchestvo Nauchno-proizvodstvennoe predprijatie "TRINITA"".
76. Isaev, V.; Rudenskaja, G.; Kostin, N.; Sergeeva, E.; Statsenko, I.; Kolesnikova, T. Method of Production of Recombinant Serine Protease Inhibitor of King Crab. RU 2560264 C1, Filed 17 April 2014, and Issued 20 August 2015. Patent Holder "Obshchestvo s ogranichennoj otvetstvennost'ju Nauchno-proizvodstvennoe predprijatie "TRINITA" (OOO NPP "TRINITA")".
77. Zikeeva, B. *Processing of Water Non-Fish Raw Materials*; Pishchepromizda: Moscow, Russia, 1950; p. 277.

8. Boeva, N.; Petrova, M.; Makarova, A. Production Method of Crab Fat. RU 2390274 C1, Filed 18 December 2008, and Issued 27 May 2010. Patent Holder "FGUP "Vserossijskij nauchno-issledovatel'skij institut rybnogo khozjajstva i okeanografii" (VNIRO)".
9. Klimova, O.; Borukhov, S.; Solovyeva, N.; Balaevskaya, T.; Strongin, A. The isolation and properties of collagenolytic proteases from crab hepatopancreas. *Biochem. Biophys. Res. Commun.* **1990**, *166*, 1411–1420. [CrossRef]
10. Klimova, O.; Vedishcheva, I.; Strongin, A. Isolation and characteristics of collagenolytic enzymes from the hepatopancreas of the crab *Chionoecetes opilio*. *Dokl. Akad. Nauk SSSR* **1991**, *317*, 482.
11. Shkuratova, E.; Shokina, Y.; Mukhin, V. Development of technology for gourmet smoked products from cod species using enzyme preparation from hepatopancreas of snow crab *Chionoecetes opilio*. *Proc. Voronezh State Univ. Eng. Technol.* **2017**, *79*, 126–137. [CrossRef]
12. The Food and Agriculture Organization (FAO). Fishery and Aquaculture Statistics: B-44. Available online: http://www.fao.org/fishery/static/Yearbook/YB2017_USBcard/root/capture/b44.pdf (accessed on 5 May 2020).
13. National Oceanic and Atmospheric Administration (NOAA). Fisheries Catch and Landings Reports: BSAI Crab. Available online: https://www.fisheries.noaa.gov/alaska/commercial-fishing/fisheries-catch-and-landings-reports (accessed on 5 May 2020).
14. The Food and Agriculture Organization (FAO). Fishery and Aquaculture Statistics: B-42. Available online: http://www.fao.org/fishery/static/Yearbook/YB2017_USBcard/root/capture/b42.pdf (accessed on 5 May 2020).

Article

Organic Material for Clean Production in the Batik Industry: A Case Study of Natural Batik Semarang, Indonesia

Nana Kariada Tri Martuti [1,*] , **Isti Hidayah** [2], **Margunani Margunani** [3] and **Radhitya Bayu Alafima** [4]

[1] Biology Department, Faculty of Mathematic and Natural Sciences, Universitas Negeri Semarang, 1st Floor, D6 Building Kampus Sekaran Gunungpati, Semarang 50229, Indonesia
[2] Mathematics Department, Faculty of Mathematic and Natural Sciences, Universitas Negeri Semarang, 1st Floor, D5 Building Kampus Sekaran Gunungpati, Semarang 50229, Indonesia; isti.hidayah@mail.unnes.ac.id
[3] Economic Department, Faculty of Economi, Universitas Negeri Semarang, C6 Building, Kampus Sekaran. Gunungpati, Semarang 50229, Indonesia; margunani@mail.unnes.ac.id
[4] Department of Industrial Technology of Agriculture, Faculty of Agricultural Technology, Institut Pertanian Bogor, Bogor 16680, Indonesia; radbaya123@gmail.com
* Correspondence: nanakariada@mail.unnes.ac.id

Received: 15 July 2020; Accepted: 16 October 2020; Published: 19 October 2020

Abstract: Batik has become more desirable in the current fashion mode within the global market, but the environmental damage induced by this fabric's synthetic dye practices is a matter of concern. This study aimed to discuss the application of organic materials as natural dyes in the clean production of textiles to maintain the environment. The research was a case study from the community services program in Kampung Malon, Gunungpati, Semarang City, Indonesia, focused on the batik home industry of the Zie Batik fabric. Furthermore, natural pigments from various plant organs (stem, leaves, wood, bark, and fruit) of diverse species, including *Caesalpinia sappan*, *Ceriops candolleana*, *Maclura cochinchinensis*, *Indigofera tinctorial*, *I. arrecta*, *Rhizopora* spp., *Strobilantes cusia*, and *Terminalia bellirica* were used for this type of material. These pigments are more biodegradable, relatively safe, and easily obtained with zero liquid waste compared to the synthetic variants. The leftover wastewater from the coloring stages was further utilized for other processes. Subsequently, the remaining organic waste from the whole procedure was employed as compost and/or timber for batik production, although a large amount of the wastewater containing sodium carbonate (Na_2CO_3), alum ($KAl(SO4)_2 \cdot 12H_2O$), and fixatives ($Ca(OH)_2$ and $FeSO_4$) were discharged into the environment during the process of mordanting and fixating, with the requirement of additional treatment.

Keywords: batik; clean production; natural; organic materials; Semarang City

---

1. Introduction

The Indonesian traditional batik is recognized as a masterpiece within the oral and intangible heritage of humanity by UNESCO [1,2]. This material's uniqueness is seen from the pattern and variety of motifs illustrating nature, diversity of fauna and flora, folktales, as well as weapons [1–4]. The fabric global market was dominated by this material in 2017, with its export value reaching USD 58.46 million, where the main destinations included Japan, the United States, and European countries. Furthermore, issues of synthetic dye practices have generated a decline in the selling demand of this fabric in several destination countries. This is because the usage of manmade variants of synthetic dyes potentially generate serious problems for animal and human health [5], including cancer [6,7], as well as polluting water sources [8] and disturbing organisms and ecosystem balance [9,10]. Furthermore, various studies have shown an extremely high content of dangerous heavy metals, including Cd, Cr [11], Cu, and Pb, around areas of the batik industry [12,13].

The poor awareness of batik makers on environmental sustainability is the reason synthetic dyes are massively used without the regulation of standardized waste management [14]. In some cases, including the synthetic batik industry in Pekalongan [8,9] and Surakarta [15], wastewater was produced with dangerous amounts of heavy metals above the environmentally permitted standard.

The momentum of the Indonesian batik export growth is simultaneously a challenge and an opportunity. Furthermore, global consumer demand for environmentally friendly (eco-friendly) products emerged as a response to the green lifestyle and environmental awareness movement. The use of these types of materials is a universal consumption trend, adopted as an effort to create a harmonious life between nature (green lifestyle) and the batik industry [16,17]. This sector has a great opportunity to manufacture eco-friendly products through the application of organic materials in batik dyes, which would be a cleaner production method.

Zie Batik is a small and medium-scale enterprise (SME) actively producing fabrics with natural dyes. This company is located in Gunungpati Subdistrict, Semarang City, with total revenue of IDR 30–40 million or USD 2100.00–2800.00/month, and is the iconic landmark of this region. Furthermore, production commenced by developing authentic batik motifs in 2006. The Zie Batik is also the first SME developing and introducing organic dyes in Semarang City, and the pioneer for standardized batik in Indonesia. Capturing the information from Zie Batik's production process is essential for disseminating proper eco-friendly batik business.

It cannot be denied that many have researched natural dyes in fabric, but their application in several places should be invented to gain a comprehensive understanding of batik dye based on the local potential. Then, even though small to medium-scale enterprises (SMEs) are well known as profit-oriented businesses, we explain this new perspective of the batik natural dye industry related to cleaner production in SMEs, especially for Semarang City as an industrial and metropolitan city. The process commences with organic pigment application and waste treatments. This form of dye prevents and integrates environmental management strategies without ignoring aspects of economic and cultural development as a sustainable development approach. The aim of this study was to comprehensively describe the clean production application of natural dyes produced from local value chain organic material as an alternative solution in this sector to maintain the environment. We also propose and depict how the SME in this study developed a batik industry with a sustainable development approach.

2. Results

2.1. The Batik Motif and Organic Material for Natural Dye

Zie Batik developed more than 10 authentic motifs or characters for their batik signature from Semarang City, with the famous one being the Javanese Puppet batik, which tells Ramayana's story. The city's iconic landmarks, portraying the historical buildings, including Tugu Muda, Lawang Sewu, Pagoda, and Warak Ngendog—the city mascot—was developed by Zie Batik to introduce natural, high-quality batik using clean production, widely used in the batik home industry in Semarang City (Figure 1). The growing demand for natural fabrics as well as the intense competition has led Zie Batik to develop other motifs illustrating fauna, flora, and the legend of the Javanese Puppet stories, as depicted in Figure 1B.

A   B

Figure 1. The most popular motifs of the Zie Batik products by themes. The iconic landmark of Semarang City or more popular with Semarangan motifs (**A**) and the legend of the Javanese Puppet stories (**B**). In the Semarangan motifs, the natural dye batik is fully decorated with the iconic buildings of Semarang City, such as Tugu Muda (TM), Lawang Sewu (LS), and Gereja Blendug (GB), as well as the city mascot Warak Ngendok (WN), a well-known mythological animal in Semarang City

Furthermore, various parts of the plant used as natural dyes, including the bark, stems, leaves, roots, seeds, fruits, flowers, and plant sap, to make the colors as in Figure 1. Table 1 shows most of the natural dye sources in Zie Batik are derived from several plant species, including *Rhizophora* spp., *Indigofera tinctoria*, *Maclura cochinchinensis*, *Pelthophorum ferruginum*, *Terminalia belerica*, and *Ceriops condolleana*.

Table 1. Dye plant resources for natural batik in the Zie Batik home industry.

Natural Dyes Plant	Vern Name	Colour	Plant Part	Used Product
Caesalpinia sappan	Secang	Red	Wood	Dried wood
Ceriops candolleana	Tingi	Brown	Fruit skin	Boiled *
Maclura cochinchinensis	Tegeran	Yellow	Bark	Dried bark
Pelthophorum ferruginum	Soga Jambal	Brown	Bark	Dried bark
Indigofera tinctoria	Tarum/Indigo	Blue indigo	Leaves, stem	Paste **
Indigofera arrecta	Indigo	Blue indigo	Leaves, stem	Paste **
Rhizopora spp.	Bakau	Brown	Fruit skin	Boiled *
Strobilantes cusia	Indigo	Blue indigo	Leaves	Paste **
Terminalia bellirica	Jelawe	Yellow	Bark	Dried bark

Note: The star mark (*) shows application of processing techniques for natural dye used before: * heat processing only; and ** representing microbial fermentation and emulsion processes.

Figure 2 depicts the various organic compounds of the natural dyes with different colors applied to the fabric. The complexity of the coloring technique depends on the design and amount of pigments used to produce the batik.

Zie Batik also continues to develop innovations in natural dye application through their direct (short processing) and indirect usage (long processing). The direct usage was practiced on several dried dye plants, including *C. sappan* wood used to produce a red color, *C. candolleana* fruits in the manufacturing of brown, and the dried bark of *M cochinchinensis* directly produced a yellow. Furthermore, leaves of *I. tinctorial*, *I. arrecta*, and or *S. cusia* were employed after a 3–5-day fermentation and emulsification process to generate an indigo paste; therefore, dissolution in a water ratio of 1:1 (m:v) before usage in the coloring stage is required. The propagule of *Rhizopora* spp. is another material used that must be boiled for over two hours before utilizing as a brown-light tint. In addition to the single-use, Zie Batik is created by mixing natural dyes to obtain new shades. The combination of these pigments, including *P. ferruginum* and *I. tinctorial*, was employed in the manufacturing of black while *C. sappan* and *C. candolleana* are used for brick-red. This organic pigment mixing relies on basic colors, including red, blue (indigo), and yellow. The addition of these paints produce a dark tone and is used to adjust the brightness level [18].

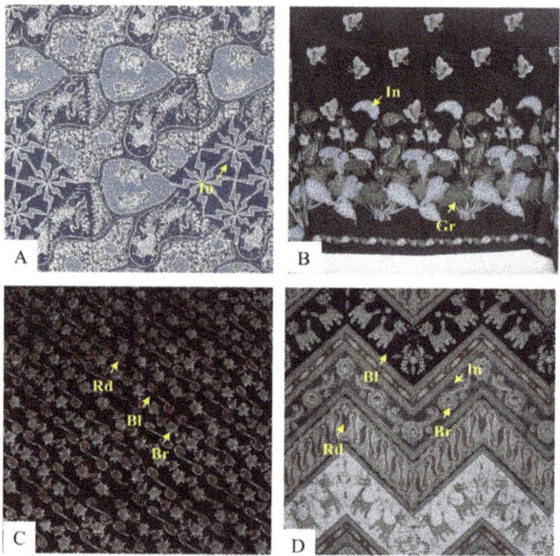

Figure 2. Various natural batik of the Zie Batik products by organic material-based natural dyes: (**A**) batik mina (fish motif), one natural dye; (**B**) caladium leaves and butterfly motifs, three natural dyes; (**C**) parang mangrove motif, three natural dyes; and (**D**) Warak Ngendhog motif, four natural dyes. In each batik, the main colors were indigo (In), green (Gr), red (Rd), black (Bl), and brown (Br).

Furthermore, Zie Batik has established a collaboration work with local farmers to build independent crop cultivations because of the natural dye demand to reduce the dependency on coloring plants from other places. This is a system of value chain utilization to reduce production costs from the energy-use sector. The simplification of the business line increases production efficiency and community involvement in actively contributing to the environment's protection [19–21].

2.2. Clean Production Management Scheme

The SME developed by Zie Batik is an ideal example of a suitable production implementation during natural dye provision for the batik coloring process. Figure 3 displays the natural dye supply stage performed by the treatment of the required organic materials through drying, boiling, fermentation, and emulsification into ready-to-use coloring materials.

Generally, the waste produced from the process was managed simply, through natural decomposition; also, the timber acquired from the hard wood natural dyer was employed as the fuel, and the resultant ash was used as a fertilizer for the indigo plant (Table 2).

However, the products were not polluted by these methods but by other processes, e.g., the utilization of a mordant followed by fixation produced the bulk of the wastewater, subsequently composed of sodium carbonate (Na_2CO_3) and alum ($KAl(SO_4)_2·12H_2O$). This challenge is possibly managed through repeated use, while home-scale water waste management systems (WMS) was used for purification, as seen in Figure 4.

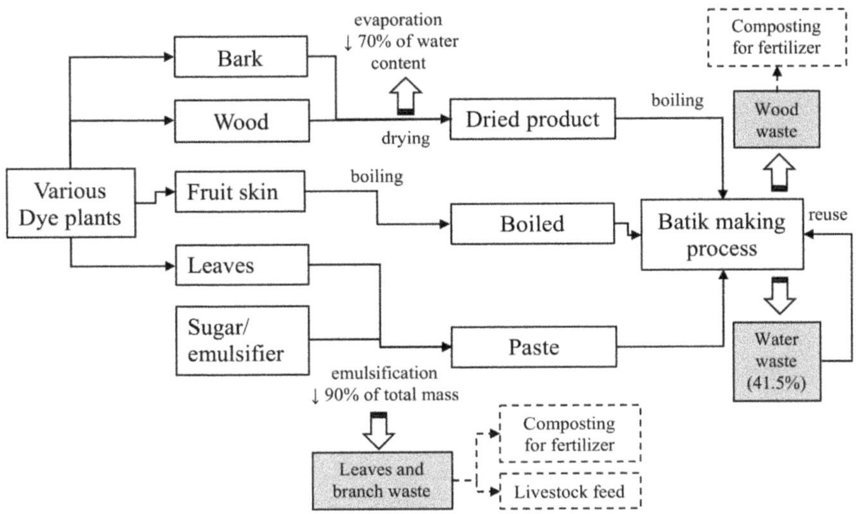

Figure 3. Natural dye-producing process and waste management during the coloring stages of the cleaner batik production method implemented by the Zie Batik SME.

Table 2. Dye plant sources and waste management.

Natural Dyes Plant	Sources	Cultivation	By-Product Waste	Waste Management
Caesalpinia sappan	Semarang, Central Java	Wide distributed but does not cultivated	Wood shavings	Decomposed or firewood
Ceriops candolleana/ C. tagal	Coastal area of Semarang City	-	No waste	-
Maclura cochinchinensis	Surakarta, Central Java, originally from Sumatra and Kalimantan	-	Wood pulp	Decomposed
Pelthophorum ferruginum *	Surakarta, Central Java	-	-	-
Indigofera tinctoria	-	Local farmers	No waste	-
Indigofera arecta	-	Local farmers	No waste	-
Rhizopora spp.	Coastal area of Semarang City	-	No waste	-
Strobilantes cusia	Japan (was introduced in 2016 for dye plant diversification in Malon)	Local farmers	No waste	-
Terminalia bellirica	Semarang	-	Wood shavings	Decomposed or timber

Note: The star mark (*) indicates a rare material that is difficult to get.

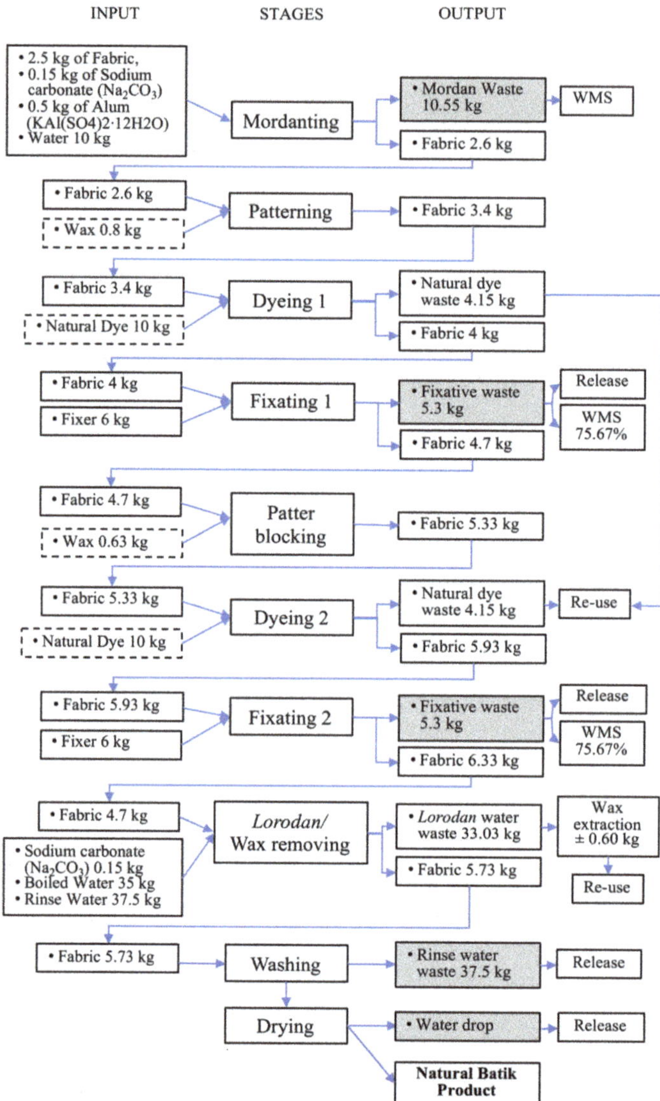

Figure 4. Natural dye batik-producing process in Zie Batik. The boxes with dotted lines represent a biodegradable component, and the grey boxes represent a water waste by-product. WMS: water waste management systems.

2.3. Clean Production Improvement Strategy

The application of proper maintenance to facilitate clean production is useful as an alternative strategy to surmount inefficiencies. The draining and retention procedures performed at each stage of mordant use, fixation, and desiccation are approaches to reduce environmental pollution and water contamination (Table 3).

Table 3. Existing waste management in Zie Batik.

Step	Problems	Existing Waste Management	Alternative Housekeeping and Waste Management
Mordanting	Wastewater contain Na_2CO_3 and $KAl(SO_4)_2 \cdot 12H_2O$ and material of fabric	Reused and deposited in the waste container	Precipitation process to separate solid waste (microfiber) and water purification using a phytoremediation technique
Patterning	-	-	-
Dyeing 1 and 2	Natural dye waste	Wood-based dye, the residual product was composted in open areasPaste-based dye was reused until no remaining compound was left	Composting process for wood in a closed and localized place
Fixating 1 and 2	Wastewater contain $CaCO_3$ and $FeSO_4$	Almost 75% was drained to the environment, and the remaining waste was deposited in the anaerobe-waste container	Streamlining the existing wastewater management installation by increasing the capacity and maintaining the installation
Pattern Blocking/waxing	Scattered and evaporated wax	Collected and reused	-
Lorodan/Wax removing	Wastewater contain wax	Removing process to collect the wax on the surface, and reused for next pattern blocking/waxing	Draining the remaining wastewater to the wastewater management installation
Washing and Drying	Water drop on the floor	Discharged into the environment	Applying a water drain collector

Source: Field observation.

Another problem encountered by Zie Batik in the implementation of clean production was the unreliable production frequency of the natural batik supply. Despite the ease of environmental damage estimation through fixed production agendas, a large magnitude of manufacturing regularity remains dependent on consumer orders. This had the potential of affecting the continuity and availability of the natural batik products in the market. Manufacturing schedules were therefore presented as a solution to preserve the longevity and sustainability of the industry. Furthermore, the execution of planning and appropriate storage and control systems for the dye materials are useful to curb fluctuations and irregular product demands.

3. Discussion

The creation of batik with natural dyes is more expensive than with synthetic compounds, as a longer time and advanced skills are required [22]. In addition, the natural dye color is dependent on many factors as a result of the biological origin of the materials. Therefore, batik textiles prepared from organic substances possess different color patterns, bestowing an exclusivity and uniqueness in comparison with artificial materials.

A heightened awareness towards environmental harm has encouraged Zie Batik to implement modifications on products from merely being profit-oriented to performing eco-friendly production activities. The use of organic materials as natural dyes was propelled by concerns of the owner regarding ecological destruction caused by synthetic dye practices. The innovations were executed on the fabric motifs, as well as the types of coloring ingredients used. Initially, organic dyes were exclusively developed from propagules (mangrove fruits) to replace synthetic brown and other dark tones. However, the textile vendor explored dye plants in Semarang and other cities to discover potential sources useful in batik production, although the majority of the necessary organic components are supplied from other districts or provinces in Java Island.

The colorants are essential items in the batik industry and of high quality; neatly executed and attractive motifs executed on comfortable fabrics are expected by clients. [23]. Therefore, the use of eco-friendly organic dye chemicals is currently preferred internationally. These compounds are obtainable from microorganisms including fungi [24], as well as plants, insects, and minerals either directly used or through extraction processes [22]. The natural pigments produced from organic matter are distinctive, as different hues are capable of being generated from a single source [25,26]. However, the use of these starting materials are not devoid of drawbacks and these include difficulty in acquisition due to market unavailability, rapid discoloration, limited color choices [27], minimal stability, pallid appearance, and greater costs in contrast with synthetic dyes [28].

The use of organic materials as an approach to ideal production operations is expected to increase the environmental consciousness of the manufacturer, reduce energy and resource consumption, improve raw material and waste management, and therefore decrease pollution [29]. Moreover, the materials employed in the dry state, including tree barks and woody parts, were dried under sunlight for 5–7 days, and then boiled to extract the concentrated pigments before use. These steps assist the achievement of dye compounds from organic materials, and activate the colors simultaneously, promoting the ease and durability of the dye attachment onto cloth fibers [30,31]. The dehydration of the bark, wood, and fruit is capable of depleting over 70% of the water content, and also increases the length of storage [32]. Meanwhile, during indigo fermentation, an approximately equivalent weight of sucrose was added to generate a denser tint. Sucrose represented the carbon source required for the yeast and bacteria fermentation in the indigo paste preparation process [33]. The organisms involved were fungi from the *Saccharomyces*, *Aspergillus*, *Penicillium*, *Pleurotus*, and *Trametes* genera, [34], as well as several bacteria from the *Alcaligenes*, *Alkalibacterium*, *Amphibacillus*, *Bacillus*, *Corynebacterium*, *Halomonas*, and *Tissierella* genera [33,35,36].

The manufacturing process utilized produced organic waste comprising leaves and stems from the indigo plants. The solid waste had the capacity of being developed into several by-products, including fertilizers and animal feed. Currently, the waste management installation constructed in the Zie Batik industry areas handles solely liquid waste, while the solid remnants are discharged into open areas to facilitate natural decomposition. The composted waste increases the soil nutrient content essential for plant growth [37], and these practices reveal natural dye use as a portion of pollution control requiring zero waste management [38,39].

Mordanting involves the boiling of textiles with chemicals to increase penetration and strengthen the natural dyes' adhesion to the fabric fibers. The use of this solution can sharpens several colors from natural dyes [40–42]. This equally increases the attractiveness of the natural dyes to textile materials and is also useful for producing good color sharpness. Furthermore, swelling increases the affinity of dye, produces a broader color spectrum, and also has better fastness [40]. However, the boiling process in the mordanting stage is useful to remove residual impurities from the weaving process and improve fabric quality. The next stage is fixating, characterized by use of chemical compounds, especially the limestone fixation process ($CaCO_3$) and/or ferrous sulfates ($FeSO_4$). This also has a greater chance of causing environmental pollution, about 75.67% of waste are discharged into the environment and were drained into a management system, as shown in Figure 4.

Furthermore, based on observations, one batch of natural batik in Zie Batik production causes liquid waste worth 95.18 kg per 10 textiles or about 9.518 kg/sheet of batik fabric on average. The amount of pollutants from this industry is higher than the synthetic batik craftsmen's production of water waste in Bogor City [43]. This is probably caused by the inefficient implementation of the clean production process. Meanwhile, based on the concept of clean production, Zie Batik is believed to be inefficient in water usage, while it serves as a main resource in the batik manufacturing process. This outcome is in line with the research of Handayani et al., [44] emphasizing considerable water inefficiency during production of natural dye batik. These colorants are used only once a day, therefore limiting the chances of storage [22]. This is due to the fact that storage of these dyes fades the color and also produces unclear highlighting. In addition, natural dyes are basically organic materials easily degraded either by temperature, oxidation or bacterial activity [45]. Meanwhile, environment conditions and microbial activities affect the transformation of organic material, and therefore are more environmentally friendly than synthetic dyes.

This study, however, answered the hesitation question of SMEs' ability to develop a core business based on environmental sustainability without ignoring economic and social aspects. However, this hesitation arose because most practitioners assume nature and business aspects conflict when sustainable developments are established. Furthermore, Zie Batik successfully proved to increase the ability of SMEs to act as a driving force in integrating sustainable development aspects through application of clean production. These efforts were made by business actors to improve the quality of this product, and equally attract buyers with creativity, uniqueness, and quality preference. The adaptation eco-friendly practices by this industry promotes more positive employee development and also increase the SMEs' productivity in addition to long-term profitability [46].

Furthermore, the dyeing process of Zie Batik was carried out 3 to 5 times to get the expected color; this is, however, different from synthetic dyes done only 1–2 times [47]. This process equally produces an exothermic reaction, therefore causing the natural dye in the solution to migrate to the fabric to have a balanced concentration [48]; this also increases the concentration of the dye on the fabric when done more often. This technique was used by Zie Batik to produce color gradations and produce more diverse variants in the form of motifs. The rinsing and drying process has not leveled up with the housekeeping standards. Furthermore, some problems, especially inadequate water reservoirs, droplets, and spills, have led to increased liquid waste, in addition to the potential danger caused by slippery floors. The production of this material is not easy to schedule, it being highly dependent on the weather and availability of the organic and fabric materials, in addition to the motif creation. Furthermore, several studies have proved natural batik contains pollutants, especially remaining wax, coloring agents, salts, and fixators; therefore, waste treatment is required before it being discharged into the rivers or drainages [49,50]. This, therefore, outlines the importance of wastewater treatment to decrease environmental pollution [51]

Furthermore, low environmental enforcement and high initial capital costs is the main obstacles towards the implementation of clean production. However, in overcoming these problems, Batik Zie received support from the relevant stakeholders, including local governments, universities, and the private sector, in addition to cooperation built between the SME actors. The campaign of this product through exhibition also continues to serve as promotion by raising the tagline "Batik by natural color".

4. Materials and Methods

This investigation was carried out as a case study research [52] with a natural observational explanatory model to describe the clean production process in batik with natural dyes. The data collection method involves a 3-year (2016–2019) observation as part of batik's small and medium-sized enterprises (SMEs) community services programs in Malon Village, Gunungpati, Semarang City, Indonesia. Furthermore, the locus of this study was in Malon Village, with a focus on the Zie Batik SME, due to the following reasons: (1) Zie Batik has been developing clean production since 2006; (2) it was the first SME to develop natural dyes as its core business in Semarang City; (3) it served as the city governments reference for eco-friendly lessons on natural batik; and (4) it provides resource personnel to act as trainers for eco-friendly process learning in natural batik production.

The data was collected by directly observing the dye- and batik-producing process and waste management in Zie Batik's working space. Furthermore, to confirm the data used in this investigation, in-depth interviews were held with 15 respondents—dye plant farmers; 3 respondents—the Zie Batik owners; 5 respondents—the Zie Batik workers; and 15 interim students at Zie Batik. This information was then arranged through a matching process to get the completeness and reliability. The reliable information was however reshaped to match an analyzable data format and was then reduced to eliminate unnecessary data. In addition, the final information was transformed into data and analyzed through categorizing, labeling, and annotating. The final result was interpreted with a narrative analysis technique. Meanwhile, the data obtained include the raw materials used, mass balance of the batik production process, the produced waste, the obstacles encountered, and the good housekeeping methods applied.

The clean manufacturing process of this industry was formulated through an analysis of the natural dyes' production from organic material. Moreover, the net production of this product includes organic material resource management of the coloring pastes narrowed down normatively. Based on the collected information, this research was, however, focused on the alternative management procedure of water waste in batik production.

5. Conclusions

Zie Batik is a small and medium-scale enterprise known to run an eco-friendly batik production business. Furthermore, use of this product serves as an alternative to reduce the impact of environmental pollution caused by synthetic dyes in textile dyeing. This occurs mainly due to the biodegradability, harmlessness, ease of acquisition, and non-toxic liquid waste characteristics of natural dyes. In supporting natural batik production, we found that the color is well-extracted from various plants, especially *C. sappan*, *C. candolleana*, *M. cochinchinensis*, *P. ferruginum*, *I. tinctorial*, *I. arrecta*, *Rhizopora spp.*, *S. cusia*, and *T. bellirica*. Meanwhile, some plant parts, including the bark, stems, leaves, roots, seeds, and sap, are used as natural coloring agents. Furthermore, Zie Batik's net production emphasizes more on the activity of producing dyes and other coloring processes. The remaining natural dyes are used for coloring activities while the organic waste produced was composted naturally and used as timber. Even though the consumption of chemical dyes can be avoided, Zie Batik's use of water in the natural batik-producing process is still wasteful and not well managed. Several improvements should be conducted to minimize the wasting of water and reducing liquid waste discharging into the environment. In addition, regarding Zie Batik's daily production process, they need to enhance the water container volume and build the water treatment plant properly.

In addition, the natural dyes involved as part of the clean production strategy by the Zie Batik SME is capable of manufacturing high-quality, unique, and profitable batik, with a high business value, thus also supporting the social realms beside that of the environmental aspects. However, the use of this product is not left without any disadvantages as it is limited by low color stability, homogeneity, and limited raw material supply compared to synthetic dyes. These limitations affect the continuity of daily production, circumvented by adjusting the manufacturing to meet the consumer's demand.

Author Contributions: This research was constructed by a cooperating program. Conceptualization, N.K.T.M. and I.H.; instrument and methodology, I.H., M.M. and R.B.A.; validation, N.K.T.M.; formal analysis, N.K.T.M.; investigation, R.B.A.; resources, N.K.T.M.; data curation, I.H.; writing—original draft preparation, N.K.T.M. and R.B.A.; writing—review and editing, I.H. and M.M.; visualization, R.B.A.; supervision, N.K.T.M.; project administration, M.M.; funding acquisition, N.K.T.M. All authors have read and agreed to the published version of the manuscript.

Funding: This study was supported by a project that has received funding from the Deputy of Strengthening Research and Technology, Ministry of Research and Technology of Republic of Indonesia through Community Service Program in the scheme of the Regional Excellence Product Development Program (PPPUD), Research and Community Service Fund (Dana Riset dan Pengabdian Masyarakat—DRPM) for 2018-2020, Grant agreement number 071/SP2H/PPM/DRPM/2020,

Acknowledgments: The authors acknowledge Marheno, Zazilah, and Sasi Syifaurohmi as the owners of Zie Batik who have provided permission, assistance, and a place for the authors to conduct this research, and then allowed to publish the paper.

Conflicts of Interest: The authors declare no conflict of interest. The funders had no role in the design of the study; in the collection, analyses, or interpretation of data; in the writing of the manuscript, or in the decision to publish the results.

References

1.   Steelyana, E.; Batik, A. Beautiful Cultural Heritage that Preserve Culture and Supporteconomic Development in Indonesia. *Binus Bus. Rev.* **2012**, *3*, 116. [CrossRef]

2. Krisnawati, E.; Sunarni, N.; Indrayani, L.M.; Sofyan, A.N.; Nur, T. Identity Exhibition in Batik Motifs of Ebeg and Pataruman. *SAGE Open* **2019**, *9*, 2158244019846686. [CrossRef]

3. Tresnadi, C.; Sachari, A. Identification of Values of Ornaments in Indonesian Batik in Visual Content of Nitiki Game. *J. Arts Humanit.* **2015**, *4*, 25–39.

4. Rudianto, M.; Sulistyati, A.; Rizali, N.; Faizin, A. *Batik Sudagaran as Cultural Creation of the Laweyan's Merchants*; Seword Press: Surakarta, Indonesia, 2019.

5. Lellis, B.; Fávaro-Polonio, C.Z.; Pamphile, J.A.; Polonio, J.C. Effects of textile dyes on health and the environment and bioremediation potential of living organisms. *Biotechnol. Res. Innov.* **2019**, *3*, 275–290. [CrossRef]

6. Thiripuranthagan, S.; Rupa, V. Detoxification of Carcinogenic Dyes by Noble Metal (Ag, Au, Pt) Impregnated Titania Photocatalysts. In *Gold Nanoparticles-Reaching New Heights*; IntechOpen: Rijeka, Croatia, 2019; pp. 1–26.

7. Chung, K.T. Azo dyes and human health: A review. *J. Environ. Sci. Heal. Part C Environ. Carcinog. Ecotoxicol. Rev.* **2016**, *34*, 233–261. [CrossRef]

8. Budiyanto, S.; Anies; Purnaweni, H.; Sunoko, H.R. Environmental Analysis of the Impacts of Batik Waste Water Polution on the Quality of Dug Well Water in the Batik Industrial Center of Jenggot Pekalongan City. *E3S Web Conf.* **2018**, *31*, 09008. [CrossRef]

9. Naqsyabandi, S.; Riani, E.; Suprihatin, S. Impact of batik wastewater pollution on macrobenthic community in Pekalongan River. *AIP Conf. Proc.* **2018**, *2023*, 020128.

10. Lestari, S.; Sudarmadji; Tandjung, S.D.; Santosa, S.J. Lethal toxicity of Batik waste water bio-sorption results in Tilapia (*Oreochromis niloticus*). *Adv. Sci. Lett.* **2017**, *23*, 2611–2613. [CrossRef]

11. Syauqiah, I.; Nurandini, D.; Lestari, R. Study of Potential Pollution of Sasirangan Liquid Waste with Biological and Chemical Parameters. *BIO Web Conf.* **2020**, *20*, 02003. [CrossRef]

12. Joko, T.; Dangiran, H.L.; Astorina, N.; Dewanti, Y. The Effectiveness of Plant Pistia Stratiotes Weight to Reduction of Heavy Metal Content Chromium (Cr) Waste at Batik Home Industry in Regency of Pekalongan. *Int. J. Sci. Basic Appl. Res.* **2015**, *24*, 45–54.

13. Arifan, F.; Winarni, S. Analysis of copper and lead on coastal environment a case study of Kaliprau coastal, Pemalang. *Adv. Sci. Lett.* **2017**, *23*, 3534–3536. [CrossRef]

14. Yaacob, M.R.; Ismail, M.; Zakaria, M.N.; Zainol, F.A.; Zain, N.F.M. Environmental Awareness of Batik Entrepreneurs in Kelantan, Malaysia—An Early Insight. *Int. J. Acad. Res. Bus. Soc. Sci.* **2015**, *5*, 338–347. [CrossRef]

15. Rezagama, A.; Sutrisno, E.; Handayani, D.S. Pollution Model of Batik and Domestic Wastewater on River Water Quality. *IOP Conf. Ser. Earth Environ. Sci.* **2020**, *448*, 012074. [CrossRef]

16. Kasiri, M.B.; Safapour, S. Natural dyes and antimicrobials for green treatment of textiles. *Environ. Chem. Lett.* **2014**, *12*, 1–13. [CrossRef]

17. Basak, S.; Senthilkumar, T.; Krishnaprasad, G.; Jagajanantha, P. Sustainable Development in Textile Processing. In *Sustainable Green Chemical Processes and their Allied Applications*; Inamuddin Asiri, A., Ed.; Springer Nature: Mumbai, India, 2020; pp. 560–571. ISBN 978-3-030-42283-7.

18. Sinaga, A.S.R. Color-based Segmentation of Batik Using the L*a*b Color Space. *SinkrOn* **2019**, *3*, 175–179. [CrossRef]

19. Fessehaie, J. *Leveraging the Services Sector for Inclusive Value Chains in Developing Countries*; International Centre for Trade and Sustainable Development (ICTSD): Geneva, Switzerland, 2017.

20. Nandamudi, S.K.; Sen, A. Landscape Restoration and Community Involvement in Biodiversity Conservation. In *Corporate Biodiversity Management for Sustainable Growth*; Sharma, R., Watve, A., Pandey, A., Eds.; Springer Nature: Pune, India, 2020; pp. 127–150. ISBN 9783030427023.

21. Baik, Y.; Park, Y.R. Managing legitimacy through corporate community involvement: The effects of subsidiary ownership and host country experience in China. *Asia Pac. J. Manag.* **2019**, *36*, 971–993. [CrossRef]

22. Zerin, I.; Farzana, N.; Sayem, A.S.M.; Anang, D.M.; Haider, J. Potentials of Natural Dyes for Textile Applications. *Encycl. Renew. Sustain. Mater.* **2020**, 873–883. [CrossRef]

23. Rahadi, R.A. The Analysis of Consumers' Preferences for Batik Products in Indonesia. *Rev. Integr. Bus. Econ. Res.* **2020**, *9*, 278–287.

24. Sánchez-Muñoz, S.; Mariano-Silva, G.; Leite, M.O.; Mura, F.B.; Verma, M.L.; Da Silva, S.S.; Chandel, A.K. *Production of Fungal and Bacterial Pigments and Their Applications*; Elsevier, B.V.: Oxford, UK, 2019; ISBN 9780444643230.

25. Verma, S.; Gupta, G. Natural dyes and its applications: A brief review. *Int. J. Res. Anal. Rev.* **2017**, *4*, 57–60.

26. Kusumawati, N.; Samik, S.; Kristyanto, A. Quality Improvement of ABBS Hand Writing Batik Production through Standardization of Natural Dyeing using Water Guava and Mango Leaves. *Atl. Highlights Chem. Pharm. Sci.* **2020**, *1*, 29–32.

27. Indrianingsih, A.W.; Darsih, C. Natural Dyes from Plants Extract and Its Applications in Indonesian Textile Small Medium Scale Enterprise. *Eksergi* **2014**, *11*, 16. [CrossRef]

28. Bhuiyan, M.A.R.; Islam, A.; Ali, A.; Islam, M.N. Color and chemical constitution of natural dye henna (Lawsonia inermis L) and its application in the coloration of textiles. *J. Clean. Prod.* **2017**, *167*, 14–22. [CrossRef]

29. Hasibuan, S.; Hidayati, J. The Integration of Cleaner Production Innovation and Creativity for Supply Chain Sustainability of Bogor Batik SMEs. *Int. J. Ind. Manuf. Eng.* **2018**, *12*, 679–684.

30. Fatima, A.; Raja, G.K.; Shah, M.A.; Gul, I.; Tabassum, S.; Batool, S.; Ahmed, W.; Ahmad, M.S. Extraction and Evaluation of a Natural Dye From Bistorta Amplexicaule. *Pak. J. Sci.* **2017**, *69*, 195.

31. Najera, F.; Dippold, M.A.; Boy, J.; Seguel, O.; Koester, M.; Stock, S.; Merino, C.; Kuzyakov, Y.; Matus, F. Effects of drying/rewetting on soil aggregate dynamics and implications for organic matter turnover. *Biol. Fertil. Soils* **2020**. [CrossRef]

32. Alfiyanti, I.L.; Fatoni, R.; Fatimah, S. The Application of Mahogany Bark (Swietenia Mahagony L.) for Natural Dyeing. *Adv. Sustain. Sci. Eng. Technol.* **2020**, *2*, 1–7.

33. Aino, K.; Narihiro, T.; Minamida, K.; Kamagata, Y.; Yoshimune, K.; Yumoto, I. Bacterial community characterization and dynamics of indigo fermentation. *FEMS Microbiol. Ecol.* **2010**, *74*, 174–183. [CrossRef]

34. De Oliveira Rodrigues, P.; Gurgel, L.V.A.; Pasquini, D.; Badotti, F.; Góes-Neto, A.; Baffi, M.A. Lignocellulose-degrading enzymes production by solid-state fermentation through fungal consortium among Ascomycetes and Basidiomycetes. *Renew. Energy* **2020**, *145*, 2683–2693. [CrossRef]

35. Okamoto, T.; Aino, K.; Narihiro, T.; Matsuyama, H.; Yumoto, I. Analysis of microbiota involved in the aged natural fermentation of indigo. *World J. Microbiol. Biotechnol.* **2017**, *33*, 1–10. [CrossRef]

36. Aino, K.; Hirota, K.; Okamoto, T.; Tu, Z.; Matsuyama, H.; Yumoto, I. Microbial communities associated with indigo fermentation that thrive in anaerobic alkaline environments. *Front. Microbiol.* **2018**, *9*, 1–16. [CrossRef]

37. Pattanaik, L.; Duraivadivel, P.; Hariprasad, P.; Naik, S.N. Utilization and re-use of solid and liquid waste generated from the natural indigo dye production process – A zero waste approach. *Bioresour. Technol.* **2020**, *301*, 122721. [CrossRef]

38. Sharma, K.; Garg, V.K. *Vermicomposting of Waste*; Elsevier, B.V.: Oxford, UK, 2019; ISBN 9780444642004.

39. Marín, M.; Artola, A.; Sánchez, A. Production of proteases from organic wastes by solid-state fermentation: Downstream and zero waste strategies. *3 Biotech* **2018**, *8*, 205. [CrossRef] [PubMed]

40. Shabbir, M.; Rather, L.J.; Bukhari, M.N.; Ul-Islam, S.; Shahid, M.; Khan, M.A.; Mohammad, F. Light Fastness and Shade Variability of Tannin Colorant Dyed Wool with the Effect of Mordanting Methods. *J. Nat. Fibers* **2019**, *16*, 100–113. [CrossRef]

41. Moniruzzaman, M.; Mondal, M.S.; Hossain, M.N. The Influence of Mordant and Mordanting Techniques on Ecofriendly Dyeing of Cotton Fabric By Extracted Used Tea. *J. Eng. Sci.* **2018**, *9*, 111–117.

42. Manian, A.P.; Paul, R.; Bechtold, T. Metal mordanting in dyeing with natural colourants. *Color. Technol.* **2016**, *132*, 107–113. [CrossRef]

43. Fauzi, A.M.; Defianisa, R.L. Analysis for cleaner production implementation strategy in batik industry in Bogor. *IOP Conf. Ser. Earth Environ. Sci.* **2019**, *325*, 012005. [CrossRef]

44. Handayani, W.; Kristijanto, A.I.; Hunga, A.I.R. A water footprint case study in Jarum village, Klaten, Indonesia: The production of natural-colored batik. *Environ. Dev. Sustain.* **2019**, *21*, 1919–1932. [CrossRef]

45. Jain, M.S.; Daga, M.; Kalamdhad, A.S. Composting physics: A degradation process-determining tool for industrial sludge. *Ecol. Eng.* **2018**, *116*, 14–20. [CrossRef]

46. Musa, H.; Chinniah, M. Malaysian SMEs Development: Future and Challenges on Going Green. *Procedia—Soc. Behav. Sci.* **2016**, *224*, 254–262. [CrossRef]

47. Richards, P.R. *Fabric Finishing: Dyeing and Colouring*; Elsevier Ltd.: New York, NY, USA, 2015; ISBN 9780857095619.

48. Indarti Rahayu, I.A.T.; Peng, L.H. Sustainable Batik Production: Review and Research Framework. *Adv. Soc. Sci. Educ. Humanit. Res.* **2020**, *390*, 66–72.
49. Birgani, P.M.; Ranjbar, N.; Abdullah, R.C.; Wong, K.T.; Lee, G.; Ibrahim, S.; Park, C.; Yoon, Y.; Jang, M. An efficient and economical treatment for batik textile wastewater containing high levels of silicate and organic pollutants using a sequential process of acidification, magnesium oxide, and palm shell-based activated carbon application. *J. Environ. Manag.* **2016**, *184*, 229–239. [CrossRef] [PubMed]
50. Sirait, M. Cleaner production options for reducing industrial waste: The case of batik industry in Malang, East Java-Indonesia. *Jkt. IOP Conf. Ser. Earth Environ. Sci.* **2018**, *106*, 012069. [CrossRef]
51. Buthiyappan, A.; Abdul Raman, A.A.; Daud, W.M.A.W. Development of an advanced chemical oxidation wastewater treatment system for the batik industry in Malaysia. *RSC Adv.* **2016**, *6*, 25222–25241. [CrossRef]
52. Harrison, H.; Birks, M.; Franklin, R.; Mills, J. Case study research: Foundations and methodological orientations. *Forum Qual. Sozialforsch.* **2017**, *18*. [CrossRef]

Publisher's Note: MDPI stays neutral with regard to jurisdictional claims in published maps and institutional affiliations.

© 2020 by the authors. Licensee MDPI, Basel, Switzerland. This article is an open access article distributed under the terms and conditions of the Creative Commons Attribution (CC BY) license (http://creativecommons.org/licenses/by/4.0/).

Communication

On the Production of Potassium Carbonate from Cocoa Pod Husks

Kouwelton Kone [1][ID]**, Karl Akueson** [2] **and Graeme Norval** [3,*][ID]

1   DFR Génie Chimique et Agro-Alimentaire, Institut National Polytechnique Félix Houphouët-Boigny de Yamoussoukro, Yamoussoukro BP 1093, Cote D'Ivoire; kouwelton.kone@inphb.ci

2   La Financière de l'Eléphant, 1er Etage, Immeuble Cormoran, Deux Plateaux Vallon, Cocody, Abidjan, Cote D'Ivoire; karl@lafinele.com

3   Department of Chemical Engineering and Applied Chemistry, University of Toronto, Toronto, ON M5S 3E5, Canada

*   Correspondence: Graeme.norval@utoronto.ca

Received: 27 July 2020; Accepted: 11 September 2020; Published: 17 September 2020

Abstract: Cocoa beans are found inside an outer husk; 60% of the cocoa fruit is the outer husk, which is a waste biomass. The husk cannot be used directly as a soil amendment as it promotes the fungal black pod disease, which reduces crop yield. The pods are segregated from the trees, and their plant nutrient value is wasted. This is particularly true for the small acreage farmers in West Africa. Cocoa pod husk is well suited to be used as a biomass source for electricity production. The waste ash is rich in potassium, which can be converted in various chemical products, most notably, high-purity potassium carbonate. This study reviews the information known about cocoa and cocoa pod husk, and considers the socio-economic implications of creating a local economy based on collecting the cocoa pod husk for electricity production, coupled with the processing of the waste ash into various products. The study demonstrates that the concept is feasible, and also identifies the local conditions required to create this sustainable economic process.

Keywords: potassium carbonate; cocoa pod husk; biomass ash

---

1. Introduction

Cocoa is native to central America, where it normally grows beneath the canopy of taller trees. The yield of cocoa is lower when cocoa is grown in this fashion. Clearing the shade trees increases the amount of direct sunlight reaching the cocoa trees, and the yield increases. The increase in yield is short-lived; the cocoa yield begins to be limited by soil conditions, and in particular, the potassium content of the soil [1]. A typical practice is to add fertilizers; these include the standard NPK (Nitrogen, Phosphorus, Potassium) mixtures, as well as calcium and magnesium supplements, depending on the soil conditions. Typical fertilization rates are in the range of 100–500 kg/annum per ha.

Cocoa beans are rich in inorganic matter; typical analyses are 2.5% K in the dry beans. At a cocoa production rate of 600 kg dry beans/ha, the potassium lost with the cocoa product is of the order of 15 kg K/ha [2]. When cocoa is harvested, the seeds are taken from the pods; 60% of the mass of the cocoa pod is the husk. Fresh cocoa pod husk contains ~1.6% K and 0.7% P [3]. The leaves contain a further ~1.5 wt% K.

In principle, the leaf litter and cocoa pod husk should be spread underneath the trees, in order to boost the organic content of the soil, as well as to provide K and P, as well as Ca and Mg. Black pod disease is a fungal rot of the cocoa pod, which reduces the cocoa yield, with yield losses of up to 30% [4]. As with other fungal diseases, affected pods must be segregated from the rest of the farm and destroyed, so that the spores are not able to infect subsequent crops.

If cocoa is grown as part of a mixed agricultural system, the soil nutrients could be replaced by the organic matter from the other trees [5]. Unfortunately, this reduces the acreage available for cocoa. Most cocoa farms are small, a few acres, and the farmers rely on increased cocoa yields [6].

It has long been known that the ash produced when cocoa pod husks are burned is rich in potash [7]. The ashes are leached, giving an alkaline solution, rich in potassium hydroxide and potassium carbonate. This solution has been used as the alkali source for local soap production.

Further, when the cocoa pod husks are burned, the black pod fungus is destroyed, greatly reducing the likelihood of future black pod outbreaks. It follows that the spoiled cocoa pods as well as the cocoa husks should be burned because ashing will improve the control of black pod disease, as well as provide a valuable mineral source.

A sustainable chemical business is presented in which the cocoa pod husks are burned, generating electricity in the remote farm regions, and the biomass ash is converted in a suite of chemical products for use. The issue of the potassium balance is considered, and techno-commercial considerations are presented.

2. Results

2.1. Analyses of the Husk and Ash

The cocoa pod husk had a higher heaving value of 18.1 MJ/kg (dry husk) and a lower heating value of 17.1 MJ/kg (dry husk), with a 12.7% ash content. The major constituents of the ash were K (320 g/kg), Ca (65 g/kg), Mg (42 g/kg), and Si (9.1 g/kg). The other elements were reported as S (7100 mg/kg), P (5600 mg/kg), Al (1100 mg/kg), and Mo (1100 mg/kg), followed by Ba (250 mg/kg), Cd (<3 mg/kg), Co (< 10 mg/kg), Cr (<10 mg/kg), Cu (150 mg/kg), Fe (940 mg/kg), Ni (33 mg/kg), Pb (<10 mg/kg), Ti (69 mg/kg), V (<25 mg/kg), and Zn (530 mg/kg).

2.2. The Farmer's Considerations

For the purpose of this analysis, a small cocoa farm (5 ha) will be considered. Farms of this size are family run; they tend to have older trees, and also have lower yields linked to the low use of chemical fertilizer [6]. For this farm, we can consider the balance loss to both the cocoa bean and the cocoa pod. For this calculation, the ratio of wet pod to dry beans is 10:1; 60% of the wet pod is wet husk, of which 1.6% is K and 0.7% is P.

The 5-ha farm generating 250 kg/ha of dry beans will produce 12.5 tons/annum of wet pod. After ashing, the farm would have 825 kg/annum of ash. This would be split into 500 kg/annum of potassium carbonate and 325 kg/annum of calcium/magnesium solid.

The farm would generate roughly 2500 $/annum through the sale of dry beans, at a price of 2000 US$ per ton. There would be an avoided cost of 620 $/annum, based on a fertilizer value of 750 US$/ton. The value of the ash is a significant improvement in the annual income of the farm.

3. Discussion

The ash contains the inorganic elements present in the cocoa husk. The predominant cations species are potassium, calcium, and magnesium (as well as sodium) and will be present as carbonates (K_2CO_3, $CaCO_3$, $MgCO_3$, Na_2CO_3). The phosphorus will be present as calcium phosphate, $Ca_3(PO_4)_2$.

Chemical analysis of cocoa pod husks from a variety of countries has been reported (3) and is consistent with these results. The average ash content was 10.7 wt%. The average potassium content was 1.63 wt%. This gives a K_2CO_3 yield of 5.76 wt% of the starting cocoa pod husks. The average calcium, magnesium, and phosphate contents are 0.33 wt%, 0.93 wt%, and 0.69 wt%. The insoluble calcium carbonate and magnesium hydroxide contents would be 0.82 wt% and 2.23 wt% of the starting cocoa pod husk.

The implication is that for every ton of cocoa pod husk, one will create ~60 kg of potassium carbonate and ~30 kg of a calcium/magnesium solid. The calcium/magnesium solid is best returned to the fields. These are necessary nutrients, and working the ash under the cocoa trees is part of an effective fertilizer program. Indeed, this would replace the import of calcium-based fertilizers.

The raw ash is a mixture of inorganic carbonates, oxides, and phosphates. Given that these inorganic species originated in the soil, it would seem straightforward to ash the cocoa husk and cocoa wood, to prevent the spread of black pod and other fungal diseases, and then to spread the ashes under the trees. Two specific issues prevent the direct application of ashes onto the soil. The first is that the ashes are dry and powdery and tend to blow away when dry. More importantly, the ash is hydroscopic (it attracts water); indeed, the oxides react with water, giving a hydroxide, which is strongly alkaline. It is not pleasant to work with the hydroscopic and strongly alkaline ash powder.

The better approach is to add the ash to water, with agitation. The potassium (and sodium) salts dissolve readily, giving a solution of potassium carbonate and potassium hydroxide, which is strongly alkaline. The concentration of the solution depends on the ratio of ash to water, and the amount of potassium in the ash, limited only by solubility.

The potassium solution is an effective fertilizer that will replace the import of potassium-based fertilizers. The solution could be evaporated to dryness, giving a crystal product, or the solution fertilizer could be applied as is. Applying the potassium as solution is the lowest cost option.

The recovery of potassium carbonate from wood ashes has been a long history [8]. Indeed, the first patent granted in the United States covered an improvement in the process for the recovery of potassium carbonate from wood ashes [8]. The ash should be near white in color; poor burning leaves carbonaceous residue in the ash, giving it a red to brown to black color. Darker-colored ashes should be burned a second time, in order to remove the residual organic matter, followed by dissolution of the ash in water, giving a calcium/magnesium/phosphate containing solid, and a potassium-rich solution. The cocoa pod husk is an excellent source of potassium salts.

3.1. Individual Farm Analysis

The data in Table 1 lead to two conclusions. The first is that the sale of cocoa beans and the loss of the husk from the farms leads to soil depletion of potassium and phosphorus along with other inorganic minerals, such as calcium and magnesium. The biomass accumulates various inorganic elements, more of which are present in the husk than in the beans. Farms that have high yields also have higher losses of inorganic elements from the soil. This leads to depletion of these elements from the soil, thereby causing a decrease in crop yields over time. The data in the table also represents the fertilizer requirement (kg/a/ha) just to make up for the losses in the biomass; but, since soluble potassium and phosphate fertilizers are applied, additional fertilizer is required to make up for the material that dissolves into the groundwater, and escapes the farm. It is noted that there is a significant amount of potassium held in the cocoa wood, which is a further source of cocoa biomass.

Table 1. Potassium balance of a 5-ha farm.

Yield (Dried Bean)	K (kg/a/ha in Dry Bean)	Wet Cocoa Pod Husk (kg/a)	K (kg/a/ha in Wet Husk)	P (kg/a/ha in Wet Husk)	K Loss (kg/a/ha)
250 kg/ha	6.25	1500	24	10.5	30.25
500 kg/ha	12.5	3000	48	21	60.5
750 kg/ha	18.75	4500	72	31.5	90.75

The mineral loss values in Table 1 are consistent with those of other reports [8]. Further, the recommended fertilizer additions [9] are significantly higher than the estimated loss of mineral; the recommended dosage rates are upwards of 350 kg/ha of K, based on a plantation of 1075 trees/ha. The dosage rate considers the K that is lost with the beans, and with the husk, as well as the potassium that accumulates in the growing wood; further, it recognizes that not all of the granular inorganic potassium ends up in the cocoa: A portion dissolves and drains into the ground water.

Finally, we note that it has long been known that wood ashes are rich in potassium; indeed, potash production was always located where forests were being cleared [7,8]. Cocoa plantations often have mixtures of other crops, such as coconut and plantain for shade. When these trees are harvested and burned, the ash will be similar in composition to that derived from the cocoa pod husk [10]. The ash can be mixed with that from the cocoa pod husk, thereby supplementing the yield of the potassium solution. Plantain is also rich in potassium, and the peelings and wood can be ashed with the ash being applied as a soil supplement.

As the farm returns the mineral content to the soil, there will be an increase in the yield of cocoa; the revenues and the value of the avoided costs both increase. This is especially true for the farm that does not use fertilizer supplements: The use of the cocoa pod husk ash allows them to increase the yield, using materials that they have on hand.

Additionally, many of the remote farming villages are not connected to the national electrical grid as they are too remote. The installation of a biomass to electricity plant will allow for the farming community to be electrified, leading to many subsequent improvements in living conditions.

3.2. Distribution of Ashing Sites

For the 5-ha farm, producing 250 kg/ha dry beans, the mass of dry pod is ~1875 kg. The 5 ha covers an area of 0.05 km^2. A circular radius of 1 km has 79 ha, or roughly 14 individual farms, and a dry pod yield of 29.6 ton.

The lower heat of combustion of dry cocoa pod husk is ~17 MJ/kg. The cocoa pod husk would be collected at harvest, and would partially dry. The heat of combustion of a wet cocoa pod husk is ~14 MJ/kg. A biomass to energy facility producing 1 MWe, and running at a 20% thermal efficiency, requires 5 MW of heat of combustion, or a supply of 1286 kg/h of wet cocoa pod husk. The annual requirement is ~8900 tons of dry cocoa pod husk. This requires a farming area of ~240 km^2, or a transportation radius of 17 km.

The implication of the thermodynamics is that the business model is better suited for multiple small facilities, rather than trying to build one very large production facility. The cocoa bean production in Cote d'Ivoire is ~2 million ton/annum, with an equivalent amount of dry cocoa pod husk produced. If it all could be used for electricity generation, the potential production capacity would be 225 MWe, which is roughly 10% of the current installed electrical capacity (2200 MWe). A small number of local producing sites will not impact the national electrical grid but will help in electrifying the remote farming communities that are not connected to the grid.

3.3. Project Scaling Issues

It is important to note that with an annual production of ~3 million ton of dry cocoa beans between Cote d'Ivoire and Ghana, there is sufficient cocoa pod husk available to saturate the global potassium carbonate market (~450,000 ton/annum). The difficulty is developing the infrastructure to collect all of the cocoa pod husks, to convert them into ash, and then to process the ash. The lack of transportation infrastructure and the low energy density of cocoa pod husk limit the scale of production that can be achieved.

The potassium carbonate product from the cocoa pod husk can readily displace some of the imported potassium carbonate. It also can be used to grow the production of palm oil-based soaps. These soaps are locally produced but not exported due to the requirement to import the alkali. The lower-cost biomass-derived alkali enables the production of soaps for export.

4. Materials and Methods

Samples of cocoa pod husk were obtained from farms in various growing regions of Cote d'Ivoire. The husk samples were analyzed using standard methods for % dry matter, % organic matter, and the elements C, N, P, K, Na, Ca, and Mg, along with higher and lower heating values. The ash produced was analyzed for the elements Al, Ba, Ca, Cd, Cl, Co, Cr, Cu, F, Fe, K, Mg, Mn, Mo, Na, Ni, P, Pb, S, Si, Ti, V, and Zn.

The information from these analyses was compared with values in the open literature. Material and energy balance calculations were performed at various scales, using EXCEL as the calculation tool. Next, the material and energy balance calculations were balanced against a farm-scale consideration, with the issue being to determine the supply for a local farm community, and the electrical generation capacity that could be obtained for that community.

The results of the calculations were compared with results of pilot plant data, obtained from a 10s of kg/h pilot plant, operated by Organic Potash Ltd., in Tema, Ghana. The calculations presented herein discuss the broader socio-economic aspects of the project.

5. Conclusions

The production of potassium carbonate from biomass-derived ash is not a new process. Cocoa has a high potassium content, and this enables the use of cocoa biomass ash as a raw material for both fertilizer and for chemical production. Normally, one would compost the biomass as a means of improving soil quality; this would return the inorganic matter to the soil as well as returning organic matter. Unfortunately, the prevalence of viral diseases precludes the use of composting as a means of returning the cocoa biomass to the soil.

The use of cocoa biomass for electricity production allows for a circular and more sustainable economy to be developed. The remote farming villages become electrified through connection to a microgrid; the standard of living of the farmers is improved without the need to import fossil fuels. The areas where the cocoa biomass is currently left to rot are cleared, which expands the fraction of arable land. The farmers have additional sources of income due to the sale of the cocoa biomass. Additionally, the filter cake is provided as a low-cost calcium, magnesium, and phosphate fertilizer, which reduces the need to import calcium-based fertilizers. Further, a small potassium chemical business will be created; the business would allow for a reduction in the imports of potassium chemicals, as well as enabling the creation of a lower-cost soap export business. Further, this decentralized business approach could be applied to some of the cooperatives, the local community organizations that consolidate the cocoa from individual farmers.

Author Contributions: Conceptualization, K.A., G.N.; methodology, K.A., G.N.; validation, K.K., K.A., G.N.; formal analysis, G.N.; investigation, K.K., K.A., G.N.; resources, K.K., K.A., G.N.; data curation, G.N.; writing—original draft preparation, G.N.; writing—review and editing, K.K., K.A., G.N.; project administration, G.N. All authors have read and agreed to the published version of the manuscript.

Funding: This research received no external funding.

Acknowledgments: The authors acknowledge the support of Organic Potash Ltd. for granting permission to publish this article.

Conflicts of Interest: The authors declare no conflict of interest.

References

1. van Vliet, J.A.; Slingerland, M.; Giller, K.E. *Mineral Nutrition of Cocoa, A Review*; Wageningen University and Research Centre: Wageningen, The Netherlands, 2015.

2. von Uexküll, H.R.; Cohen, A. Potassium Requirements of Some Tropical Tree Crops. *Potassium Requir. Crops* **1980**, 71–104. Available online: https://www.ipipotash.org/uploads/udocs/ipi_research_topics_no_7_potassium_requirements_of_crops.pdf (accessed on 25 July 2017).

3. Bonvehi, J.S.; Jorda, R.E. Constituents of Cocoa Husks. *Z. Für Nat. C* **1998**, *53*, 785–792. [CrossRef]

4. Acebo-Guerrero, Y.; Hernández-Rodríguez, A.; Heydrich-Pérez, M.; El Jaziri, M.; Hernández-Lauzardo, A.N. Management of BlackPod Rot in Cacoa, A Review. *Fruits* **2012**, *67*, 41–48. [CrossRef]

5. Asare, R. *Cocoa Agroforests in West Africa: A Look at Activities on Preferred Trees in the Farming Systems*; Forest & Landscape Denmark: Hørsholm, Denmark, 2005.

6. Wessel, M.; Quist-Wessel, P.M.F. Cocoa Production in West Africa, a Review and Analysis of Recent Developments. *NJAS Wagen. J. Life Sci.* **2015**, *74–75*, 1–7. [CrossRef]
7. Babayemi, J.O.; Adewuyi, G.O.; Dauda, K.T.; Kayode, A.A.A. The Ancient Alkali Production Technology and the Modern Improvement. *Asian J. Appl. Sci.* **2011**, *4*, 22–29. [CrossRef]
8. Maxey, D.W. Samuel Hopkins, the Holder of the First U.S. Patent: A Study of Failure. *Penn. Mag. Hist. Biog.* **1998**, *122*, 3–37.
9. Ghana, Y. Available online: http://www.yara.com.gh/crop-nutrition/crops/cocoa/key-facts/nutritional-summary/ (accessed on 2 August 2017).
10. Baon, J.B. Use of Plant Derived Ash as Potassium Fertilizer and Its Effects on Soil Nutrient Status and Cocoa Growth. *J. Tanah Trop.* **2009**, *14*, 185–193.

© 2020 by the authors. Licensee MDPI, Basel, Switzerland. This article is an open access article distributed under the terms and conditions of the Creative Commons Attribution (CC BY) license (http://creativecommons.org/licenses/by/4.0/).

 *recycling*

Article

Exploring Biogas and Biofertilizer Production from Abattoir Wastes in Nigeria Using a Multi-Criteria Assessment Approach

Idi Guga Audu [1,2,*], Abraham Barde [3], Othniel Mintang Yila [4], Peter Azikiwe Onwualu [5] and Buga Mohammed Lawal [1]

1. Raw Materials Research and Development Council, 17 Aguiyi Ironsi St, Maitama District, Abuja 900211, Nigeria; mblawal@hotmail.com
2. Department of Food Engineeering, Technical University Dresden, Bergstraße 120, 01069 Dresden, Germany
3. Armament Engineering Department, Air Force Institute of Technology, Kaduna 800282, Nigeria; abraham-barde@nda.edu.ng
4. Ministry of Meteorology, Energy, Information, Disaster Management, Environment, Climate Change and Communications (MEIDECC), Nuku'alofa 00100, Tonga; othnielyila@gmail.com
5. Materials Science and Engineering Programme, African University of Science and Technology, Abuja 900211, Nigeria; aonwualu@aust.edu.ng
* Correspondence: idig.audu@gmail.com

Received: 2 June 2019; Accepted: 12 August 2020; Published: 19 August 2020

Abstract: Management of waste streams from abattoirs is a major challenge in developing countries. Harnessing these wastes as resources for the production of biogas and biofertilizer could contribute to curbing the environmental menace and to addressing the problems of energy and food deficits in Nigeria. However, large scale uptake of the technology is faced with techno-socio-economic and the lack of data required for effective investment decisions. In this study, the potential use of waste generated in the north central region of Nigerian abattoirs, representing approximately 12% of the land and 6% of the population, were evaluated for suitability for biogas and biofertilizer production. Data acquired from the study sites were used for computational estimation and integrated into strengths, weaknesses, opportunities, and threats (SWOT) analysis to give a detailed overview of the prospects and the limiting factors. The study revealed that high investment costs and public subsidies for fossil fuels are the key limiting factors while the prospects of tapping into the unexploited carbon markets and multiple socio-economic and environmental benefits favors investment. Public supports in the form of national policy reforms leading to intervention programs are required for progress.

Keywords: abattoir wastes; biogas; biofertilizer; anaerobic digestion; environmental pollution

1. Introduction

Due to increasing population and standards of living, energy production in Nigeria and CO_2 emission increased from 146.3 million tons of oil equivalent (Mtoe) and 28.06 Mt of CO_2 in 1990 to 239.77 Mtoe and 85.09 Mt of CO_2 in 2016 respectively [1], Figure 1. The greatest part of the primary energy mix is from biomass and wastes contributing 82.2% while oil, natural gas and hydropower contributes meager values of 10.6%, 6.8% and 0.4% respectively [2]. On the other hand, land use and forestry contributed the most CO_2 emission (38.2%) followed by energy (32.6%) with wastes, agriculture and industrial processes contributing 14%, 13% and 2.1% respectively [3]. Like other African countries, where only about 36% of the population has access to electricity [4], electricity deficit is a challenge in Nigeria. Although the total installed capacity for electricity generation from power plants is 12,067 MW, actual generation was only 3941 MW in 2015 [3]. Some of the challenges facing the electricity supply sector that leads to this under-delivery include shortage of gas, poor maintenance, inadequate regulations or inability to enforce them and vandalization of energy infrastructure [3]. In the absence of electricity and high cost of fossil fuels like cooking gas (LNG) and kerosene, most people turn to wood fuel and agricultural residues as energy sources for cooking. Additionally, the level of food insecurity in Nigeria has steadily increased since the 1980s, rising from 18% in 1986 to about 41% in 2004 [5]. Sub-optimal supply of agricultural inputs, especially fertilizers, is one of the key challenges leading to persisting food insecurity [6]. Unfortunately, the current major source of fertilizers in Nigeria is fossil based. Application of synthetic fertilizers contributes significantly to greenhouse gas (GHG) emissions [7,8]. Wastes from abattoirs contribute significantly to possible foodborne disease hazards and have adverse effects on air quality, agriculture, potable water, and aquatic life [9]. Uncontrolled disposal of the wastes results in methane (CH_4) emissions from the manure, untreated organic wastes and wastewater. The current waste disposal practices in most abattoirs in Nigeria is to dump rumen content and other solid wastes at nearby designated sites while the liquid phase resulting from washings are allowed to flow into the drains [9–12]. However, animal by-products from slaughterhouses could be processed by anaerobic digestion (AD) to simultaneously produce methane for energy and nutrient rich effluent as biofertilizer [13], which may be a way of curbing the menace of such wastes in many cities in Nigeria [14]. Biogas and biofertilizer production from abattoir waste streams could contribute to lowering fossil fuel and inorganic fertilizer applications. This will also provide an efficient waste recycling method, thereby lowering GHG emissions from open manure storage [15]. Biogas is a valuable source of energy with varying applications such as electricity generation, lighting, cooking, and transportation [16]. The spent substrate is a very useful biofertilizer and can be used to offset the financial as well as the environmental costs associated with the use of mineral fertilizers [15,16]. Sustainable and efficient biogas and biofertilizer production from such waste recycling is a promising method towards attaining a circular economy. These facts support the need to invest in efficient recycling method like AD, to process wastes streams generated in abattoirs as a means to curb the environmental hazards as well as provide valuable materials from renewable resources to contribute in meeting energy and food needs.

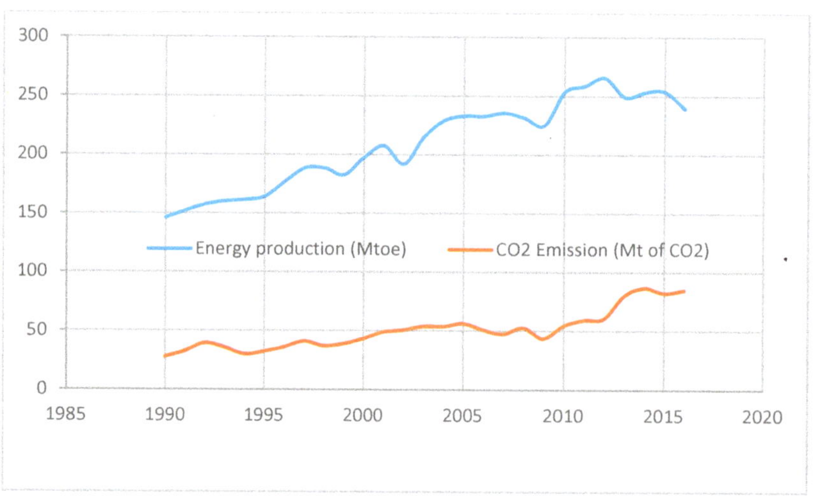

Figure 1. Energy production and CO_2 emissions between 1990 to 2016 in Nigeria (source: [1]).

Several experiments have demonstrated success in converting abattoir wastes by AD into biogas and biofertilizer [17,18], as well as its potency in reducing environmental pollution problems associated with abattoir waste disposal. The GHG emission reduction potential could be determined using mathematical computation given by the Intergovernmental Panel on Climate Change (IPCC) [19], Joint Global Change Research Institute (JGCRI) [20], and Tolera and Alemu [21]. The AD process could also be improved by applying enzymes to produce animal feed from fibrous agricultural wastes like sugarcane bagasse and wheat straw for improved bio-utilization of nutrient contents and the biodegradation of fibrous feeds [22]. In a study by Klintenberg et al. [23] in Namibia, an AD process using a combination of manure, blood, and stomach contents similar to the proportions in abattoir waste generated the highest biogas cumulative output compared to other sampled combinations. Ware and Power [24] demonstrated in Ireland that mixed waste streams from cattle slaughterhouses, consisting of 28.4%, 41.0%, and 27.3% fats, carbohydrates, and proteins, respectively are viable for producing high levels of CH_4. Other studies in Nigeria revealed that abattoir substrates have a low concentration of toxic substances, which would inhibit the process, ultimately resulting in high CH_4 output, suggesting that the substrates are suitable for biogas production [25,26].

The idea of converting abattoir wastes to produce biogas and biofertilizer has long been pursued and is increasingly relevant in sustainability. For example, TekniskaVerken in Linkoping, Sweden has long term experience in handling large volume of slaughterhouse wastes for biogas production [27]. Through sustained efforts on research and development by this company, several improvements have been recorded including better odor control, higher organic wastes loading rate, higher biogas quality and yields [27]. Svensk Biogas AB (SvB) operates a co-digestion plant for slaughterhouse waste for biogas production in which the input from slaughterhouse waste ranges from 35% to 75% [27]. The plant which began in 1996 increased from annual capacity of 55,000 tons to 100,000 tons in 2010. Underpinning the success of this company is the continuing research on biogas production by the company in association with Linkoping University and the Swedish University of Agricultural Science [28]. Despite the several benefits and demonstrated experiences for biogas and biofertilizer production from abattoir wastes, large-scale development of the technology in Nigeria is still at a nascent stage [29]. The reasons for the present lack of uptake of this technology have not been analyzed, and therefore, require investigation. The question around using "slaughterhouse waste for biogas and biofertilizer production" is a multi-dimensional one and requires a methodology that allows systematic investigation of such multi-dimensional issues. The issues include questions regarding the local situation on energy access, resource depletion, climate change mitigation, feedstock availability and terms of accessibility, current use of feedstock that may compete with the proposed biogas plant and space to accommodate biogas plant in close proximity to the source of wastes. Further to these, is the economic demand side for the products (fertilizer and electricity) All these need to be investigated. In addition, socio-economic and environmental considerations are critical in understanding the barriers against uptake of the technology. Thus, the research questions this study attempts to address are:

i.   What is the potential for biogas and biofertilizer production using abattoir wastes in Nigeria?

ii.   What are the socio-economic and environmental merits and demerits of adopting this technology?

iii.   Why the very low level of the development and adoption of the technology in Nigeria?

Consequently, the objective of this study was, therefore, to evaluate the possibility of biogas and biofertilizer production from waste streams in Nigerian abattoirs. The adoption of this technology could be limited by economic, environmental, social, and ethical constraints. Due to complexities in implementing biogas projects, wide range of factors related to biogas system analysis, site-specific properties and the local community situations must be considered to provide workable information for decision-making in specific localities. Suitable tools are required to ensure a well-thoughtout analysis of the factors that affect biogas systems, to ensure informed decision-making. Several tools have been used for such analyses. For instance, the Political, Economic, Social, Technological, Legal and Environmental (PESTLE); the Technique for Order of Preference by Similarity to Ideal Situations (TOPSIS); the Analytic Hierarchy Process (AHP); and the Strengths, Weaknesses, Opportunities and Threats (SWOT). Among these, SWOT has the advantage of combining some elements of the other methods, making it more comprehensive. Therefore, in this study, strengths, weaknesses, opportunities, and threats (SWOT) analysis was selected as the tool for initial assessment of the techno-socio-economic benefits and risks of biogas production from abattoirs in Nigeria. The analysis is targeted at identifying internal strengths and weaknesses as well as the external opportunities and threats that can endanger biogas production from abattoirs in the selected case study in Nigeria. To avoid the limitations of SWOT, which only provides qualitative descriptions [30,31], we incorporated tangible and measurable data from the selected case study sites to provide a comprehensive view of the possibility of biogas and biofertilizer production from abattoirs in Nigeria. The case study sites were the abattoirs in the Federal Capital Territory (FCT)-Abuja, and Niger and Nassarawa States.

2. Materials and Methods

2.1. Study Location, Data Collection Method and Quantifications

Description of study sites: The study sites were in the north central region of Nigeria. The selected sites included Karu abattoir in the FCT-Abuja, Minna and Suleja abattoirs in Niger State, and Lafia abattoir in Nassarawa State (Figure 2). The total land area of the study location is 110,795 km^2, which represents 12% of Nigeria's land area. The population in 2015, projected from the 2006 census, was 11 million, thus representing 6% of Nigeria's population. The region produces livestock (cattle, sheep, and goats), and is among the leading Nigerian producers of some crops such as cassava, rice, yam, melon, plantain, and banana.

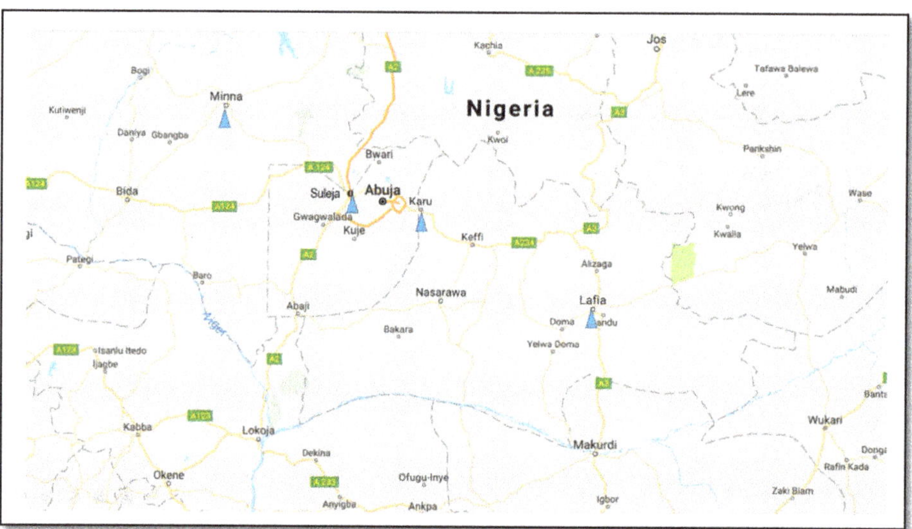

Figure 2. Map of the study region showing the location of the sites (blue rectangle). Source—Google maps.

Data collection/collation approach: We initially developed and distributed a questionnaire to identify the types and number of animals slaughtered per day. We then visited each study site to gain information on their production and administration. In addition, the local situation such as current energy sources, current use of the wastes, available facilities for waste disposal, waste disposal methods and effects on the environment were assessed for each study site. The volume of wastes generated and, hence, other parameters were calculated from the data of the number of animal slaughter acquired from the survey. Generally, the mathematical computations used in this study were in accordance with acceptable standard coefficients and measurements.

Feedstock generation potential/estimation methods: Based on the primary data of the average daily slaughters acquired from the respective study sites, the annual slaughters, total wastes and potentials were determined. For estimating the quantity of waste from abattoir which could be utilized for AD, a coefficient of 35% of the body weight of the animal slaughtered has been used by the World Bank 1998 and Akinbomi et al. 2014 [32]. The waste generated by each animal was therefore taken as 35% of its body weight; for cattle, the average body mass used was 353 kg, while the corresponding numbers for sheep and goat were 33 and 23 kg, respectively [32]. Waren and Power (2016) used 193.37 kg/head of cattle [24], a value higher than 35% of 353 kg used for cattle in this study. The value 35% of the body weight was deliberately used in this study to avoid overestimation. In this study, the data obtained for sheep and goats were combined values, as the number of sheep and goats were not segregated, therefore, an average of 28 kg was used in the estimates for sheep and goats.

Biogas Production Potential and Energy Potential Estimation Methods: Some coefficients used in estimating biogas and electricity generation potential include: dry matter (DM) which was taken as 15% of total wastes [33], volatile solid (VS) as 96.7% of DM [24], and the biomethane potential (BMP) was 700 m^3/t VS [34]. Provision was made for downtime, to cover the possibilities for maintenance, unforeseen stoppages, and incubation time before biogas generation began. Based on this, 75% of the BMP was applied to obtain the overall electricity potential. The electricity potential was based on 3.73 kWh/m^3 CH$_4$ [32], using the adjusted BMP.

Bio-Fertilizers Yield Potential Estimation Method: For the biofertilizer potential, only about 60% VS is converted to biogas [35]. Similarly, VS reductions to a range of 40% to 46% after about 80 days of AD of some organic substrates were observed by Schirmer et al. [36]. We therefore applied 40% of VS as the dry matter remaining after the AD process. Thus, potential biofertilizer yield (dry) was by Equation (1) [29].

$$\text{PBF (dry)} = (\text{DM} - \text{VS}) + (40\%\text{VS}) \tag{1}$$

where: DM = dry mass, VS = volatile solids, i.e., portion of DM that is potentially converted to biogas.

The optimal total solid of the digester was 15% which was used to estimate the bulk density by applying Equation (2), proposed by Chen [37], from which the volume of the slurry was obtained.

$$\varrho = 998/(1 - 0.00345\text{TS}) \tag{2}$$

where: ϱ = bulk density (kg/m^3), TS = total solids (%); TS ranges from 0–16%. Taking TS = 15%, ϱ = 1052 kg/m^3 was obtained and applied.

Estimation of Possible Reduction of Green House Gas (GHG) Emissions: The decrease in GHGs by the production of biogas from abattoir waste was calculated using the mathematical computational method developed by IPCC and applied by Tolera and Alemu [21], expressed as GHG reduction potential of AD equal to the estimate of GHG emissions from dumping sites minus the estimate of GHG emissions from AD:

$$GHG\ Emission = \lfloor((Q \times DOC \times DOCF \times F1 \times 1.336) - R) \times (1 - OX) \times 25\rfloor \tag{3}$$

Estimation of cutback of GHGs using AD for biogas production:

$$GHGr = \lfloor((Q \times DOC \times DOCF \times F1 \times 1.336) - R) \times (1 - OX) \times 25\rfloor - \sum(Qj \times EFj) \tag{4}$$

whereby ([($Q \times DOC \times DOCF \times F1 \times 1.336$) – R] × (1 – OX) × 25) is the *GHG* emission potential of the dumping sites and $\sum(Qj \times EFj)$ is the *GHG* emission potential of the abattoir waste AD production plants for biogas factoring in the CH$_4$ global warming potential used to convert the quantity of methane emitted to the carbon (iv) oxide in equivalence (CO$_2$ eq) from the quantity of abattoir waste produced.

The variables are defined as follows:

Q = the quantity of slaughterhouse waste from waste records (t/kg);
DOC = the degradable organic carbon expressed as a proportion of abattoir waste with default value (DV) = 0.12;
$DOCF$ = the fraction of degradable organic carbon dissimilated for the abattoir waste whose DV = 0.7;
$F1$ = the fraction of methane produced from dumping sites, DV = 0.50;
The value 1.336 is the rate that carbon is being converted to methane;
R = annual recovery of methane, quantified in tons (here no recovered methane);
OX = the oxidation factor, DV = 0.1 for well-managed and DV = 0 for unmanaged);
The value 25 is the CH_4 global warming potential;
Qj = the amount of the given type of waste j (here is only abattoir waste);
EFj = the biogas emission factor of the given waste type j, DV = 0.02 kg CO_2 eq.

Estimation of Biogas Equivalence of Fossil Fuels: Amigun and Blottnitz [18] and B-Sustain's [38] provided a comparative energy value estimation which showed that 1 m^3 of biogas is equivalent to coefficient factor of 0.45 kg liquefied petroleum gas (LPG), 0.6 kg kerosene (K), 0.4 kg furnace oil (FO), 0.7 kg petrol (P), 0.5 kg diesel (D), or 3.50 kg firewood(FW) in the same activities.

Biogas Equivalent of a given

$$\text{Fossil Fuel} = \sum \text{CFF} \times \text{BV}$$

where CFF is coefficient factor of any fuel as given above; BV is the biogas volume produced.

2.2. SWOT Analysis

Due to complexities in implementing biogas projects, various assessment methodologies have been applied [39]. In view of the wide range of factors related to biogas system analysis, site-specific properties and the local community situations must be considered to provide workable information for decision-making in specific localities. Suitable tools are required to ensure a thorough analysis of the factors that affect biogas systems for informed decision-making. Several tools have been used for such analyses. For instance, the Political, Economic, Social, Technological, Legal and Environmental (PESTLE) tool is widely used to analyze the prospects and potential risks and screen the external marketing environment of an investment or a company [40]; the Technique for Order of Preference by Similarity to Ideal Situations (TOPSIS) grades the alternatives with respect to their geometric distance from the positive and negative ideal solutions [41]; the Analytic Hierarchy Process (AHP) applies mathematics and psychology to organize and analyze complex decisions [42]; and the Strengths, Weaknesses, Opportunities and Threats (SWOT) is a tool for auditing a system and its internal and external environment by building on the strengths and opportunities, correcting the weaknesses and protecting the threats [43]. Among these, SWOT seems to be the most popular, with its advantage of combining some elements of the other methods. For example, several aspects of PESTLE are well captured in SWOT. SWOT is a strategic planning tool, which originated from business management, and is applied to detect and assess the strengths (S), weaknesses (W), opportunities (O) and threats (T) of a project or product that is being evaluated. The structure of the SWOT matrix (Figure 3) defines the strengths and weaknesses of the assessed system as internal characteristics, while the opportunities and threats are external factors that influence the success or failure of the system [44].

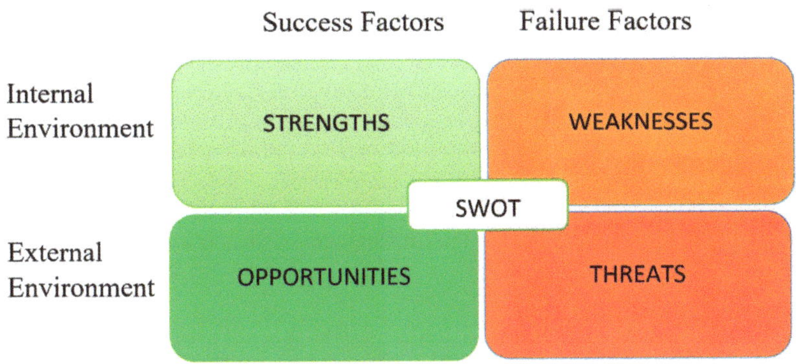

Figure 3. SWOT matrix structure: [44].

2.3. SWOT Matrix Development and Identification of Factors

Based on the SWOT matrix framework in Figure 3 and results of the initial site surveys, the various SWOT factors were identified (Table 1). The steps were as follows:

- Identification of all internal aspects of biogas and biofertilizer production from the abattoirs, which might influence the project, followed by classification of favorable factors as strengths and unfavorable ones as weaknesses.
- Identification of external factors that may influence the project, looking at the global and local scenarios in Nigeria, and classifying the negative factors as threats and the positive ones as opportunities.

Table 1. Identified SWOT factors for biogas and biofertilizer production using abattoir wastes in Nigeria.

	Positive	Negative
Internal Environment	**Strengths** • Feedstock availability • Suitable climate • Provision of clean energy and biofertilizer alternative • Kills pathogenic organisms • Solves waste disposal problems • Reduction of GHG emission and useful for closing carbon cycle • Existing market for products • Flexibility for small, medium and large plant	Weaknesses • High investment cost • Lack of continuity in developing technical proficiency • Relative novelty, adoption may face resistance, sensitization needed • Lack of access to water • High protein in abattoir wastes • Pathogens from contaminated materials • Oversimplification of the biogas system
External Environment	Opportunities • Energy deficits and rural settings, favors decentralization • Food insecurity and calls for diversifying Nigeria's economy • Improved public health • Job opportunities • Increased economic activities • Synergizes global goals of climate mitigation • Public Logistic support	Threats • Competitive product undercuts prices • Low level of understanding of environmental problems among citizens • Too expensive and high lending rate • Public subsidies for fossil-based energy and fertilizer

3. Results

3.1. State of the Abattoirs Assessed—Quantitative and Qualitative Assessment

A total of four abattoirs, namely, Minna, Suleja, Karu, and Lafia, were evaluated. Based on the survey, Table 2 shows the raw data of daily slaughter and types of animal slaughtered from which annual feedstock values were calculated and further analyses were conducted to obtain the electricity and biofertilizer potentials. Animals slaughtered included cattle, sheep, and goats. However, at Minna abattoir, occasionally, a few camels were also slaughtered. Suleja had the highest number of cattle slaughtered followed by Karu, while Karu had the highest number of sheep and goats slaughtered followed by Minna.

Table 2. Average daily number of animal slaughters from the study sites.

Abattoir Assessed	Suleja	Minna	Lafia	Karu
Location/State	Niger	Niger	Nassarawa	FCT
Slaughter (Cattle)	180	60	45	135
Slaughter (Sheep and Goats)	19	95	61	650

In each abattoir, there was an animal holding area to keep a few animals for some time, either to fatten them before they were slaughtered or to house them when the number of available animals exceeded the number slated for slaughter for the day. Therefore, some animal dung was generated in the animal holding area during the night. Additionally, fruit and vegetable markets adjoin the abattoirs at Suleja and Karu, and a cattle market is located near the Lafia abattoir. On the other hand, the Minna abattoir is strictly isolated for abattoir operations. Thus, besides the wastes generated in the abattoirs, additional wastes were obtainable from fruit and vegetable markets and the animal holding areas near the abattoirs.

None of the abattoirs had automated systems for meat and waste processing. The rumen contents were conveyed using wheelbarrows to the designated dumping sites (Figure 4). There were obvious stinking smells in the areas. Two of the rumen dump sites were close to streams near the Karu and Suleja abattoirs. These show a more pronounced danger of surface water pollution, resulting in possible public health and environmental hazards, especially to the users of water residing downstream.

(a)   (b)

Figure 4. Waste from Suleja abattoir dumped at a nearby stream. (**a**) Waste dumped at nearby stream; (**b**) Dumping of waste.

In all the study sites, wastes streams are currently not put into use, the wastes could therefore be made available for AD plants following appropriate discussions and agreements. The authorities concerned in supervising the activities of the abattoirs in all the sites expressed similar interest on the idea of AD plant as means for wastes management and efficient resource utilization. While the abattoirs in Minna, Suleja and Lafia have abundant land area that could accommodate AD plants near the slaughterhouses, Karu abattoir has limited land area for this project in the current site. The current energy source for cooking in the communities in the study sites is not any different to most part of Nigeria. These are fossil fuel and more often wood fuel that is hauled without replenishing, thereby leading to deforestation, erosion and degradation of soil quality.

3.2. Estimate of Waste Generated, Biomethane, Electricity, Biofertilizer and Methane Emission Mitigation Potentials

The raw data obtained from each abattoir were used to estimate the biogas, electricity, and biofertilizer potentials, and the values are presented in Table 3. In general, as expected, potentials derivable corresponds to the wastes generated which depends on the type and number of animals slaughtered per day. As can be seen in the data, wastes contributed by sheep and goats were each less than 12% at Suleja, Minna and Lafia abattoirs. Only Karu site had a relatively higher contribution from sheep and goats valued at 28% of the total wastes. Estimate of the total solids and volatile materials were based on contributions from cattle wastes being the dominant sources, as well as those from sheep and goats.

Table 3. Estimates of total wastes generated, biomethane, electricity and mitigated methane emission potentials from the primary data.

Item/Abattoir	Suleja	Minna	Lafia	Karu	Total
Total no. of annual slaughter (Cattle)	65,700	21,900	16,425	49,275	153,300
Total no. of annual slaughter (Sheep and Goats)	6935	34,675	22,265	237,250	301,125
[32]TBW—353 kg/animal (10^3 t/y) (Cattle)	23.19	7.73	5.80	17.39	
[32]TBW—28 kg/animal (10^3 t/y) (Sheep and Goats)	0.19	0.97	0.62	6.64	
[32]Waste—35% TBW (10^3 t/y) (Cattle)	8.12	2.71	2.03	6.09	
[32]Waste—35% TBW (10^3 t/y) (Sheep and Goats)	0.07	0.34	0.22	2.33	
Total waste (10^3 t/y)	8.19	3.05	2.25	8.41	21.89
[33]DM, 15% of total waste (10^3 t/y)	1.23	0.46	0.34	1.26	3.28
[24]VS, 96.7% of DM (10^3 t/y)	1.19	0.44	0.33	1.22	3.18
[34]BMP @ 700 m^3/t vs. (10^3 m^3)	831.08	309.23	228.20	854.21	
75% factor BMP (10^3 m^3)	623.31	231.92	171.15	640.66	1667
[32]PE @ 3.73 kWh/m^3CH$_4$ (kW)	265.23	98.69	72.83	272.61	709.36
Spent slurry in (10^3 m^3)	117	44	33	122	
PBF dry, (10^3 t/y)	0.515	0.192	0.142	0.530	1.378
Vol. of slurry added daily (m^3)	320	121	89	334	
[26]Digester Capacity, 14 days HRT (m^3)	4500	1700	1250	4700	

TBW total body weight, TS total solids vs. volatile solids, BMP biomethane potential, PE potential electricity based on 75% BMP, PBF DM potential biofertilizer dry matter. [32]Akinbomi et al. 2014, [24]Ware and Power 2016, [33]Deublein and Steinhauser 2008, [34]Schnurer and Jarvis 2010), [26]Rabah et al. 2011.

Suleja and Karu that are near Abuja had higher number of slaughtered animals which is likely to be attributed to higher human populations given the influx of people to Abuja. Therefore, it is expected that big cities like Lagos, Onitsha, Kano and Port-Harcourt would have very high number of animals slaughtered. Although detailed estimates of wastes generated in the proximity of the study sites are beyond the scope of this study, the idea of combining other biodegradable wastes with the wastes from the abattoirs may prove useful. Besides increasing the volume of materials for processing for the benefits of economy of scale, using a variety of substrates is known to increase biogas yields and profitability [45].

Estimation of Possible Reduction of GHG Emissions: A model for computation proposed by the IPCC with default values for the various coefficients and used by Tolera and Alemu [21] was applied to estimate the values of GHG reduction potential in terms of t CO$_2$ eq, Figure 5. The aggregate GHG reduction potential by installation of AD plants at the 4 sites is 30.71 t CO$_2$ eq.

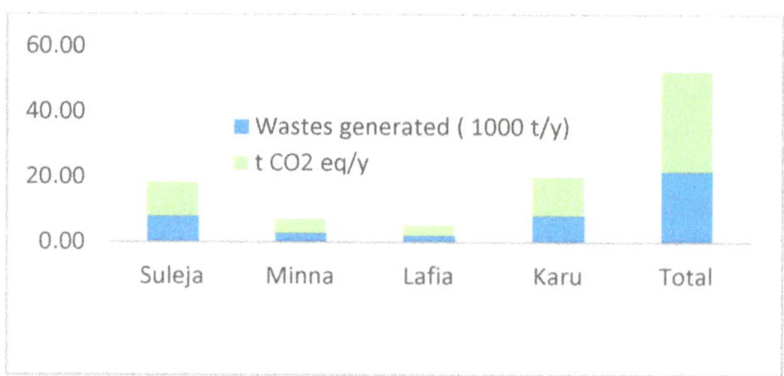

Figure 5. Annual reduction of GHGs (t CO_2 eq) using anaerobic digester (AD).

Estimation of Biogas Equivalence of Fossil Fuels: The computation shows that the 1.667×10^6 m^3/year of biogas estimated is equivalent to 667 t of furnace oil, 750 t of liquefied petroleum gas, 834 t of diesel, 1000 t of kerosene, 1167 t of petrol, and 5835 t of charcoal/firewood per year in the same functions as presented in the Table 4.

Table 4. Estimation of equivalence of biogas potentials with some fossil fuels in the same function.

Abattoir	Waste	BMP	FO 0.4 kg	LPG 0.45 kg	D 0.5 kg	K 0.6 kg	P 0.7 kg	FW 3.5 kg
	(10^3 t/y)	(10^3 m^3)			t			
Suleja	8.19	623	249	280	312	374	436	2182
Minna	3.05	232	93	104	116	139	162	812
Lafia	2.25	171	69	77	86	103	120	599
Karu	8.41	640	256	288	320	384	448	2242
Aggregate	21.89	1667	667	750	834	1000	1167	5835

BMB is biomethane potentials, FO is furnace oil, LPG is liquefied petroleum gas, D is diesel, K is kerosene, P is petrol, and FW is firewood.

3.3. SWOT Assessment and Analyses

In general, the quantitative data and qualitative description of the state of the abattoirs evaluated as presented in Sections 3.1 and 3.2 supports the SWOT factors. A concise presentation of findings from assessment and analyses of the identified SWOT factors are given below (Tables 5–8) and further discussed in the Section 4 using the results of our findings and relevant literatures.

Table 5. SWOT factors—Strengths.

Identified Factor	Assessment/Finding
Feedstock availability	Karu abattoir had the highest feedstock value (8400 t/y), followed closely by Suleja (8200 t/y), while the abattoirs at Minna and Lafia had lower values (3000 and 2200 t/y, respectively). The corresponding digester capacities based on optimal 15% total solids were 4500, 1700, 1250, and 4700 m^3. The total available feedstock which could be used for production of biogas and biofertilizer for all the study sites amounts to 21,900 t/y.
Suitable climate	The region, i.e., study sites have characteristic optimum climatic condition for suitable anaerobic digestion processes. Operating temperature for AD range between 10 and 55 °C, with 35 and 55 °C being optimal for mesophilic and thermophilic digestion respectively [46]. Nigeria has a tropical climate with temperature ranges between 27–40 °C, suitable for the optimal performance of the digester, with no requirements to use the produced gas for heating the reactor, unlike in Europe [47].

Table 5. Cont.

Identified Factor	Assessment/Finding
Combined provision of better alternative energy and biofertilizer	Capable of generating a combined total of 710 kW. The electricity generation potentials for Karu, Suleja, Minna, and Lafia abattoirs were found to be 273, 265, 99, and 73 kW capacities, respectively. At the current average rate of 13.5 kg fertilizer per hectare in Nigeria, the 4 study sites have a combined potential to provide fertilizer for about 100 hectares.
Kills pathogenic organisms	It has been demonstrated that a number of pathogenic organisms like *S. enterica* and *M. paratuberculosis* are reduced and inactivated in anaerobic environments [48,49].
Contributes in solving waste disposal problems	In all the abattoirs, the waste streams exist as a nuisance and managing them is a key challenge. The wastes could therefore be dedicated to AD in a way that takes care of interest groups. Harnessing these wastes as resources for production of biogas for energy and biofertilizer for improved soil fertility could contribute to curbing the environmental menace and addressing the problems of energy and food deficits in Nigeria.
Reduction of GHG emission and useful for closing carbon cycle	Contributes to reducing GHG emissions emanating from direct disposal to the fields. Installation of ADs at the 4 sites depicts GHG reduction potential of 30 t CO_2 eq.
Existing market for products	Due to massive deforestation, there is limited forest resources and soil degradation. The traditional direct use of biomass for fuel is not sustainable. Moreover, there is increasing interest in more modern options such as cooking gas.
Flexibility for small, medium and large plant	The possibility and capability to set up and run a biogas generating plant in small, medium, and large scales in particular Karu and Suleja abattoirs is advantageous for efficient use of resources in a sustainable and environment friendly manner. Minna and Lafia sites are more suited for small and medium plants to avoid the challenge of feedstock shortage. Yet, there is the benefits of economy of scale in biogas plant operation where larger capacity plant is more viable economically [50–52].

Table 6. SWOT factors—Weaknesses.

Identified Factor	Assessment/Finding
High investment cost	Chukwuma et al. [51] demonstrated higher value of profitability index for AD plant with bigger capacity. The cost of investment for AD plant is relatively high, several millions of Naira [51]. When built as small scales, such as petite backyard operations, biogas systems tend to be too costly, are hardly profitable, and rarely make significant contribution to the family or community [50,52].
Lack of equipment fabrication facilities for making the digesters and accessories	There are no refabricated digesters. For diverse applications ranging from cooking to electricity generation, there is need to have compatible equipment and accessories such as gas holder, gas bottles, pressure regulator, water trap, burner stove and lamp, biogas generating sets and biogas stoves and their accessories as well as installation materials, user training and after sales services. Unfortunately, these are lacking in Nigeria.
Lack of continuity in developing technical proficiency	No strategic, sustained, and substantial research, development, and training on building robust technological capacity to set up and run such plants efficiently. No technical standards and codes for AD installation and maintenance and no established testing methodologies.

Table 6. *Cont.*

Identified Factor	Assessment/Finding
Relative novelty, adoption may face dislikes, sensitization needed	Adopting, adapting, and advancing a new technology often requires proper sensitization, reorientation, and commitment from all stakeholders. Demonstration of the technology through pilot programs and marketing may be necessary in case there is the initial reluctance in adopting and adapting to new techniques and products like these.
Limited of access to water	In all the study sites evaluated, tap water was available only at Karu site but regular water flow is usually interrupted by incessant power outage. This inadequate water supply has been noted as one of the challenges grabbled with by the abattoirs and the neighboring residents.
High protein in abattoir wastes	The blood and meat trimmings are part of wastes which contribute to high protein content of the wastes.
Pathogens from contaminated materials	Pathogens are present in the waste and can also arise from production processes [53], posing hazards while handling waste inputs to the digester. Spent substrates such as biofertilizer could also contain pathogens depending on the incidence of viable pathogenic organisms. in the input and spent substrates.
Oversimplification of the biogas system	Considering it simply as a receptacle for wastes and a provider of gas and fertilizer may likely cause such failures, eventually resulting in deficient performance, leading to abandonment of the plant.

Table 7. SWOT factors—Opportunities.

Identified Factor	Assessment/Finding
Energy deficits and rural settings, favors decentralization	Frequent power nationwide and most rural settings do not have access to electricity and conventional cooking facilities such as LPG (cooking gas), kerosene, and electricity.
Food insecurity and calls for diversifying Nigeria's economy	Biofertilizer availability could contribute to providing a sustainable solution to the current food insecurity in Nigeria. Crop yields higher by 11–20% compared to controls have been reported after the application of spent digester effluent [54].
Improved public health	Some aerobic organisms are killed by the fermentation process in an anaerobic environment. Biogas systems could also serve as a better alternative for management of abattoir waste, which could otherwise be disposed in open fields forming breeding grounds for pathogenic organisms, therefore enhancing public health.
Job opportunities	Going by Arnott's [52] projection, job opportunities for about 790 people could be generated from AD plants producing a total of 1390 t of dry biofertilizers at the four study sites.
Increased economic activities	The time spent for collecting and carrying wood by women and children could be swapped for education, more productive activities, or simply recreation and leisure time [55].
Synergizes global goals of climate mitigation	Estimates by this study shows the aggregate GHG reduction potential by installation of AD plants at the 4 sites is 30.71 t CO_2 eq. Thus, installation of AD systems could contribute in GHG mitigation by preventing disposal to the open fields. The use of biogas in place of fossil-based alternatives further provides avenues for reduction of GHG emissions.
Public logistic support	Gaining public support might be easy owing to the socio-economic benefits associated with AD systems.

Table 8. SWOT factors—Threats.

Identified Factor	Assessment/Finding
High lending/loan rat	Bank lending rates in Nigeria range from 16.91% to 29.26% and include stringent collateral requirements. This financial predicament may not be favorable for investing in AD, making it difficult for willing investors to start such a project despite its prospects.
Public subsidies for fossil-based energy and fertilizer	In Nigeria, there are public subsidies for fossil-based energy and chemical fertilizers. This is a threat to the competitiveness of the AD system.

4. Discussion

4.1. Site Specific Conditions and Effects on Products

For all the sites investigated, wastes from cattle had the largest proportion contributing 99%, 89%, 90%, 72% of the total wastes corresponding to Suleja, Minna, Lafia, and Karu sites respectively. Although wastes are largely from cattle in all the sites, differences in waste composition due to contributions from varieties of the animals may influence the chemistry of the digestion processes and final products. Greater impact due to the differences could be expected from the Karu site which had the largest contribution from sheep and goats by a value of 28%.

In the following sections, quantitative and qualitative data acquired from the case study are incorporated into the SWOT analysis to provide insights regarding the possibility of biogas and biofertilizer production from abattoir wastes in Nigeria.

4.2. Strengths

4.2.1. Feedstock Availability

A vital component of AD is the feedstock, which is readily available in the evaluated abattoirs. In all the abattoirs, the waste streams exist as a nuisance, and therefore, could be dedicated to AD following business agreement with the authorities and other interest groups. It is important to secure official access and also sort out any tax or fees that may be needed to afford uninterrupted access to the wastes and, as well, gain the logistic support of those concerned. The ministry of agriculture in each state is the authority in charge. Karu abattoir had the highest feedstock value (8400 t/y), followed closely by Suleja (8200 t/y), while the abattoirs at Minna and Lafia had lower values (3000 and 2200 t/y, respectively). The corresponding digester capacities based on optimal 15% total solids were 4500, 1700, 1250, and 4700 m^3 corresponding to 265, 99, 73- and 273-kW capacity in terms of electricity potentials for Suleja, Minna, Lafia and Karu abattoirs, respectively. The specific investment cost per kW or MW capacity is higher for smaller plants and lower for bigger plants [56], and this favors investment in Karu and Suleja abattoirs, where higher values of wastes are obtained compared to Lafia and Minna. Although digester capacities between 20–60 m^3 are deemed good enough for small-to-medium scale business enterprises going by the assertion that a family would consume a minimum of 0.8 m^3/d [52,57], thus, each of the evaluated abattoirs meets the required waste volumes to operate as medium-scale business enterprises. However, Carlini et al. [50], demonstrated an economy of scale in AD plants businesses where a larger (1000 kW capacity) plant yielded break-point in 4 years' time compared to smaller (100 kW) plant that took 10 years to reach break-point. Similarly, in a study of the economic viability of AD in 3 selected sites in Anambra state—Nigeria, Chukwuma et al. [51] demonstrated higher value of profitability index for AD plant with bigger capacity. When built as small backyard operations, biogas systems tend to be too costly, are hardly profitable, and rarely make significant contribution to the family or community [52]. The bulk of the abattoir wastes (in slurry form) is water, with just 15% TS content [24]. In view of this, decentralization of biogas plants to localities generating the biomass has more advantages than hauling the biomass to a central location [56]. Thus, there is no need to transport waste from one abattoir to another. However, in view of the benefit of economy of scale, possibilities of increasing the economic viability of the system could be explored by sourcing feedstock from nearby sources, like fruits and vegetable markets, animal holding places, and livestock farms e.g., poultry, piggery.

4.2.2. Better Alternative Energy Source

Compared to the current biomass, which is inefficiently burnt for cooking by over 81% of Nigerians dwelling in rural areas, biogas is a better source of energy. At the global level, about three billion people use biomass for cooking and heating. The biomass is burnt in inefficient cook stoves or used to fuel open fires [58]. Annually, over 4.3 million premature deaths are caused by illnesses attributable to household air pollution resulting from this practice [58]. The data of the four study sites indicate that a total of 1.667×10^6 m^3 of CH$_4$ could be generated annually as a clean source of energy. If it's used for electricity generation, it will provide a total of 709 kW (6220 MWh/y). A family could consume a minimum of 0.8 m^3/d of biogas for cooking [57], therefore, if the generated CH$_4$ is used as cooking gas, it could provide for 5709 families at the rate of 0.8 m^3/d per family. Thus, there is immense potential in the use of biogas from abattoir wastes which could contribute to averting the public health burden attributable to inefficient use of biomass that causes household air pollution. At household level, the accessories commonly used in places like India and Nepal where biogas is widely utilized include gas holder, gas bottles, pressure regulator, water trap, burner stove and lamp. However, these are not available in Nigeria since wide application of the technology is yet to pick-up. With a good supporting national program, such accessories and pre-fabricated digesters should be part of the tools for successful dissemination and uptake of the technology.

4.2.3. Ability to Kill Pathogenic Organisms

Improper organic waste disposal could provide breeding grounds for disease-causing organisms like *Salmonella* sp., *E. coli*, and *Shigella* sp., therefore, proper disposal is advantageous as it improves public health. Many of these pathogenic organisms are aerobic in nature and are killed by the fermentation process in an anaerobic environment. It has been demonstrated that a number of pathogenic organisms like *S. enterica* and *M. paratuberculosis* are reduced and inactivated in anaerobic environments [48,49]. However, AD cannot remove all the pathogens; hence biogas plants in Sweden that use slaughterhouse waste are required by law to pasteurize their substrate prior to feeding it into the digester [27]. Although desirable, it is worth noting that pasteurization could increase the cost of production.

4.2.4. Solution to Waste Disposal Problems

Use of the abattoir wastes for biogas systems provides a sustainable solution to the problem of waste disposal [59]. Thus, rather than spending money for waste disposal, the waste could be converted into valuable products (including biogas for energy and biofertilizer for improved soil fertility). Multiple products harvesting could improve the economy of AD operations. It is worth noting here, that treatment of digestate from biogas production can also be costly, especially if transported from far distances. Indeed, several researches have been going on in developing cost-effective processes for nutrients recovery and components separation from biogas digestate such as nutrient recovery from digestate and production of tailor-made chemicals, production of organic chemicals by treatment of effluents, etc. [60]. Separation techniques that have been explored include solid-liquid separation of digestate, using for example, membrane purification, screw press or centrifuge [60]. Nevertheless, market leading technology is still to evolve [60].

4.2.5. Suitability of the Climate

Operating temperature for AD range between 10 and 55 °C, with 35 and 55 °C being optimal for mesophilic and thermophilic digestion respectively [46]. Nigeria has a tropical climate with temperature ranges between 27–40 °C, which is perfect for the optimal performance of the digester. Thus, the extra cost of heating the reactor could be done away with in most parts of Nigeria throughout the year, unlike temperate regions such as Europe, where 20–30% of gas production is used for heating [47]. To improve the economy of AD operation, the option of cogeneration or combined heat and power (CHP) is an appropriate method to be considered. In this method, beside electricity, the heat generated could be harnessed. This is a well-known technology widely applied in some countries in the temperate regions like the Sweden, Germany, Italy, etc. where the process heat is converted to meet household heating needs or provide for the heating needs of some operations/processes like AD. In the case of a tropical country like Nigeria, the prospects of channeling the heat for industrial drying could be the future in incorporating the concept of CHP in AD technology development.

4.2.6. Reduction of GHG Emissions and the Use for Closing the Carbon Cycle

AD treatment of wastes to obtain biogas prevents GHG emissions emanating from direct disposal to the fields, as well, the use of biogas as alternative to fossil-based fuels further contributes in reducing the release of GHG emissions, thus a climate smart alternative [61]. The additional environmental benefits in installing ADs are corroborated by the greenhouse gas (GHG) reduction potential, an aggregate value of 30 ton of CO_2 equivalents per year for the study sites. As well, in terms of energy output, the aggregate value of 1667×10^3 m^3 of biomethane potential for the study sites could substitute an average of 750 t of fossil-based fuels (LPG, kerosene, petrol, diesel) or 5835 t of firewood. The use of digestate can lead to enhancement of soil carbon content. The biomass used to produce biogas is indirectly derived from photosynthesis. When burnt, the released CO_2 is reabsorbed in subsequent photosynthetic processes. This presents a closed carbon cycle loop, and thus presents an environment-friendly source of energy.

4.2.7. Existing Market for Products

The increasing population and growing wealth in Nigeria have resulted in a growing demand for both energy and biofertilizer products. Moreover, there is a deficit with regard to these two products, and therefore, they are likely to find buyers. In addition to these readily accessible market opportunities, the feed-in tariff to the national gas grid presents another prospective market outlet. When chemical fertilizer was introduced in Nigeria some farmers were reluctant in patronizing it, however it became a sought-after product when positive outcome from initial testers were revealed. Similarly, it is possible that some farmers may be reluctant to use the digestate, however, positive outcome from initial testers would likely attract more customers. It is also worth noting that the use of animal dung as source of biofertilizer is common among farmers in Nigeria especially by mixed farming practitioners, one of the widely practiced systems in Nigeria. Poultry droppings, cattle, sheep and goats' dung are already highly valued fertilizers by many farmers.

4.3. Weaknesses

4.3.1. High Investment Costs

Recently, Chukwuma et al. [51] evaluated the economic viability of biogas plant in 3 selected sites in Anambra state of Nigeria to be fed using cattle and chicken dung. Based on the estimated volumes of feedstock to be hauled to central location of each community, plant capacities of 3661.35, 3969.45- and 3546.18-kW electricity for Onitsha North, Njikoka and Dunukofia were projected respectively. The total cost of investment consisting of fixed cost and variable or annual costs (maintenance, transportation and operational cost) were 3.895, 4.155 and 3.797 billion Nigerian Naira equivalent to 10.760, 11.478 and 10.488 million USD in that order. In another study of Italian scenario aimed at analyzing the economic performance of co-digestion plants fed with agro-industrial wastes as a function of installed capacity in kW, the study revealed the benefits of economy of scale in biogas plant operation where larger capacity plant is more viable economically [50]. The study revealed total investment costs of 758,000€, 2.147 million € and 3.217 million € and breakpoints of 10, 5 and 4 years for plant capacities of 100, 500 and 1000 kW respectively [50]. Although the cost estimate for Nigeria's scenario by Chukwuma et al. [51] that indicates 2892 USD/kW for the most viable option seems on the high side, it is however cheaper compared to Italian scenario by Carlini et al. [50] which gave a value 3667 USD/kW. From these examples the investment cost for medium-scale biogas systems (which could be operated profitably), as against a small backyard system (which is hardly profitable), is extremely difficult for direct investment without some funding facilities to stimulate investment. Unfortunately, investing in AD does not seem attractive because most people prefer investments with clear financial returns than the overall environmental and socio-economic benefits. The idea of pro-bono investment is also unfortunately not common in Nigeria. Thus, the onus lies with stakeholders in the public sector to develop programs and policies that should stimulate and facilitate investments in this sector. The public sector should embrace a national biogas program as service offerings of energy provision, water provision and recycling, waste treatment, and nutrient capture for improved food production which are valuable inputs to socio-economic development in Nigeria. Leveraging the feed-in tariff policy in Nigeria, it may be possible to raise the value of AD products so that they could favorably become competitive with prices of alternative products in the markets.

4.3.2. High Protein in Abattoir Wastes

The blood and meat trimmings in abattoir wastes are high protein content waste products [24,62]. This content could be a source of sulfide formation during the AD process. Accumulation of sulfides in the digester results in higher concentrations of corrosive hydrogen sulfide (H_2S) in the biogas, ultimately leading to sulfide inhibition of the methanogens [63,64]. In addition to sulfide formation due to high protein content, ammonia is formed when protein degrades. The presence of ammonia increases the pH in the digester. If the proportion of wastes from the abattoir is very high, the pH is likely to increase beyond 8.0 which can be growth-limiting for some volatile fatty acid (VFA)-consuming methanogens [65]. At such high pH values, the high fermentation rates of proteins and fats are accompanied by fatty acid accumulation. Additionally, the presence of neutral NH_3 in the digester could be toxic to some beneficial microorganisms as NH_3 can easily pass through the cell membranes of bacteria, disrupting intercellular pH and concentrations of other ions [64]. It is thus desirable to lower NH_3 levels in the high-protein substrate materials during AD. This is an area of active research for recovery of the NH_3 or providing alternative pathways to circumvent the problems [66,67]. Furthermore, at increased pH and temperature, the equilibrium shifts towards toxic NH_3 levels, giving rise to undesired synergistic relationships among the three parameters [68]. Despite these problems, results have shown that an alternative CH_4-producing pathway is activated at elevated levels of NH_3 [69]. Acetate is converted into H_2 and CO_2 in this pathway by syntrophic acetate oxidizers (SAOs), followed by the subsequent reduction of CO_2 to CH_4 by hydrogen-utilizing methanogens. Thus, through this pathway, methane is produced by selective inhibition of acetate-utilizing methanogens, as NH_3 released during protein degradation [69]. The potentially low C/N ratio of slaughterhouse waste can be improved by co-digestion with some other feedstock such as manure, sewage sludge, food waste or straw [34]. Generally, co-digestion with other substrates help by diluting the toxic components, improving C/N ratio and imparting pH and moisture content adjustments [70,71].

4.3.3. Pathogens from Contaminated Materials

Pathogens are present in the waste and can also arise from production processes [53], posing hazards while handling waste inputs to the digester. Spent substrates such as biofertilizer could also contain pathogens depending on the incidence of viable pathogenic organisms in the input and spent substrates. The survival rate of these organisms in the sludge could determine the need for additional treatment of the spent substrates. While precautionary measures may be needed in handling the wastes, laboratory analysis to test for the presence of pathogenic organisms and ensure process quality control in the input and output substrates is necessary. Alternative solution to this problem is to pasteurize the feedstock before AD treatment as done in Sweden and probably some other countries.

4.3.4. Lack of Continuity in Developing Technical Proficiency

Long term success in operating an AD system requires continuity on research and development—making observations on the go and resolving problems, and policy creation and implementation regarding hygiene and environmental standards. These are lacking in Nigeria, for example, there are no technical standards and codes for AD installation and maintenance and no established testing methodologies. Contributing to these weaknesses is that public institutions and the legal system are often unable to monitor and enforce the law over the long term [72]. A government administration that could ensure sanity by enforcing rule of law could support AD to thrive.

4.3.5. Relative Novelty

Adapting to a new technology generally requires sensitization, reorientation, and commitment from all stakeholders. In some cases, demonstration of the technology through pilot programs and marketing could solve the initial reluctance in adopting and adapting to new techniques. Such adaptation necessitates the establishment and implementation of awareness programs, which entail an additional cost, like releasing a new product in the market. One circumstance that favors adopting and adapting to AD systems is the decreasing forest resources in which in some cases people travel several kilometers to hunt for wood fuel. AD system would therefore present an amenable alternative that could lift off the burden of the hardship faced in scouting for wood fuel plus the negative impacts on health and environment in its use.

4.3.6. Limited Access to Water

Water availability is one of the conditions that could potentially limit the uptake of AD. In all the study sites evaluated, tap water was available only at Karu site but regular water flow is usually interrupted by incessant power outage. This is a depiction of most situations of dilapidated social amenity infrastructure in Nigeria that need to be addressed to become more investment friendly. Under normal conditions, the public sector provides such facilities while firms like AD company only need to pay utility bills. Although each of the study site is situated at municipal areas, with pipe borne water supply from existing rivers or dams, the water supply is, however, inconsistent. For example, the nearest dam to Karu is Usman Dam that supply Abuja municipal areas including Karu. The inconsistency in water supply may be due to power failure, poor maintenance and poor budget administration of funds allocated for the purpose.

4.3.7. Oversimplification of the Biogas System

A sizable number of AD pilot plants in Nigeria have failed to deliver the desired goals in the past. The goals of the pilot plant are to demonstrate the process, learn from its working and move to the stage of application. Oversimplification of the biogas system by considering it simply as a receptacle for wastes and a provider of gas and fertilizer was likely the cause for such failures, as it eventually resulted in deficient performance, leading to abandonment of the technology. Most of the biogas plants established in the past were pilot-scale demonstrations using public funds, lacking the required follow-up to make it work. For the system to work, there is need for both the contractors (suppliers) and beneficiaries to work together with a mindset of providing solution. For example, following instructions, feedback, record keeping, keenly making observations, fixing problems, trying options etc. are necessary in ensuring that a system works, giving room for further improvement. In case of any pilot demonstration plants, beneficiaries should be made to have some stakes for better ownership and commitment. Private sector investment could contribute to resolving this threat.

4.4. Opportunities

4.4.1. Energy Deficits and Rural Settings Favor Decentralization

About 70% of Nigerians dwell in rural areas, working as peasant farmers. They have limited access to modern cooking facilities such as LPG (cooking gas), kerosene, and electricity. Moreover, only about 36% of Nigerians have access to electricity and power outages are common. These factors favor decentralization of biogas and fertilizer production in communities where feedstocks are more readily available, as well, the products generated from AD can be used by the community.

4.4.2. Food Insecurity and Calls for Diversifying Nigeria's Economy

At the current average rate of 13.5 kg fertilizer per hectare in Nigeria, the 4 study sites have a combined potential to provide fertilizer for about 100 hectares. Biofertilizer availability could contribute to providing a sustainable solution to the current food insecurity in Nigeria. Crop yields higher by 11–20% compared to controls have been reported after the application of spent digester effluent [54]. In the same vein, the application of spent digester effluent has healing effects on soil structure, countering the detrimental influence of increasing soil acidity, topsoil erosion, and micro-nutrient depletion due to long-term usage of inorganic fertilizer [73]. On the other hand, despite the huge agricultural potential, Nigeria's economy largely depends on oil. Therefore, there have been several calls for diversification of Nigeria's economy. Combining these two scenarios presents great opportunities for the development of biogas systems.

4.4.3. Improved Public Health

Prevention of organic waste disposal in open fields which could otherwise be breeding grounds for pathogenic organisms such as *Salmonella* sp., *E. coli*, and *Shigella* sp., by employing AD systems leads to improved public health. Some aerobic organisms are killed by the fermentation process in an anaerobic environment. Biogas systems could also serve as a better alternative for management of human excreta, therefore enhancing public health.

4.4.4. Job Opportunities

Biogas plant construction is labor-intensive and could therefore provide many job opportunities for business and technical managers, construction masons, plant operators for waste loading, etc. For example, engineers and technicians with skills in design and construction of biogas facilities and accessories could become actively employed in this sector were it to be functional. In countries like Germany and Sweden, biogas is an active sector providing employment for several people. In Nepal, the Biogas Support Program provides opportunities to people by systematically improving their skills via training and providing employment to at least 9000 people [55]. There are, therefore, several untapped job opportunities in this sector in Nigeria.

4.4.5. Increased Economic Activity

The generation of both biogas and biofertilizer will have a positive economic impact. Energy is the driver of economic activities, nearly every product and service require energy inputs. Availability of biogas as an alternative for cooking would help women and children who typically spend hours to hunt and collect wood for fuel. The time spent for collecting and carrying wood by women and children could be swapped for education, more productive activities, or simply leisure time [55].

4.4.6. Synergies with Global Climate Change Mitigation Goals

The dangers posed by climate change point to the urgency of creating and maintaining sustainable environment. For this reason, international and national leaders have set targets for climate change mitigation. For example, the Nigerian Government though the Electricity Regulatory Commission

(NERC) has put in place a Feed in Tariff (FIT) policy aimed at promoting investment in renewable energy for power generation to achieve 10% of total energy mix [51]. AD system is one of the options that could contribute to these goals. Installation of AD systems in the various abattoirs in Nigeria could contribute in CH_4 emissions mitigation by preventing disposal to the open fields. In addition to this, mitigations of other emissions like CO_2 and N_2O are also possible. The use of biogas in place of fossil-based alternatives further provides avenues for mitigation of GHG emissions. Significant impact could be made in mitigating GHG emissions by a national program that could widen the scope to cover the states of the federation and by sustaining efforts long into the future for cumulative effect. Investing in this sector is a step in the right direction especially as it synergizes global climate change mitigation goals.

4.4.7. Logistic Support from the Public

Given the socio-economic benefits associated with AD systems, securing public support is likely. For example, the top priorities for most government and development agencies include job creation, improved access to power, food security, and improving the quality of life of the citizens. These goals fit well with the benefits accruable by adopting AD technology. This could be a powerful tool in reaching out to citizens who are in desperate need for solutions to improve the quality of life. In most cases, these are some of the campaign promises that politicians make to get voted into power, therefore AD system could be a tool to deliver dividends of democracy to the citizens.

4.5. Threats

4.5.1. High Lending Rate

As the investment cost for AD is prohibitive, bank loans could serve as a possible funding facility. However, in Nigeria bank lending rates range from 16.91 to 29.26% and include stringent collateral requirements. This financial predicament may not be favorable for investing in AD, making it difficult for willing investors to start such a project despite its prospects. However, since investment in this sector shares synergies with the global goals of food and energy security as well as environmental sustainability, appropriate policy planning could help to resolve this threat in favor of the technology's uptake. Higher rates of AD application experienced in China, India, Germany and Nepal are due to government supporting policies and financial incentives.

4.5.2. Public Subsidies for Fossil-Based Energy and Fertilizers

Public subsidies for fossil-based energy and chemical fertilizers may threaten the competitiveness of the AD system. The system does not enjoy a large public support currently, and thus, there is no level playing ground in financial terms for investing in the technology, rendering it a risky venture. Besides the subsidies, carbon emissions and soil degradation resulting from the use of conventional fossil energy sources and synthetic fertilizers respectively, are not accounted for. Thus, conventional technologies appear superior to cleaner energy options in purely financial terms [72]. Attaching a price to carbon as an incentive in the event of carbon reduction, as is the case in the Clean Development Mechanism, can help to tackle part of this problem. In a policy brief by Bassi et al. [74], it was pointed out that the EU should focus on carbon market as a better strategy than subsidies for low-carbon renewable energy to achieve further reductions in emissions from power sectors. Further, it was stated that renewable sources of electricity are becoming cost-competitive with fossil fuels and will soon no longer need subsidies [74]. Indeed, climate policy incentivizes renewable energy technologies growth more compared to conventional businesses. As of 2011, no African country has placed a price on carbon and thus there is no truly level playing field in terms of cost and benefits between conventional and renewable technologies [72]. However, in 2015, there was a dialogue on adopting carbon pricing in Lagos, Nigeria, during which the stakeholders endorsed the idea and explored essential modalities.

The key outcome was the formation of the Carbon Pricing Leadership Coalition Nigeria [75]. Therefore, this threat can be tackled by keying into global carbon markets.

5. Conclusions

In this case study, a multi-criteria approach was used where quantitative data from the study sites were incorporated into the SWOT analysis to assess the prospects of AD systems for abattoir wastes management in Nigeria as a cleaner technology approach. The study established more strengths and opportunities than weaknesses and threats in favor of uptake of AD technology. The wastes generated in each abattoir in the study sites meets the required volume to operate as small to medium scale business enterprise. Furthermore, if AD systems are installed in the four study sites, combined potentials of about 1.66×10^6 m^3/y, 1380 t/y, 709 kW and 30.71 t CO_2 eq corresponding to biomethane, dry biofertilizer, electricity and GHG reduction potentials respectively, are attainable. Larger cities, such as Lagos, Kano, Kaduna, Ibadan, Enugu, Onitsha, and Port Harcourt, which record more animal slaughters, will have corresponding higher potentials. Moreover, the prospects of sourcing more feedstock near any proposed plant, such as chicken droppings, cattle dung, pig droppings etc., from nearby farmers could enable bigger capacity plants which has better economic viability than smaller plants. Despite these benefits, there are negative factors in the SWOT analysis such as high investment cost and the relative novelty of the technology (in the internal environment), as well as high lending rates and public subsidies for fossil-based energy and fertilizer (in the external environment). Thus, to make these benefits a reality, the negative factors in the SWOT analysis must be adequately addressed using political instruments and public policies. The high propensities of surmounting these negative elements are predicated on the positive factors in the SWOT analysis. The positive factors include provision of clean energy and spent slurry as biofertilizer, destruction of some pathogenic organisms, solving waste disposal problems, and existing needs for both biogas and biofertilizer (in the internal environment). Other factors in the external environment that could incentivize uptake of AD systems include persisting energy deficits and food insecurity in Nigeria, the prospects for improved public health, job opportunities, and the fact that the AD system shares synergies with global climate change mitigation goals, like mitigating GHG emissions. In general, although the study revealed much potential for investment in AD system to harness the wastes generated in Nigerian abattoirs, a lot needs to be done to tackle the hindering factors. Based on these findings, the study recommends private sector investment supported by public policies. Public support in form of national programs with the mandate of providing coordinated research and development, marketing, technical, facilitating access to climate funding opportunities, financial supports and framework as well as setting out standards, codes and regulation for the private sector participation, are required for progress.

Author Contributions: I.G.A. conceived and designed the study; I.G.A. performed the site evaluation survey; I.G.A., B.M.L., O.M.T., P.A.O. and A.B. analyzed the data; I.G.A. and A.B. wrote the paper; B.M.L., O.M.Y., P.A.O. and A.B. revised the paper critically for intellectual inputs. All authors have read and agreed to the published version of the manuscript.

Funding: This research was funded by the Raw Materials Research and Development Council, Abuja—Nigeria; we are grateful.

Acknowledgments: The authors wish to thank the Raw Materials Research and Development Council, for funding this study. The authors also thank Wolfgang Glasser and Kemjika Ajoku who read the manuscript and made useful comments. Mike Hobbs edited the manuscript for English gramma errors. We are grateful. Maria Obi also supported in some aspects of the site evaluation; we are grateful.

Conflicts of Interest: The authors declare no conflict of interest.

References

1. IEA. International Energy Agency. *France*. 2019. Available online: https://www.iea.org/countries/Nigeria/ (accessed on 10 December 2019).
2. Energypedia. 2019. Available online: https://energypedia.info/wiki/Nigeria_Energy_Situation#cite_ref-JICA.2C_February_2007.2C_The_Master_Plan_Study_for_Utilization_of_Solar_Energy_in_the_Federal_Republic_of_Nigeria.2C_Final_Report.2C_Volume_1.2C_p._3-20_8-0 (accessed on 22 December 2019).
3. World Resources Institute Climate Analysis Indicators Tool (WRI CAIT 4.0, 2017). *GHG Emissions are Reported in Units of Carbon Dioxide Equivalents. Global Warming Potentials (GWPs) are the 100-Year GWPs from the Intergovernmental Panel on Climate Change (IPCC) Second Assessment Report (SAR)*; World Resources Institute: Washington, DC, USA, 2017.
4. Knee, A. *The Nature of Energy*; Magazine of the World Conservation Union; International Union for Conservation of Nature and Natural Resources: Gland, Switzerland, 2007; pp. 1–44.
5. Sanusi, R.A.; Badejo, C.A.; Yusuf, B.O. Measuring household food insecurity in selected local government areas of Lagos and Ibadan, Nigeria. *Pak. J. Nutr.* **2006**, *5*, 62–67.
6. Abu, O. Food security in Nigeria and South Africa: Policies and challenges. *J. Hum. Ecol.* **2012**, *38*, 31–35. [CrossRef]
7. IPCC. *Climate Change 2014: Mitigation of Climate Change, Contribution of Working Group III to the Fifth Assessment Report of the Intergovernmental Panel on Climate Change*; Cambridge University Press: Cambridge, UK; New York, NY, USA, 2014.
8. Zhang, J.; Jiang, J.; Tian, G. The potential of fertilizer management for reducing nitrous oxide emissions in the cleaner production of bamboo in China. *J. Clean. Prod.* **2016**, *112*, 2536–2544. [CrossRef]
9. Adeyemi, I.G.; Adeyemo, O.K. Waste management practices at the Bodija abattoir, Nigeria. *Int. J. Environ. Stud.* **2007**, *64*, 71–82. [CrossRef]
10. Adeyemo, O.K.; Adeyemi, I.G.; Awosanya, E.J. Cattle cruelty and risks of meat contamination at Akinyele cattle market and slaughter slab in Oyo State, Nigeria. *Trop. Anim. Heal. Prod.* **2009**, *41*, 1715–1721. [CrossRef]
11. Chukwu, O.; Adeoye, P.A.; Chidiebere, I. Abattoir Waste Generation, Management and the Environment: Minna, Nigeria. *Int. J. Biosci.* **2011**, *1*, 100–109.
12. Oruonye, E.D. Challenges of Abattoir Waste Management in Jalingo Metropolis, Nigeria. *Int. J. Resour. Geogr.* **2015**, *1*, 22–23.
13. Salminen, E.; Rintala, J. Anaerobic digestion of organic solid poultry slaughterhouse waste—A review. *Bioresour. Technol.* **2002**, *83*, 13–26. [CrossRef]
14. Akinbami, J.-F.; Ilori, M.; Oyebisi, T.; Akinwumi, I.; Adeoti, O. Biogas energy use in Nigeria: Current status, future prospects and policy implications. *Renew. Sustain. Energy Rev.* **2001**, *5*, 97–112. [CrossRef]
15. Lukehurst, C.T.; Frost, P.; Seadi, T.A. Utilization of digestate from biogas plants and biofertilizer, task 37. *IEA Bioenergy* **2010**, *24*.
16. Amon, T.; Bredow, H.V.; Gromke, J.D.; Dohle, H.; Fischer, E.; Fischer, E.; Friehe, J.; Gattermann, H.; Grebe, S.; Grope, J.; et al. *Guide to Biogas-from Production to Use*; Fachagentur Nachwachsende Rohstoffe e. V. (FNR): Eschborn, Germany, 2010.
17. Cvetkovic, S.; Radoičić, T.K.; Vukadinović, B.; Kijevčanin, M.L. Potentials and status of biogas as energy source in the Republic of Serbia. *Renew. Sustain. Energy Rev.* **2014**, *31*, 407–416. [CrossRef]
18. Amigun, B.; Von Blottnitz, H. Capacity-cost and location-cost analyses for biogas plants in Africa. *Resour. Conserv. Recycl.* **2010**, *55*, 63–73. [CrossRef]
19. IPCC/Intergovernmental Panel on Climate Change, National Greenhouse Gas Inventories and Uncertainty Management in National Greenhouse Gas Inventories, IPCC National Greenhouse Gas Inventory Program. 2000. Available online: http://www.ipcc-nggip.iges.or.jp/public/gp/gpgaum.htm. (accessed on 28 March 2019).
20. JGCRI/Joint Global Change Research Institute, GCAM v4.4 Documentation: Global Change Assessment Model (GCAM), Documentation for GCAM. 2018. Available online: http://jgcri.github.io/gcam-doc/. (accessed on 27 February 2019).
21. Tolera, S.T.; Alemu, F.K. Potential of Abattoir Waste for Bioenergy as Sustainable Management, Eastern Ethiopia, 2019. *J. Energy* **2020**, *2020*, 6761328. [CrossRef]

22. Kholif, A.; Elghandour, M.M.M.Y.; Rodríguez, G.; Olafadehan, O.; Salem, A.Z.M. Anaerobic ensiling of raw agricultural waste with a fibrolytic enzyme cocktail as a cleaner and sustainable biological product. *J. Clean. Prod.* **2017**, *142*, 2649–2655. [CrossRef]

23. Klintenberg, P.; Jamieson, M.; Kinyaga, V.; Odlare, M. Assessing Biogas Potential of Slaughter Waste: Can Biogas Production Solve a Serious Waste Problem at Abattoirs? *Energy Procedia* **2014**, *61*, 2600–2603. [CrossRef]

24. Ware, A.; Power, N. Biogas from cattle slaughterhouse waste: Energy recovery towards an energy self-sufficient industry in Ireland. *Renew. Energy* **2016**, *97*, 541–549. [CrossRef]

25. Adamu, A.A.; Mohammed-Dabo, I.A.; Hamza, A.; Ado, S.A. Predicting rate of biogas production from abattoir waste using empirical models. *Int. J. Sci. Eng. Res.* **2017**, *8*, 1238–1245.

26. Rabah, A.B.; Baki, A.S.; Hassan, L.G.; Musa, M.; Ibrahim, A.D. Production of biogas using abattoir waste at different retention time. *Sci. World J.* **2010**, *5*, 23–26.

27. Ek, A.E.W.; Hallin, S.; Vallin, L.; Schnürer, A.; Karlsson, M. Slaughterhouse waste co-digestion—Experiences from 15 years of full-scale operation. World renewable energy congress 2011-Sweden. *Bioenergy Technol.* **2011**, *9*, 64–71.

28. Hellman, J.; Ek, A.E.W.; Sundberg, C.; Johansson, M.; Svensson, B.H.; Karlsson, M. Mechanisms of increased methane production through re-circulation of magnetic biomass carriers in an experimental continuously stirred tank reactor. In Proceedings of the 12th World Congress on Anaerobic Digestion, Guadalajara, Mexico, 31 October–4 November 2010.

29. Ngumah, C.; Ogbulie, J.N.; Orji, J.C.; Amadi, E.S. Biogas potential of organic waste in Nigeria. *J. Urban Environ. Eng.* **2013**, *7*, 110–116. [CrossRef]

30. Kurttila, M.; Pesonen, M.; Kangas, J.; Kajanus, M. Utilizing the analytic hierarchy process (AHP) in SWOT analysis—A hybrid method and its application to a forest-certification case. *Policy Econ.* **2000**, *1*, 41–52. [CrossRef]

31. Okello, C.; Pindozzi, S.; Faugno, S.; Boccia, L. Appraising Bioenergy Alternatives in Uganda Using Strengths, Weaknesses, Opportunities and Threats (SWOT)-Analytical Hierarchy Process (AHP) and a Desirability Functions Approach. *Energies* **2014**, *7*, 1171–1192. [CrossRef]

32. Akinbomi, J.; Brandberg, T.; Sanni, S.A.; Taherzadeh, M.J. Development and dissemination strategies for accelerating biogas production in Nigeria. *Bioresources* **2014**, *9*, 5707–5737.

33. Deublein, D.; Steinhauser, A. *Biogas from Waste and Renewable Resources*; Wiley-VCH Verlag GmbH & Co. KGaA: Hoboken, NJ, USA, 2008; pp. 27–83.

34. Schnurer, A.; Jarvis, A. Microbiological Handbook for Biogas Plants, 1-74. Swedish Waste Management U2009:03. *Swed. Gas Cent. Rep.* **2010**, *207*, 1–74.

35. Burke, D. *Dairy Waste Anaerobic Digestion Handbook*; Environmental Energy Company: Olympia, WA, USA, 2001; pp. 1–57.

36. Schirmer, W.N.; Jucá, J.F.T.; Schuler, A.R.P.; Holanda, S.; Jesus, L.L. Methane production in anaerobic digestion of organic waste from Recife (Brazil) landfill: Evaluation in refuse of diferent ages. *Braz. J. Chem. Eng.* **2014**, *31*, 373–384. [CrossRef]

37. Chen, Y.R. *Engineering Properties of Beef-Cattle Manure*; ASAE paper no. 82-4085; ASAE: Lisbon, Portugal, 1982.

38. B-Sustain, Environmental and Social Benefits of Biogas Technology. 2013. Available online: http://www.bsustain.in/faqs.html (accessed on 22 May 2018).

39. Dzene, I.; Romagnoli, F. Comparison of different biogas use pathways for Latvia: Biogas use in CHP vs. biogas upgrading. In In Proceedings of the 9th International Conference Environmental Engineering 2014, Vilnius, Lithuania, 22–23 May 2014.

40. Kolios, A.; Read, G. A Political, Economic, Social, Technology, Legal and Environmental (PESTLE) Approach for Risk Identification of the Tidal Industry in the United Kingdom. *Energies* **2013**, *6*, 5023–5045. [CrossRef]

41. Jadidi, O.; Hong, T.S.; Firouzi, F.; Yusuff, R.M. An optimal grey based approach based on TOPSIS concepts for supplier selection problem. *Int. J. Manag. Sci. Eng. Manag.* **2009**, *4*, 104–117. [CrossRef]

42. Wind, Y.; Saaty, T.L. Marketing Applications of the Analytic Hierarchy Process. *Manag. Sci.* **1980**, *26*, 641–658. [CrossRef]

43. Liu, T.; McConkey, B.G.; Ma, Z.; Liu, Z.; Li, X.; Cheng, L. Strengths, Weaknessness, Opportunities and Threats Analysis of Bioenergy Production on Marginal Land. *Energy Procedia* **2011**, *5*, 2378–2386. [CrossRef]

44. Kretschmer, W.; Bischoff, S.; Hanebaeck, G.; Muller, H. *SWOT Analysis and Biomass Competition Analysis for SUPRABIO Biorefineries*; IUS Institute for Environmental Studies, Weibel and Ness GmbH: Heidelberg, Germany, 2014.

45. Eze, J.I. Studies on Generation of Biogas from Poultry Droppings and Rice Husk from a Locally Fabricated Biodigester. Master's Thesis, University of Nigeria, Nsukka, Nigeria, 1995; pp. 64–65.

46. Connaughton, S.; Collins, G.; O'Flaherty, V. Psychrophilic and mesophilic anaerobic digestion of brewery effluent: A comparative study. *Water Res.* **2006**, *40*, 2503–2510. [CrossRef]

47. Nijaguna, B.T. *Biogas Technology*; Book 142 New Age International (P) Limited, Publishers 4835/24; Ansari Road: Daryaganj, New Delhi, 2006.

48. Olsen, J.E.; Larsen, H.E. Bacterial decimation times in anaerobic digestions of animal slurries. *Boil. Wastes* **1987**, *21*, 153–168. [CrossRef]

49. Kearney, T.E.; Larkin, M.; Frost, J.; Levett, P. Survival of pathogenic bacteria during mesophilic anaerobic digestion of animal waste. *J. Appl. Bacteriol.* **1993**, *75*, 215–219. [CrossRef] [PubMed]

50. Carlini, M.; Mosconi, E.M.; Castellucci, S.; Villarini, M.; Colantoni, A. An Economical Evaluation of Anaerobic Digestion Plants Fed with Organic Agro-Industrial Waste. *Energies* **2017**, *10*, 1165. [CrossRef]

51. Chukwuma, C.E.; Ndrika, V.I.O.; Chukwumuanya, E.O. Economic Viability of Bio-energy Plants for Location Analysis in Parts of Anambra State of Nigeria. *JEAS* **2018**, *13*, 40–53.

52. Arnott, M. *The Biogas/Biofertilizer Business Handbook. Appropriate Technologies for Development*; Peace Corps Information Collection and Exchange Manual R-48: Washington, DC, USA, 1982; pp. 1–117.

53. World Bank. Pollution Prevention and Abatement Handbook—World Bank Documents. Available online: https://www.google.de/?gfe_rd=cr&ei=wOT_WMWpFJDVXrzyiOgE#q=meat+processing+and+rendering+pollution+and+prevention+abatement+handbook1998 (accessed on 10 January 2017.).

54. Marchaim, U. *Biogas Processes for Sustainable Development*; FAO: Rome, Italy, 1992; pp. 1–99.

55. SNV. *Biogas Support Program Fuels Rural Household Energy Supply in Nepal. A Paper for UNCTAD Expert Meeting on Green and Renewable Technologies as Energy Solutions for Rural Development*; SNV: Geneva, Switzerland, 2010.

56. WBA. *Biogas—An Important Energy Source*; Factsheets; World Bioenergy Association (WBA): Stockholm, Sweden, 2013.

57. Tumwesige, V.; Avery, L.; Austin, G.; Balana, B.; Bechtel, K.; Casson, E.; Davidson, G.; Edwards, S.; Eshete, G.; Gebreegziabher, Z.; et al. Small-Scale Biogas Digester for Sustainable Energy Production in Sub-Saharan Africa: A Review. 1st World Sustainability Forum 2011. Available online: www.wsforum.org. (accessed on 12 April 2015.).

58. WHO. Household Air Pollution and Health. Fact Sheet No. 292. Available online: http://www.who.int/mediacentre/factsheets/fs292/en/2016 (accessed on 5 August 2017).

59. Alfa1, M.I.; Wamyil, F.B.; Daffi, R.E.; Igboro, S.B. Design, Development and Evaluation of Slaughterhouse Anaerobic Digestion Plant Model. *Am. J. Eng. Res.* **2017**, *6*, 70–74.

60. Drosg, B.; Fuchs, W.; Al Seadi, T.; Madsen, M.; Linke, B. Nutrient recovery by biogas digestate processing. *IEA Bioenergy Task* **2015**, *37*, 1–40.

61. Sindibu, T.; Solomon, S.; Ermias, D. Biogas and bio-fertilizer production potential of abattoir waste as means of sustainable waste management option in Hawassa City, southern Ethiopia. *J. Appl. Sci. Environ. Manag.* **2018**, *22*, 553. [CrossRef]

62. EC (European Commission). European Parliament Commission Regulation 2009/1069/EC. Available online: https://eur-lex.europa.eu/legal-content/en/ALL/?uri=CELEX:32009R1069 (accessed on 19 August 2020).

63. Ochieng' tieno, F.A. Anaerobic digestion of wastewaters with high strength sulfates. *Discov. Innov.* **1996**, *8*, 143–150.

64. Chen, Y.; Cheng, J.J.; Creamer, K.S. Inhibition of anaerobic digestion process: A review. *Bioresour. Technol.* **2008**, *99*, 4044–4064. [CrossRef]

65. Lay, J.-J.; Li, Y.-Y.; Noike, T. Influences of pH and moisture content on the methane production in high-solids sludge digestion. *Water Res.* **1997**, *31*, 1518–1524. [CrossRef]

66. Tada, C.; Yang, Y.; Hanaoka1, T.; Sonoda, A.; Ooi, K.; Sawayama, S. Effect of natural zeolite on methane production for anaerobic digestion of ammonium rich organic sludge. *Bioresour. Technol.* **2005**, *96*, 459–464. [CrossRef] [PubMed]

67. Nordell, E.; Hallin, S.; Johansson, M.; Karlsson, M. The Diverse Response on Degradation Rate of Different Substances upon Addition of Zeolites. In Proceedings of the Third International Symposium on Energy from Biomass and Waste, Venice, Italy, 8–11 November 2010; ISBN 978-88-6265-008-3.
68. Siegrist, H.; Vogt, D.; Garcia-Heras, J.L.; Gujer, W. Mathematical Model for Meso- and Thermophilic Anaerobic Sewage Sludge Digestion. *Environ. Sci. Technol.* **2002**, *36*, 1113–1123. [CrossRef] [PubMed]
69. Schnürer, A.; Nordberg, A. Ammonia, a selective agent for methane production by syntrophic acetate oxidation at mesophilic temperature. *Water Sci. Technol.* **2008**, *57*, 735–740. [CrossRef]
70. Esposito, G.; Frunzo, L.; Liotta, F.; Panico, A.; Pirozzi, F. BMP tests to measure the biogas production from the digestion and co-digestion of Complex organic substrates. *Open J. Environ. Eng.* **2012**, *5*, 1–8. [CrossRef]
71. Patil, V.S.; Deshmukh, H.V. A review on co-digestion of vegetable waste with organic wastes for energy generation. *Int. Res. J. Biol. Sci.* **2015**, *4*, 83–86.
72. Fischer, R.; Lopez, J.; Suh, S. Barriers and drivers to renewable energy investment in Sub-Saharan Africa. *J. Environ. Invest.* **2011**, *2*, 54.
73. Sheahan, M.; Black, R.; Jayne, T.S. Are farmers underutilizing fertilizer? Evidence from Kenya. In Proceedings of the International Association of Agricultural Economists (IAAE) Triennial Conference, Foz do Iguaçu, Brazil, 18–24 August 2012; pp. 18–24.
74. Bassi, S.; Carvalho, M.; Doda, B.; Fankhauser, S. *Decarbonizing the European Union Credibly, Effectively and Acceptably*; Policy brief; Grantham Research Institute on Climate Change and the Environment and the Centre for Climate Change Economics and Policy: London, UK, 2017.
75. Envir-News. Nigeria Adopts Carbon Pricing to Curb Global Emission. 2015. Available online: http://www.environewsnigeria.com/nigeria-adopts-carbon-pricing-curb-global-emission/2015 (accessed on 24 April 2017).

© 2020 by the authors. Licensee MDPI, Basel, Switzerland. This article is an open access article distributed under the terms and conditions of the Creative Commons Attribution (CC BY) license (http://creativecommons.org/licenses/by/4.0/).

 *recycling*

Article

Utilization of Waste Cooking Oil via Recycling as Biofuel for Diesel Engines

Hoi Nguyen Xa [1], Thanh Nguyen Viet [2], Khanh Nguyen Duc [2]  **and Vinh Nguyen Duy [3],***

[1]   University of Fire Fighting and Prevention, Hanoi 100000, Vietnam; xahoinguyen@gmail.com
[2]   School of Transportation and Engineering, Hanoi University of Science and Technology,
      Hanoi 100000, Vietnam; thanh.nguyenviet@hust.edu.vn (T.N.V.); khanh.nguyenduc@hust.edu.vn (K.N.D.)
[3]   Faculty of Vehicle and Energy Engineering, Phenikaa University, Hanoi 100000, Vietnam
*   Correspondence: vinh.nguyenduy@phenikaa-uni.edu.vn

Received: 16 March 2020; Accepted: 2 June 2020; Published: 8 June 2020

Abstract: In this study, waste cooking oil (WCO) was used to successfully manufacture catalyst cracking biodiesel in the laboratory. This study aims to evaluate and compare the influence of waste cooking oil synthetic diesel (WCOSD) with that of commercial diesel (CD) fuel on an engine's operating characteristics. The second goal of this study is to compare the engine performance and temperature characteristics of cooling water and lubricant oil under various engine operating conditions of a test engine fueled by waste cooking oil and CD. The results indicated that the engine torque of the engine running with WCOSD dropped from 1.9 Nm to 5.4 Nm at all speeds, and its brake specific fuel consumption (BSFC) dropped at almost every speed. Thus, the thermal brake efficiency (BTE) of the engine fueled by WCOSD was higher at all engine speeds. Also, the engine torque of the WCOSD-fueled engine was lower than the engine torque of the CD-fueled engine at all engine speeds. The engine's power dropped sequentially through 0.3 kW, 0.4 kW, 0.6 kW, 0.9 kW, 0.8 kW, 0.9 kW, 1.0 kW and 1.9 kW.

Keywords: feedstock; waste cooking oil; engine characteristics; exhaust emissions; specific energy consumption; fuel consumption

---

1. Introduction

Compression ignition (CI) engines have been used in the industrial and agricultural sectors, and in construction, power factories, and transportation, for several decades. This broad use has resulted in a growing demand for petroleum-based diesel [1–5]. However, global fossil-fuel reserves have become limited due to rising fuel prices, depleting petroleum reserves and issues related to atmospheric pollution [4–9]. As a result, many researchers have concentrated on the discovery of renewable, carbon-neutral, and environmentally-friendly non-petroleum-based diesel in recent years [10–13]. Biodiesel is produced from various feedstocks, such as rapeseed, soybean, cottonseed oil, palm oil, and jojoba oil, and can be used to fuel internal combustion engines with no significant differences from petroleum-based fuels [14–17]. Since most biodiesel is manufactured using edible oils, the price of biodiesel is higher than conventional diesel. This is a significant barrier to the commercialization of biodiesel [8]. The use of edible feedstocks could also intensify the competition between fuel supply and food production on agricultural land, increasing the cost of food and oil [14,18–20]. As such, low-priced, inedible feedstock-based fuels, such as waste cooking oil synthetic diesel (WCOSD), should be adopted due to their competitive pricing compared to conventional diesel, and to ensure food security worldwide. The use of feedstock-based fuels can also help to lessen environmental issues by reducing waste-oil disposal.

Ahmet Necati Ozsezen et al. performed the experiment on canola oil methyl esters (COME) and waste palm oil esters (WPOME) [21], and they observed that while maximum engine torque slightly decreased, brake specific fuel consumption (BSFC) increased when compared to commercial diesel (CD) fuel. In terms of combustion characteristics, although the peak cylinder gas pressures for COME and WPOME were respectively 8.33 MPa and 8.34 MPa, at a 6.75° crankshaft angle (CA) after top dead center (ATDC), the peak cylinder gas pressure for CD fuel was 7.89 MPa at a 7° CA ATDC. In another study, K. Muralidharan explored the effects of compression ratio on the combustion characteristics of a variable compression ratio engine fueled by diesel and methyl esters of waste cooking oil (WCO) blends [15]. As a consequence, the compression ratios were changed, corresponding to the values of 18:1, 19:1, 20:1, 21:1 and 22:1. At the compression ratio of 21:1, although the BSFC of the B40 blend (a blend of 40% biodiesel) was 0.259 kg/kWh, the BSFC of the CD was 0.314 kg/kWh. Muralidharan's results also revealed that the engine fueled with WCOSD methyl ester had a lower heat release rate, a lower maximum rate of pressure rise, a longer ignition delay and a higher mass fraction, that was burnt at a higher compression ratio, than the engine that was fueled with diesel. In summary, fuel blends can increase nitrogen oxide emissions and decrease the emissions of hydrocarbon and carbon monoxide. When A. Abu-Jrai et al. [20] tested blends of treated WCOSD and CD on a naturally aspirated diesel engine, to investigate the engine's exhaust emission and combustion characteristics, they observed that the total combustion duration of B50 was longer than the total combustion duration of CD fuel. They also noted that the combustion was advanced at all engine loads and that the BSFC for B50 was slightly higher than the BSFC for conventional diesel. In addition, while the concentration of NOx emissions in the engine fueled by B50 increased by 37%, 29% and 22%, smoke emissions dropped by 42%, 31% and 30%, when compared to the emission reductions of the full load at 25%, 50% and 75%, respectively, for the engine fueled by conventional diesel.

In H. An et al.'s [22] investigation into the impacts of biodiesel that had been derived from WCO on emission characteristics, combustion characteristics and the performance of a test engine, they concluded that the use of biodiesel/blended fuels led to a higher BSFC, particularly at partial load conditions and low engine speeds. The thermal brake efficiency (BTE) of the engine fueled with biodiesel was also found to be slightly lower than the BTE of the engine fueled with conventional diesel at a 25% load. By comparison, the engine's BTE was higher, with conventional diesel loads of 50% and 100%. In addition, major emissions, such as HC and NOx, were slightly lower for biodiesel than they were for CD. During the combustion process, the ignition delay was slightly shorter, and the peak heat release rate was lower for the engine fueled by biodiesel. However, at low engine speeds, these factors adopted an opposite trend, significantly impacting the engine's emissions and combustion processes.

In addition to the research projects that have been discussed above, several other studies have investigated the use of trans-esterification biodiesel fuel, produced from WCOSD, in CI engines. Regardless, no study has investigated the impacts of synthetic biodiesel on the cooling and lubricant temperature of the conventional CI engines. As such, this research aims to discuss the technology that is used to produce biodiesel derived from WCO, and evaluate its characteristics and usability in conventional CI engines. To achieve these aims, experiments were conducted to produce WCOSD. In addition, experimental procedures were performed to measure the characteristics of test engines fueled by either diesel or WCOSD. The results of this research are the foundation for using catalyst cracking biodiesel that is derived from WCO in diesel vehicles worldwide.

2. Results and Discussions

2.1. Comparison of Engine Performance Characteristics

In the performance test, WCOSD (the researched fuel) and CD (the reference fuel) were compared at full and partial load conditions, and at engine speeds between 1200 rpm and 2400 rpm. The performance of an engine is expressed through BTE, BSFC, and engine torque or engine power output. Before investigating the effects of WCOSD on the engine's performance characteristics at full and partial load conditions, in addition to the effects of WCOSD on the engine's operating characteristics at full load conditions, the engine's experimental operation conditions, which were achieved by cooling the temperatures of the water and lubricant, needed to be considered. The engine torque at a full load condition and the engine brake power at a full load condition are shown in Figure 1.

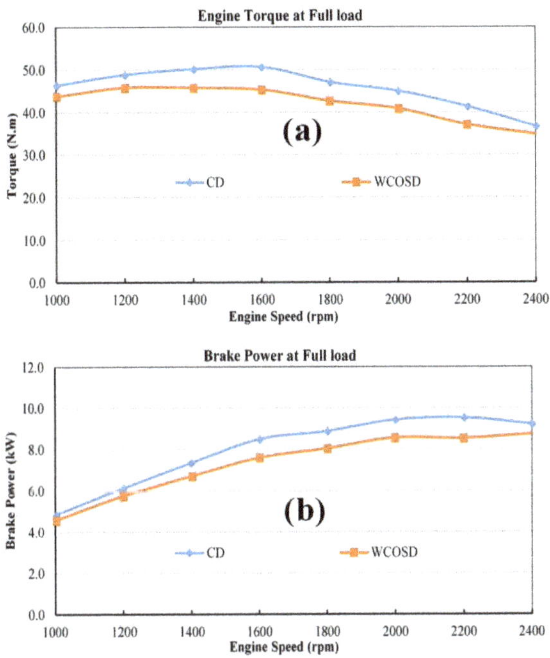

Figure 1. (**a**) engine torque and (**b**) engine brake power at full load conditions.

These results reveal that the engine torque with WCOSD was lower at all engine speeds than the engine torque with CD. The engine torque of the engine fueled with WCOSD was 2.6 Nm, 3.0 Nm, 4.4 Nm, 5.4 Nm, 4.4 Nm, 4.1 Nm, 4.3 Nm, and 1.9 Nm lower than the engine torque of the engine fueled with CD at engine speeds of 1000 rpm, 1200 rpm, 1400 rpm, 1600 rpm, 1800 rpm, 2000 rpm, 2200 rpm and 2400 rpm, respectively. Furthermore, engine power dropped sequentially from 0.3 kW to 0.4 kW, then to 0.6 kW, 0.9 kW, 0.8 kW, 0.9 kW, 1.0 kW, and finally to 1.9 kW. Reductions in engine torque and brake power were due to WCOSD's lower heating value than that of CD (Table 1). WCOSD's density is also less than CD's. As a result, WCOSD's fuel mass supply was lesser than CD's fuel mass supply under the same configurations of the fuel supply system. This trend is generally similar to the study results mentioned in [22], where the thermal brake efficiency of the engine fueled with WCO was also found to be slightly lower than that of CD at a 25% load.

Table 1. Specifications of the test engine.

Parameter	Characteristic
Type of the engine	Single-cylinder, four strokes, direct injection water-cooled, naturally aspirated.
Number of cylinders	1
Bore x Stroke	97 mm × 96 mm
Displacement	709 cm^3
Continuous rated power output	9.2 kW/2400 rpm
Starting system	Electric system
Air cleaner type	Wet/dry type
Lubricating system	Forced lubrication with a pump
Cooling system	Radiator
Combustion system	Direct injection
Max. torque	49 Nm/1600 rpm
Compression ratio	18.1
Max. output	10.3 kW/2.400 rpm
Dry weight	116 kg

BSFC is one of the most important parameters of an engine and is defined as the fuel consumption per unit of power in a unit of time. It measures how efficiently an engine uses the fuel that is supplied to produce work. Figure 2 shows the BSFC of the WCOSD and CD fuels at full load conditions. The curves illustrate that the BSFC of WCOSD decreased at all engine speeds but 2400 rpm. The BSFC of the engine running with WCOSD dropped from 1.4 g/kWh to 10.4 g/kWh, then to 14.1 g/kWh, 1.4 g/kWh, 6.5 g/kWh, 10.5 g/kWh and 5.3 g/kWh, or by values of 0.5%, 4.2%, 5.9%, 0.6%, 2.6%, 4.1% and 2.0%, respectively. In comparison, the CD-fueled engine's BSFC dropped at speeds of 1000 rpm, 1200 rpm, 1400 rpm, 1600 rpm, 1800 rpm, 2000 rpm, and 2200 rpm. At 2400 rpm, the BSCF of the engine fueled with WCOSD increased to 4.2 g/kWh, or by 1.6%. This means that the WCOSD-fueled engine was efficient in its use of the chemical energy that was supplied by the fuel.

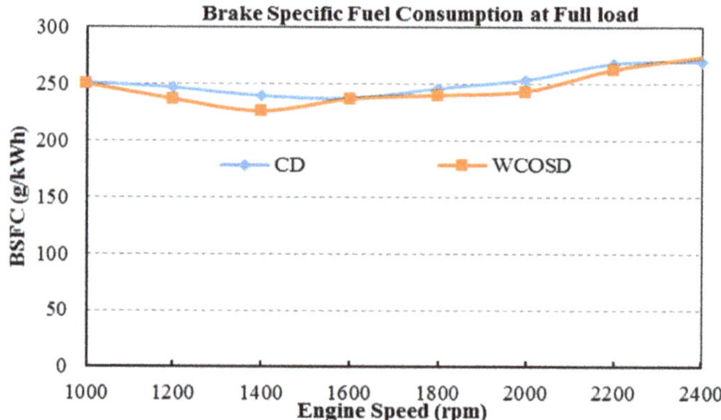

Figure 2. The BSFC at full load condition.

BTE evaluates how efficiently an engine can transform the chemical energy of a fuel into useful work. It is determined by dividing the brake power of an engine by the amount of energy input to the system. The BTE can be determined by dividing the useful work by the lower heating value of the fuel. Figure 3 shows the BTEs of the test engines that were fueled with WCOSD and CD. WCOSD exhibited a higher thermal efficiency than CD at all speeds. The BTEs of the engines fueled with WCOSD were 0.6%, 1.9%, 2.6%, 0.7%, 1.3%, 1.8% and 1.0% higher than the BTEs of the engine fueled with CD. The engine speeds of the engine fueled by CD ranged from 1000 rpm to 2200 rpm. While at lower heating values, the BTEs of the engine fueled with WCOSD were lower than the BTEs of the engine fueled with CD, the thermal efficiency of the engine fueled with WCOSD was higher than the thermal efficiency of the engine fueled with CD. This higher thermal efficiency can be explained by the reduced BSFC of the engine fueled by WCOSD, as shown in Figure 2. In other words, the conversion of the fuel's chemical energy into the engine's mechanical energy is more efficient when the engine is fueled by WCOSD, as the density and kinematic viscosity of CD is lower than the density and kinematic viscosity of WCOSD, reducing the likelihood for evaporation, decreasing the efficiency of combustion and increasing fuel consumption.

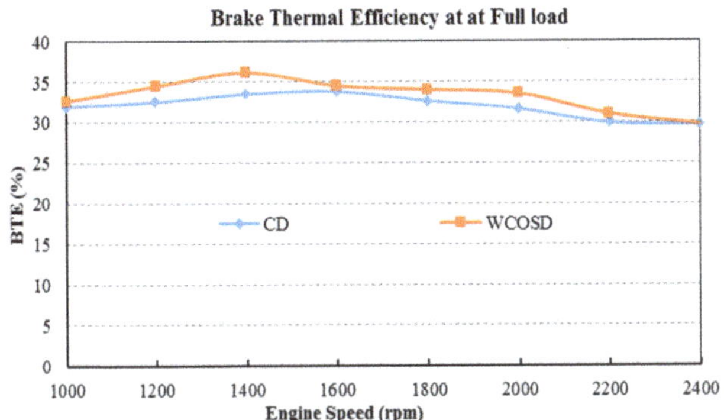

Figure 3. The brake thermal efficiency at full load condition.

2.2. Comparison of Lubricant Temperature and Cooling Water Temperature

2.2.1. Lubricant Temperature and Cooling Water Temperature at Full Load

Figure 4 shows the variations in the lubricant oil temperatures for the two fuels. The lubricant oil temperature increased as the engine speed increased, reaching a maximum speed of 1800 rpm. WCOSD's lubricant oil temperature was lower than CD's lubricant oil temperature at all speeds. To clarify, WCOSD's lubricant oil temperatures were 3.5 °C, 3.9 °C, 3.1 °C, 2.6 °C, 2.3 °C, 1.5 °C and 8.2 °C lower than CD's lubricant oil temperatures at engine speeds of 1000 rpm, 1200 rpm, 1400 rpm, 1600 rpm, 1800 rpm, 2000 rpm and 2400 rpm, respectively. At 2200 rpm, the differences between the two fuels' lubricant oil temperatures were insignificant.

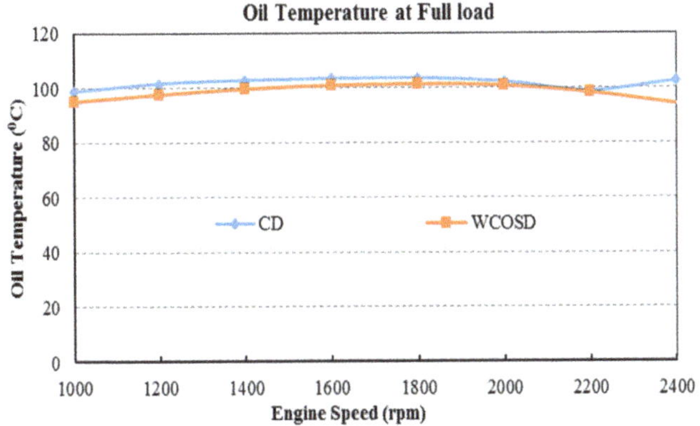

Figure 4. Lubricant oil temperature at full load condition.

Cooling water temperature is a parameter that indicates the operating state of the engine. As showed in Figure 5, The cooling water temperature of the CD-fueled engine was slightly higher than the cooling water temperature of the WCOSD-fueled engine. The highest difference in cooling water temperature was 5.6 °C at an engine speed of 1400 rpm. The cooling water temperature differences at each test point changed from 0.7 °C to 5.6 °C. The cooling water temperature of the engine fueled with WCSOD was 3.3 °C, 5.6 °C, 4.9 °C, 4.2 °C and 2.5 °C lower than the cooling water temperature of the engine fueled by CD at engine speeds of 1200 rpm, 1400 rpm, 1600 rpm, 1800 rpm and 2000 rpm, respectively. By comparison, the WCOSD-fueled engine's cooling water temperature was 1.2 °C higher than the CD-fueled engine's cooling water temperature at 2200 rpm. At 1000 rpm and 2400 rpm, the differences in the cooling water temperatures were insignificant for both engines.

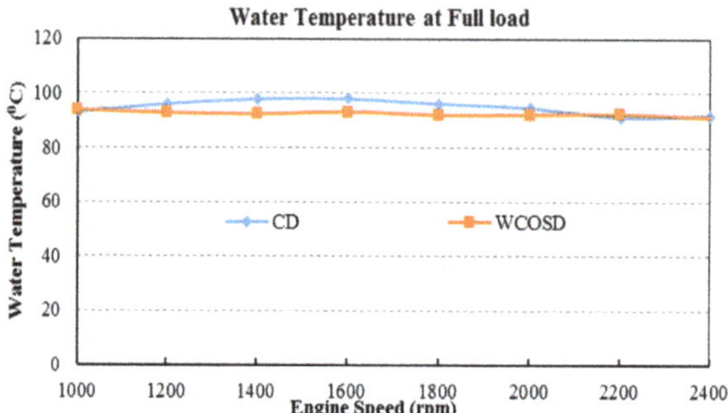

Figure 5. Cooling water temperature at the full load condition.

As illustrated above, at a full load, the same speed and when fueled by CD or WCOSD, there was a less than 5 °C difference in the engines' lubricant and cooling water temperatures, except at the speed of 2400 rpm. This can be explained by the differences in the heat values of each fuel. Indeed, the heat value of CD was higher than the heat value of WCOSD (Table 1). As a result, the higher heat release of CD led to higher cooling water and lubricant temperatures.

2.2.2. Lubricant Temperature, Cooling Water Temperature and Partial Load Conditions

Figure 6 shows the oil lubricant temperature with partial load conditions and at different speeds. In addition, Figure 6a reveals a variation of the oil lubricant temperature against an engine load of 1200 rpm. The oil lubricant temperatures of the engine fueled with CD were higher than the oil lubricant temperatures of the engine fueled with WCOSD. At engine speeds of 1400 rpm, 1800 rpm, 2000 rpm and 2200 rpm, this difference was higher at low and high loads, and smaller at medium loads. When the engine load was low, the CD-fueled engine's oil lubricant temperature was lower than the WCOSD-fueled engine's oil lubricant temperature. However, for high engine loads, this trend shifted in the opposite direction. At 1600 rpm, the differences in lubricant temperature were higher at low loads and lower at high loads.

Figure 7 presents the examination of torque and the cooling water temperatures at seven different engine speeds. According to the experiment results, the cooling water temperatures of the WCOSD-fueled engines were higher than the cooling water temperatures of the CD-fueled engines at low and medium load conditions. In contrast, the cooling water temperatures of the WCOSD-fueled engines were lower than the cooling water temperatures of the CD-fueled engines at high load conditions. For most test conditions that used the same engine speeds and load conditions, the differences between the lubricant temperatures and cooling temperatures were inadequate. Further, when the engines were tested with both fuels, the parameters of each test point were recorded when the engines reached a steady state. Thus, the parameters for the performances of the engines at the same speeds and load conditions could be fully compared and analyzed.

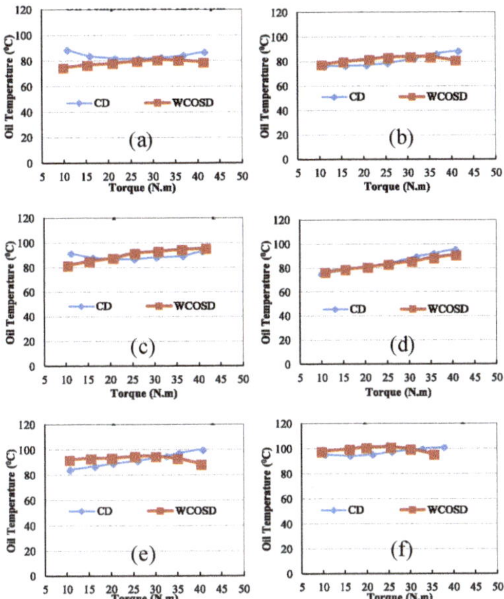

Figure 6. Lubricant oil temperatures at partials load and different engine speeds: (**a**) at 1200 rpm, (**b**) 1400 rpm, (**c**) 1600 rpm, (**d**) 1800 rpm, (**e**) 2000 rpm, (**f**) 2200 rpm.

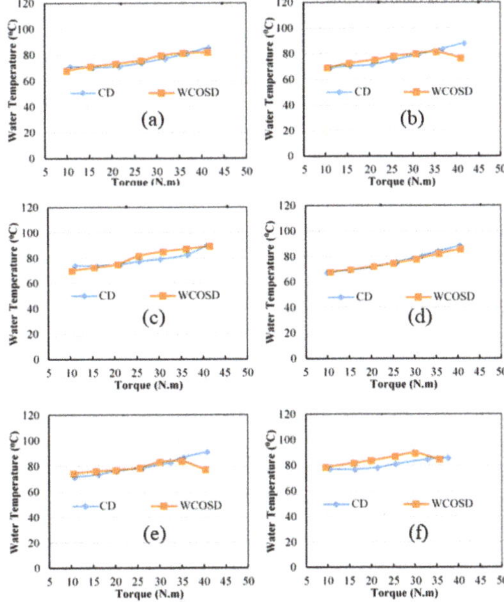

Figure 7. Cooling water temperatures at partial loads and different engine speeds: (**a**) at 1200 rpm, (**b**) 1400 rpm, (**c**) 1600 rpm, (**d**) 1800 rpm, (**e**) 2000 rpm, (**f**) 2200 rpm.

In general, when the test engine was at a full load condition, CD's higher heating value resulted in higher cooling water and lubricant temperatures. At low load conditions, CD's higher viscosity negatively affected the quality of the injection and air-fuel mixture. As a result, CD's output power was lower than WCOSD's output power. In summary, CD fuel produced lower cooling water and lubricant temperatures at lower load conditions, due to the lower released heat.

3. Testing Equipment, Experimental Setup and Test Procedure

3.1. Characteristics of Commercial Diesel Fuel and Waste Cooking Oil Synthetic Diesel Fuel

This research compared the characteristics of test engines that were fueled by CD and WCOSD. The CD fuel was bought at a fuel station in Saraburi and was used as the reference fuel for this study. The characteristics of CD and WCOSD are listed in Table 2.

Table 2. Chemical and physical properties of tested fuels.

Properties		WCOSD	CD	Method Test
Composition (%)	C	77.14	81.17	
	H	14.6005	15.285	
	N	0.077545	0.066545	
Density (kg/m^3)		820.289	827.485	ASTM D1298
Heating Value (MJ/kg)		44.245	44.864	ASTM 04-5865
Flash Point (°C)		93.5	81.5	ASTM D92
Viscosity (mm^2/s)		2.92	3.74	ASTM D445
Cetan Number		47.7	49.2	ASTM D613

For this research, catalyst cracking biodiesel was successfully manufactured from WCOSD at the laboratory. The WCOSD was pre-treated to remove solid particles and excessive water and was then used as the raw material for cracking without further treatment. During the cracking process, the free fatty acids (FFAs) in the WCOSD were decarboxylated to hydrocarbon and CO_2. FFAs cannot deactivate base catalysts that are in soap form. Therefore, before carrying out the reaction, it is beneficial to use transesterification on WCOSD by removing the FFAs from the raw material, through neutralization with an alkaline solution, esterification with glycerine, extraction with solvents and distillation or removal of the fatty acids with ion-exchange to avoid the form of soaps. Indeed, to manufacture biodiesel fuel oil, in place of diesel, with a higher stable energy output than that of the conventional fatty acid methyl esters (FAME) process (60%), the synthesizing process known as the catalytic cracking method (CCM) was used with a non-food oil at the pilot-scale level. As shown in Figure 8, this new biodiesel production process was comprised of the simple catalytic cracking of different kinds of animal fats and vegetable oils. The results showed that the product quality was similar to, or even higher than, the product quality of CD fuel.

The pathway of the cracking reaction involved the following: (1) the hydrolysis of the triglycerides to glycerine and free acid, (2) the dehydration of the glycerine to gaseous hydrocarbon, and (3) the decarboxylation of an FFA, such as carboxylic acid, hydrocarbon and CO_2. The fuel was produced via the pyrolysis system using a schematic diagram, as shown in Figure 9. An internal agitate was then installed inside the reactor to allow for the continued extraction of the overflow of the reactive catalysts, their residues and their follow-up. In addition, a back-flow device was installed to control the outflow of high boiling point elements, and to allow for the selective recovery of kerosene and diesel through multistage cooling.

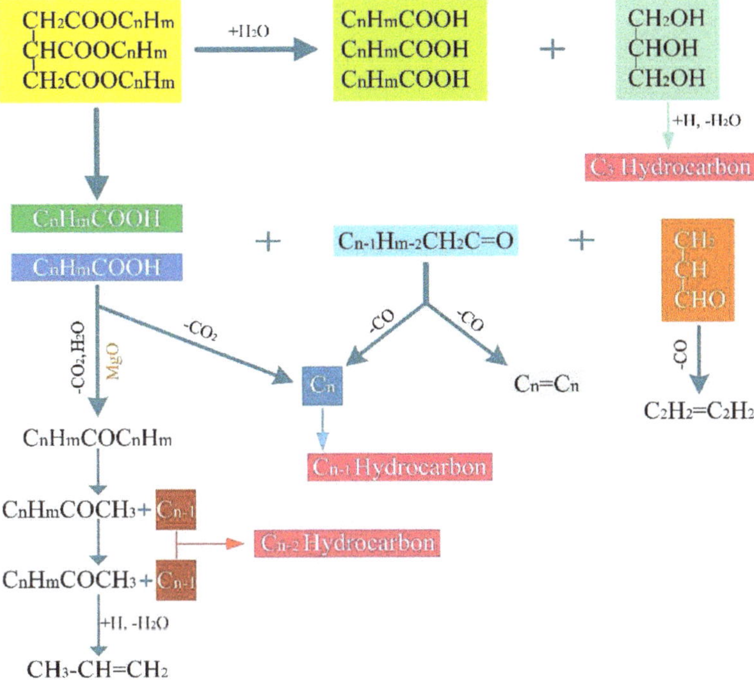

Figure 8. Reaction mechanisms of cracking reaction.

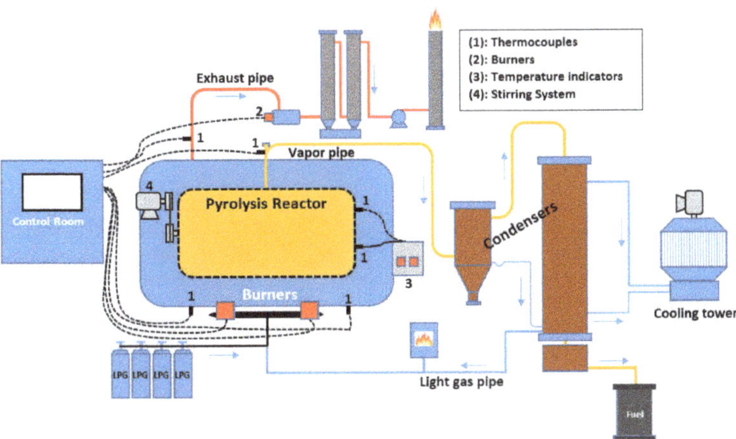

Figure 9. Schematic diagram of the pyrolysis system during the production of waste cooking oil.

The catalyst was a magnesium oxide catalyst that was supported with activated carbon. The catalyst advances the Decarboxy-cracking of WCO and other triglycerides, to generate hydrocarbons with a middle-distillate range. The composition of the catalyst was an 89.20 weight percentage of activated carbon, and a 9.75 weight percentage of magnesium oxide.

According to the results of the analysis based on the synthesized oil that was yielded from the pyrolysis system, the fuel performance of the new biodiesel showed levels that were nearly identical to those of CD. The comparison of major indicators can be summarized as follows: (1) Sulphur—<3 ppm weight (the same as FAME); (2) Cetane Index—47.7 (lower than FAME but sufficient to meet the requirements); and (3) Flow Point—−25 °C (higher and better than FAME). The chemical and physical properties of the WCOSD are given in Table 2.

For the use of non-food oils, such as used cooking oil, Jatropha, and palm oil, in producing biodiesel, the present research aimed to establish the synthesizing process known as CCM. The new biodiesel (WCOSD) production process involved the simple catalytic cracking of a variety of vegetable oils and animal fats (Figure 8). The pathway of the cracking reaction can be summarized as the hydrolysis of triglycerides to glycerine and free acid; the dehydration of glycerine to gaseous hydrocarbon and water; the decarboxylation of an FFA, such as carboxylic acid, to hydrocarbon and CO2, using MgO (Equations (1) and (2)); and the decarbonylation and reaction of hydrocarbon.

$$MgO + RCOOH \ \rightarrow \ MgCO_3 + RH \tag{1}$$

$$MgCO_3 \ \rightarrow \ MgO + CO_2 \tag{2}$$

As shown in Figure 9, the pilot plant was a type of side-on internal agitated reactor with a capacity of 200 L/day. As a consequence, an internal agitator was installed inside the reactor to allow for the extraction of the overflow of reactive catalysts and residues. In addition, a back-flow device was installed to control the outflow of high boiling point elements and to allow for the selective recovery of kerosene and diesel through multistage cooling. According to the results of the analysis based on the synthesized oil that was yielded from the pilot plant, the characteristics of the new biodiesel were nearly identical to those of CD, wherein the sulfur content was less than 3 ppm, the cetane index was 47.7, and the flow point was about −25 °C.

3.2. Experimental Apparatus

The engine that this study used was a single-cylinder, four strokes, naturally aspirated and water-cooled DI CI engine. The engine's specifications are summarized in Table 1.

The test engine was coupled with a DC regenerative dynamometer—40KW (LKA 4180-AA). The specifications of the dynamometer are presented in Table 3.

Table 3. Specifications of the 40-kW direct current generator.

Model of Dynamometer	LKA-4180
Electric Supply	3×380–420 V, 50 Hz
Maximum mechanical speed	4500 min^{-1}
Base speed (min^{-1}) at armature voltage (V)	1420 min^{-1}
Cooling system	Air with a flow rate of 1300 m^3/h
Power output	62.2 kW
Armature voltage	440 V
Exciter voltage	310 V
Rated armature current	151 A
Exciter current	2.00 A
Torque	398 Nm

3.3. Experimental Procedures

The experimental process in this study proceeded in two main phases as showed in Figure 10. The purpose of the first phase is to investigate the effects of catalyst cracking biodiesel from waste cooking oil on the performance of the tested engine. To reach this purpose, the effects of WCOSD on the emission characteristics and performance of the test engine are compared to those of commercial diesel. The effects of injection timing on the operation of the engine fueled with WCOSD are investigated in the second phase. To change injection timing, the camshaft of the engine is changed. The camshafts include the three following modes: (1) Standard camshaft (STD), with this camshaft both diesel and WCOSD are tested; (2) Modified camshaft with fuel cam advanced 2° CA (STD −2); and (3) Modified camshaft with fuel cam retarded 2° CA (STD +2).

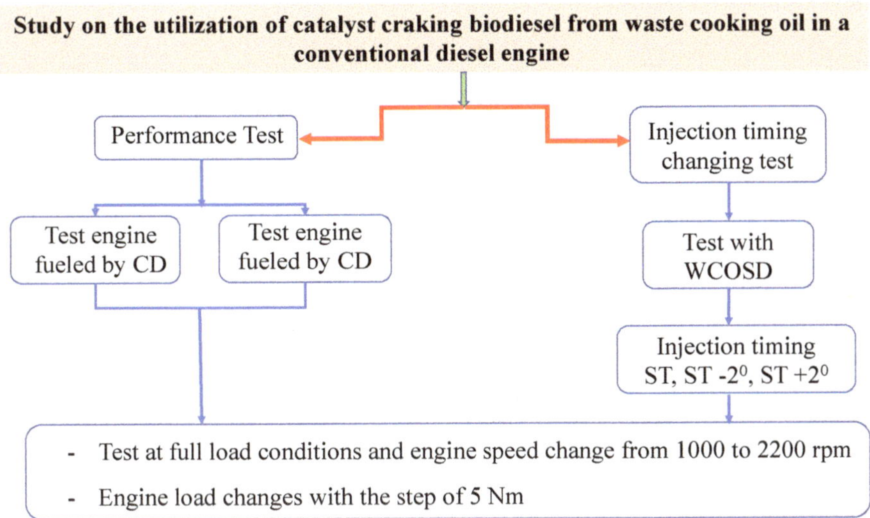

Figure 10. Experimental processes.

Consequently, the engine is tested with commercial diesel fuel as a reference fuel. Second, the engine is tested with WCOSD. The engine firstly needs to warm-up to the steady operating conditions. At each test point, after adjusting to achieve desired speed and torque (or power), we wait until the engine reaches a stable operating state, then measure all desired parameters. These parameters include engine speed, power output (or torque), fuel consumption, intake air flow rate, exhaust and intake gas temperatures, oil temperatures, cooling water temperature, fuel temperature, environmental conditions and exhaust emission. However, this research is focused on the effect of fuels on the engine characteristics of performance, lubricant and cooling temperature; thus, we only show the results relating to these issues at the standard camshaft value.

4. Conclusions

Based on the catalytic cracking method that was used in this research, WCO was used to successfully manufacture biodiesel. In addition, an experiment was conducted to evaluate the performance and temperature characteristics of the test engines' cooling water and lubricant oil after they were fueled by CD or WCOSD. When compared to the CD-fueled engine at a full load condition, the WCOSD-fueled engine's torque dropped from 1.9 Nm to 5.4 Nm at all speeds, and its BSFC dropped at almost every speed at a full load condition. The BTEs of the WCOSD-fueled engine were higher than the BTEs of the CD-fueled engine at all engine speeds but 2400 rpm, at a full load. The exhaust temperatures of the engine running with WCOSD were slightly lower. The lubricant oil temperature for the WCOSD-engine dropped from 0.5 °C to 8.2 °C, and the cooling water temperature of that engine was slightly lower as well. When compared to the BSCF of the CD-fueled engine at a partial load condition, the BSFC of the engine running with WCOSD was almost always lower. In general, WCOSD can be used to fuel conventional CI engines, because it allows the engine to work well and to operate smoothly at all operation conditions, its engine performance at a full load is comparable to the engine performance of CD, and its engine performance at a partial load is comparable to, or slightly better than, the engine performance of CD at certain points.

Author Contributions: Experimental design was performed by T.N.V. and H.N.X.; fieldwork was conducted by H.N.X.; and statistical analyses were performed by T.N.V., K.N.D. and H.N.X.; H.N.X. and V.N.D. contributed to the writing—review & editing of the paper. All authors have read and agreed to the published version of the manuscript.

Funding: This research received no external funding.

Conflicts of Interest: The authors declare no conflict of interest.

References

1. Calder, J.; Roy, M.M.; Wang, W. Performance and emissions of a diesel engine fueled by biodiesel-diesel blends with recycled expanded polystyrene and fuel stabilizing additive. *Energy* **2018**, *149*, 204–212. [CrossRef]

2. Jiaqiang, E.; Pham, M.H.; Deng, Y.; Nguyen, T.; Duy, V.N.; Le, D.H.; Zhang, Z. Effects of injection timing and injection pressure on performance and exhaust emissions of a common rail diesel engine fueled by various concentrations of fish-oil biodiesel blends. *Energy* **2018**. [CrossRef]

3. Duc, K.N.; Tien, H.N.; Duy, V.N. Performance enhancement and emission reduction of used motorcycles using flexible fuel technology. *J. Energy Inst.* **2018**, *91*, 145–152. [CrossRef]

4. Manigandan, S.; Gunasekar, P.; Poorchilamban, S.; Nithya, S.; Devipriya, J.; Vasanthkumar, G. Effect of addition of hydrogen and TiO_2 in gasoline engine in various exhaust gas recirculation ratio. *Int. J. Hydrogen Energy* **2019**, *44*, 11205–11218. [CrossRef]

5. Manigandan, S.; Gunasekar, P.; Devipriya, J.; Nithya, S. Emission and injection characteristics of corn biodiesel blends in diesel engine. *Fuel* **2019**, *235*, 723–735. [CrossRef]

6. Anawe, P.A.L.; Folayan, J.A. Data on physico-chemical, performance, combustion and emission characteristics of Persea Americana Biodiesel and its blends on direct-injection, compression-ignition engines. *Data Brief* **2018**, *21*, 1533–1540. [CrossRef]

7. Bello, E.I.; Oguntuase, B.; Osasona, A.; Mohammed, T.I. Characterization and engine testing of palm kernel oil biodiesel. *Eur. J. Eng. Technol.* **2015**, *3*, 1–14.

8. Yadav, C.; Saini, A.; Bera, M.; Maji, P.K. Thermo-analytical characterizations of biodiesel produced from edible and non-edible oils. *Fuel Process. Technol.* **2017**, *167*, 395–403. [CrossRef]

9. Nguyen, D.V.; Duy, V.N. Numerical analysis of the forces on the components of a direct diesel engine. *Appl. Sci. (Switzerland)* **2018**, *8*, 761. [CrossRef]

10. Nguyen Duc, K.; Nguyen Duy, V.; Hoang-Dinh, L.; Nguyen Viet, T.; Le-Anh, T. Performance and emission characteristics of a port fuel injected, spark ignition engine fueled by compressed natural gas. *Sustain. Energy Technol. Assess.* **2019**. [CrossRef]

11. Duc, K.N.; Duy, V.N. Study on performance enhancement and emission reduction of used fuel-injected motorcycles using bi-fuel gasoline-LPG. *Energy Sustain. Dev.* **2018**. [CrossRef]

12. Anh, T.L.; Duy, V.N.; Thi, H.K.; Xa, H.N. Experimental investigation on establishing the HCCI process fueled by n-heptane in a direct injection diesel engine at different compression ratios. *Sustainability (Switzerland)* **2018**, *10*, 3878. [CrossRef]

13. Duc, K.N.; Tien, H.N.; Duy, V.N. A Study of Operating Characteristics of Old-Generation Diesel Engines Retrofitted with Turbochargers. *Arab. J. Sci. Eng.* **2018**, *43*, 4443–4452. [CrossRef]

14. Huang, J.; Wang, Y.; Qin, J.; Roskilly, A.P. Comparative study of performance and emissions of a diesel engine using Chinese pistache and jatropha biodiesel. *Fuel Process. Technol.* **2010**. [CrossRef]

15. Muralidharan, K.; Vasudevan, D. Performance, emission and combustion characteristics of a variable compression ratio engine using methyl esters of waste cooking oil and diesel blends. *Appl. Energy* **2011**. [CrossRef]

16. Fukuda, H.; Kondo, A.; Noda, H. Biodiesel fuel production by transesterification of oils. *J. Biosci. Bioeng.* **2001**. [CrossRef]

17. Han, X.; You, K.; Tan, J.; Wang, J.; Ge, Y.; He, C. Characteristics of polycyclic aromatic hydrocarbons emissions of diesel engine fueled with biodiesel and diesel. *Fuel* **2010**, *89*, 2040–2046. [CrossRef]

18. Saravanan, S.; Nagarajan, G.; Sampath, S. International Journal of Sustainable Energy Combined effect of injection timing, EGR and injection pressure in reducing the NO x emission of a biodiesel blend Combined effect of injection timing, EGR and injection pressure in reducing the NO x emission of. *Int. J. Sustain. Energy* **2014**, *33*, 386–399. [CrossRef]

19. Meng, X.; Chen, G.; Wang, Y. Biodiesel production from waste cooking oil via alkali catalyst and its engine test. *Fuel Process. Technol.* **2008**. [CrossRef]

20. Abu-Jrai, A.; Yamin, J.A.; Ala'a, H.; Hararah, M.A. Combustion characteristics and engine emissions of a diesel engine fueled with diesel and treated waste cooking oil blends. *Chem. Eng. J.* **2011**. [CrossRef]

21. Ozsezen, A.N.; Canakci, M. Determination of performance and combustion characteristics of a diesel engine fueled with canola and waste palm oil methyl esters. *Energy Convers. Manag.* **2011**, *52*, 108–116. [CrossRef]
22. An, H.; Yang, W.M.; Maghbouli, A.; Li, J.; Chou, S.K.; Chua, K.J. Performance, combustion and emission characteristics of biodiesel derived from waste cooking oils. *Appl. Energy* **2013**. [CrossRef]

 © 2020 by the authors. Licensee MDPI, Basel, Switzerland. This article is an open access article distributed under the terms and conditions of the Creative Commons Attribution (CC BY) license (http://creativecommons.org/licenses/by/4.0/).

MDPI

St. Alban-Anlage 66

4052 Basel

Switzerland

Tel. +41 61 683 77 34

Fax +41 61 302 89 18

www.mdpi.com

Recycling Editorial Office

E-mail: recycling@mdpi.com

www.mdpi.com/journal/recycling

www.ingramcontent.com/pod-product-compliance
Lightning Source LLC
LaVergne TN
LVHW071246170726
843515LV00006BA/1343